NATO ASI Series

Advanced Science Institutes Series

A series presenting the results of activities sponsored by the NATO Science Committee, which aims at the dissemination of advanced scientific and technological knowledge, with a view to strengthening links between scientific communities.

The Series is published by an international board of publishers in conjunction with the NATO Scientific Affairs Division.

A Life Sciences	Plenum Publishing Corporation
B Physics	London and New York
C Mathematical and Physical Sciences	Kluwer Academic Publishers
D Behavioural and Social Sciences	Dordrecht, Boston and London
E Applied Sciences	
F Computer and Systems Sciences	Springer-Verlag
G Ecological Sciences	Berlin Heidelberg New York Barcelona
H Cell Biology	Budapest Hong Kong London Milan
I Global Environmental Change	Paris Santa Clara Singapore Tokyo

Partnership Sub-Series

1. Disarmament Technologies	Kluwer Academic Publishers
2. Environment	Springer-Verlag / Kluwer Academic Publishers
3. High Technology	Kluwer Academic Publishers
4. Science and Technology Policy	Kluwer Academic Publishers
5. Computer Networking	Kluwer Academic Publishers

The Partnership Sub-Series incorporates activities undertaken in collaboration with NATO's Cooperation Partners, the countries of the CIS and Central and Eastern Europe, in Priority Areas of concern to those countries.

NATO-PCO Database

The electronic index to the NATO ASI Series provides full bibliographical references (with keywords and/or abstracts) to about 50 000 contributions from international scientists published in all sections of the NATO ASI Series. Access to the NATO-PCO Database compiled by the NATO Publication Coordination Office is possible in two ways:

- via online FILE 128 (NATO-PCO DATABASE) hosted by ESRIN, Via Galileo Galilei, I-00044 Frascati, Italy.

- via CD-ROM "NATO Science & Technology Disk" with user-friendly retrieval software in English, French and German (© WTV GmbH and DATAWARE Technologies Inc. 1992).

The CD-ROM can be ordered through any member of the Board of Publishers or through NATO-PCO, B-3090 Overijse, Belgium.

Series F: Computer and Systems Sciences, Vol. 156

The NATO ASl Series F Special Programme on
ADVANCED EDUCATIONAL TECHNOLOGY

This book contains the proceedings of a NATO Advanced Research Workshop held within the activities of the NATO Special Programme on Advanced Educational Technology, running from 1988 to 1993 under the auspices of the NATO Science Committee. The books published so far in the Special Programme are listed briefly, as well as in detail together with other volumes in NATO ASI Series F, at the end of this volume.

Springer

Berlin
Heidelberg
New York
Barcelona
Budapest
Hong Kong
London
Milan
Paris
Santa Clara
Singapore
Tokyo

Microcomputer-Based Labs: Educational Research and Standards

Edited by

Robert F. Tinker

The Concord Consortium
Educational Technology Lab
37 Thoreau Street, Concord, MA 01742, USA

Springer

Published in cooperation with NATO Scientific Affairs Division

Proceedings of the NATO Advanced Research Workshop on
Microcomputer-Based Labs: Educational Research and Standards,
held in Amsterdam, The Netherlands, November 9–13, 1992

Library of Congress Cataloging-in-Publication Data

Microcomputer-based labs : educational research and standards / edited
by Robert F. Tinker.
        p.   cm. -- (NATO ASI series. Series F, Computer and systems
sciences ; vol. 156)
    "Proceedings of the NATO Advanced Research Workshop on
Microcomputer-Based Labs: Educational Research and Standards, held
in Amsterdam, the Netherlands, November 9-13, 1992"--
    Includes bibliographical references and index.
    ISBN-13: 978-3-642-64740-6   e-ISBN-13: 978-3-642-61189-6
    DOI: 10.1007/978-3-642-61189-6

    1. Laboratories--Study and teaching--Data processing.   2. Science-
-Study and teaching--Data processing.   3. Physics--Study and
teaching--Data processing.   4. Microcomputers.   5. Education--Data
processing.   I. Tinker, Robert.   II. NATO Advanced Research
Workshop on Microcomputer-Based Labs: Educational Research and
Standards (1992 : Amsterdam, Netherlands)   III. Series: NATO ASI
series.   Series F, Computer and systems sciences ; no. 156.
Q183.A1M53   1996
507'.8--dc20
                                                              96-30575
                                                                   CIP

CR Subject Classification (1991): K.3

ISBN-13: 978-3-642-64740-6 Springer-Verlag Berlin Heidelberg New York

© Springer-Verlag Berlin Heidelberg 1996
Softcover reprint of the hardcover 1st edition

Typesetting: Camera-ready by editor

SPIN: 10543911   45/3142 – 5 4 3 2 1 0

Preface

Microcomputer-based labs, the use of real-time data capture and display in teaching, represent one of the most valuable innovations microcomputers have contributed to science teaching. This technology gives the learner new possibilities to explore and understand the world and to see it represented symbolically in ways that greatly increase comprehension. As this book chronicles, the quarter-century, international effort to develop and understand microcomputer-based labs (MBL) has resulted in a rich array of innovative implementations and some of the most convincing evidence for the value of computers to improve learning.

This book is a sampler of work in MBL by an outstanding international group of scientists and educators based on papers they presented at a NATO-sponsored seminar held in Amsterdam in 1992. Since the seminar, the papers have been extensively edited and updated, so that this volume provides a unique, international view of the current state of this field. We did not try to include every possible project, interface, application, and curriculum. Instead, this volume is representative of the range of educational applications of MBL, including some very commonplace, widely-adopted applications as well as some speculative or specialized ones.

In addition to the obvious importance of MBL itself, the story told here contributes valuable policy lessons about the development and spread of educational innovations in general. The history of the way MBL has developed, the people involved, the time required, and the role of funders all contain lessons that are important for anyone who hopes to understand and contribute to the development of educational innovations.

The NATO seminar, "Microcomputer Based Labs: Educational Research and Standards," was held November 9–13, 1992, at the University of Amsterdam, the Netherlands, with funding from the NATO Advanced Research Workshops program (grant AET ARW 920492) and the University of Amsterdam. It was one of over thirty NATO seminars focused on advanced educational technologies that was managed by Luis V. da Cunha whose excellent support of this series is greatly appreciated.

The editor owes a special debt to A.L. Ellermeijer who was the host in Amsterdam and co-director of the project. His organizational skills, dedication to improved education through MBL, and good humor is without equal. Boris Berenfeld and Ard Hartsuijker, who filled out the seminar organizing committee, made tremendous contributions to the successful completion of this project. Any omissions and errors in this volume cannot be traced to the dedicated members of the organizing committee who provided continuing assistance and advice; they can only be my responsibility.

The Concord Consortium   Robert F. Tinker
Concord, Massachusetts, USA
September 1996

Table of Contents

Part I: Overview

Part II: Research

Introduction

Origins of Innovation

The idea of computers in the lab is nothing new to scientists. Almost all research has come to rely on computers to control experiments, to collect data, to represent the data in visual forms such as graphs, and to extract trends and summary information from the sometimes huge amounts of data. In addition, the same computers are often used to explore mathematical models and compare these models to the observed data. The importance of doing this in the lab as the experiment is underway — in "real time" — has obvious advantages for repeating suspect results or quickly following up unexpected leads. In fact, the first personal computer may have been the 1960's era Digital Equipment Corporation's PDP-8L, where the "L" indicates that it was optimized for laboratory use in recognition of the importance of having real-time data acquisition and analysis capacity in the lab. Many of the first users of this capacity were experimental physicists who tended to be quite comfortable with the use of electronics, sensors, and instruments in their research.

It is no surprise, then, that experimental physicists interested in education were familiar with the advantages of real-time data acquisition and, as inexpensive micro-computers appeared in the mid-1970s, immediately saw that these advantages could be brought to education. In the 1975–85 decade, this realization seems to have occurred independently in at least England, the Netherlands, Germany, and the United States. In each country, this early work resulted in a strand of research and development represented in this volume. The book does over-represent MBL applications in physics and hardly mentions the existence of a range of applications to education in the other sciences, mathematics, or technology. This somewhat accidental bias reflects the fact that physicists were instrumental in the development of MBL so applications outside physics have almost always been later to be developed and studied.

About 1975, the 6502 microprocessor was developed by a small company called MOS Technology. This was not the first 8-bit microprocessor, but it was superior in many respects to the Intel 8080 then dominant. To market their innovation to engineers, MOS Technology created the KIM, which consisted of the 6502 on a board together with a keypad, six digits of display, a 1K assembler in ROM, and 1K of

RAM memory. My team at TERC[1], then developing an NSF-funded college-level instrumentation course for physics students, built a simple 8-bit analog-to-digital and digital-to-analog interface for the KIM and applied it to a number of MBL experiments. One program measured temperature and displayed the temperature history on an oscilloscope. Another, which needed another 1K RAM on the interface, recorded sound and displayed the Fourier transform of the sound. It was clear then that this inexpensive computer could be a universal instrument with broad applications to education. Over the next decade, with the help of a group from the American Association of Physics Teachers, a selection of KIM-based applications like these were incorporated into a workshop for teachers that reached over four thousand physics teachers, mostly in colleges.

It was the development of the 6502-based Apple II computer and funding to TERC from the National Science Foundation starting in 1984 that led to the first significant pre-college use of MBL in the United States. TERC at that time was a place where progressive educators, academic psychologists, experienced project staff, college faculty, scientists, and recent college graduates worked together to address educational problems. During that grant, Jim Pengra, a college physics teacher on sabbatical at TERC, at my suggestion, first interfaced to an Apple II an ultrasonic detector developed for the focusing system of a Polaroid camera, and wrote software to display the real-time position, velocity, and acceleration of any object in front of the detector. This led to the development of the ultrasonic motion detector that figures so prominently in many papers in this book.

At approximately the same time, similar developments were underway in Europe. In Great Britain, there were centers associated with universities devoted to helping infuse microcomputers into education, several of which developed MBL applications for the UK school computers. In the Netherlands these developments happened within the part of the University of Amsterdam graduate physics program devoted to didactics. In Germany, MBL was also first developed at universities with strong commitment to education.

The significance of these histories for policy is that all MBL development seemed to need the involvement of scientists and educators working together outside the existing educational establishment. None of these developments happened in schools of education nor in institutions devoted primarily to curriculum development. It may be that all educational innovations that lead to fundamental change require a mix of talent not found within the educational establishment.

It is also significant that the first MBL was developed with NSF grant funding in excess of $1M that gave staff the opportunity to innovate. Even with this funding, it was an un-funded faculty member on sabbatical with the freedom to contemplate and take risks who made the most significant single contribution to this area. It would seem that the right mix of talented scientists, educators and researchers, generous funding, pressure to produce, and freedom to make mistakes is essential for significant innovation.

---

[1] At that time, TERC was known as the Technical Education Research Centers, Inc. It has subsequently simplified its name to TERC, Inc. It address is 2067 Mass. Ave., Cambridge, MA 02140.

The Spread of Acceptance

The MBL work described in this book has been underway for a quarter-century and still has limited but growing implementation in schools. Part of this long development period is related to the parallel revolution in computers that has resulted in CPU doubling in capacity every 18 months. The difference between the KIM and a modern RISC-based microcomputer is far greater than the difference between the first horseless carriages and modern cars. Still, the effect of these changes on MBL is surprisingly superficial; most of the new processing power goes into the user interface rather than the core computations. The Fourier transform on the KIM was adequate for many learning situations but required a fair bit of mumbo-jumbo to load and launch; now an application with similar performance runs easily inside an intuitive graphical user interface.

The more important determinant of the long times required for MBL adoption is the organizational inertia of schools. It is not usually advantageous to simply replace a traditional lab with an equivalent one using MBL. This kind of "substitution" policy is easiest for schools to implement, but the result of such a substitution is often a simple lab made more difficult and expensive by the inclusion of computers with no educational gain. The MBL context adds capacity and flexibility that, to be exploited, requires the lab to be reconceptualized, giving students more opportunity to explore and learn through investigations. This, in turn, often requires a change in teaching style that takes time and institutional commitment. This sounds as though education must change simply to accommodate this new technology. If this were the only reason to change, the logical conclusion would be to not use the technology. These changes, however, are exactly the ones most thoughtful educators are advocating and those central to the changes envisioned by the new standards and curriculum frameworks in most countries. Thus, the widespread, effective utilization of MBL is tied to large-scale institutional change that is very slow.

The adoption patterns in America and Europe are quite different in ways that reflect the degree of centralization of schooling. The highly decentralized nature of U.S. education results in the quick adoption of innovation by a few schools but the very slow adoption by the majority. The paper by deBeurs and Ellermeijer describes a very different adoption strategy in the Netherlands. After long and thoughtful discussions, two places were found for MBL in the curriculum used by the schools serving the top one-quarter of students. The objectives of this instruction were identified, and teams assigned to generate the hardware and software, model curriculum materials, and teacher professional development plans. Textbook vendors were shown the model curricula and encouraged to develop their own materials that fit the required curricula. At the senior level, the purpose of using MBL was to give students tools for their investigations; at the earlier grade MBL was introduced as part of informatics.

With a coordination that is at the same time simple and almost incomprehensible to Americans, the necessary computers were purchased for all the target schools, the MBL hardware and software was delivered, the new curriculum distributed, and all the teachers assigned to use the MBL were trained. The result was that by 1988 all the target Dutch schools were equipped with excellent MBL materials that were integrated into instruction and taught by well-prepared teachers in a thoughtful and flexible way. To MBL advocates in the Netherlands this process was painfully slow and there was a time before 1988 when the country was notably lagging in MBL use. But with their thorough and thoughtful implementation, they quickly became a leader. The papers by Szydlowski and Mioduszewska attest to that leadership. The advantages of this approach are obvious; its disadvantages are that it does take time and inhibits applications outside those officially sanctioned.

A key point of the U.S. history of MBL was coining the term "MBL" in 1983 for the proposal that led to that first NSF grant to TERC. The importance of adding another term to the overfilled educational lexicon was to capture, in one phrase, the technology and a progressive educational application of the technology that empowered the student to learn through investigations. A side benefit of creating that name is that the impact of that grant can be followed through the propagation of the MBL name. The grant started this process by leading to the first MBL products and, through commercialization, significant use of MBL[2]. Collaboration with the Voyage of the Mimi project led to the first commercial elementary grade uses of MBL, in grades 4–6[3]. A contract between TERC and IBM led to a new line of hardware and more integrated and advanced software[4]. The inclusion of MBL work in other grants at TERC led to the development of new probes, a field station, use of MBL with hand-held computers, and other related work that is ongoing at the Concord Consortium and many other institutions. There are now dozens of MBL products with no association to TERC except the use of the name[5], and even a popular calculator-based interface using the derivative initials "CBL." While no statistics are available, it is probably true that a majority of U.S. high school and college physics teachers use MBL, that some chemistry teachers use MBL, but that the idea is yet to have much impact on other disciplines or on younger students. Again, there are no statistics, but the situation is probably similar in other countries, with probably less MBL use outside physics because of the lack of local autonomy and the need for central planning.

The most obvious lesson we can learn from these histories from both sides of the Atlantic concerns the slow rate of fundamental change. To create something like MBL and have it widely implemented takes decades. The dissemination path is hard

---

[2] The company was HRM Software and the first product was "Experiments in Physiology" marketed in 1993. This used the Apple II game port as an interface to inexpensive temperature and light sensors.

[3] Sunburst now markets this as part of "Whales and their Environment." See http://www.nysunburst.com/ or call 1-800-321-7511.

[4] This is called the Personal Science Lab now available through Team Labs at 1-800-PSL-HELP.

[5] A sample of U.S. products lines, by no means complete would include those in the previous footnotes and the following: Vernier Software has a full range of inexpensive interfaces, software, and probes, including the CBL. <http://www.teleport.com/~vernier/> or call 503-297-5317. LOGAL has a good line. <http://server.logal.com/home.html> Tel-Atomic has Kis and Champ II interfaces with lots of probes. 1-800-622-2866. Pulse Metric has a cardiovascular monitoring package. 1-800-92-PULSE. Quantum Technology has a "LEAP" interface and materials coordinated with BSCS Biology. 1-303-674-9651. The Accu-Labs Products Group has a system they call SensorNet with a range of probes. 1-209-522-8874. PASCO has an excellent physics-oriented interface with lots probes. 1-800-772-7800.

to track, but, for MBL, started with receptive college and university faculty and early adopters, and only reached significant numbers when commercial suppliers picked up the approach and started marketing products that incorporated it. Stimulating this entire effort is the innovating group that needs to have support for the long term. In Europe this was done with the help of university departments or affiliates with modest funding from the central government. In the United States the schools of education are so divorced from content that they have not taken the lead in developments like MBL and non-profits like TERC have sprung up to fill the void. Grants to these non-profits need to be larger because they do not have the hidden subsidies of a university. It is, therefore, a major challenge in the United States to make a long-term commitment to a field like MBL, and put together long-term funding through a succession of grants, contracts, and licenses.

The Paucity of the Literature

The publication of this volume finally begins to correct a glaring paucity of literature about MBL. In spite of its importance, this is the first book about MBL and the first time the general reader can sample applications and research representing 25 years of work across the globe. For an innovation that is as important and widespread, one would expect there to be a far larger supporting academic and general literature. Major research and development efforts in MBL have gone completely unchronicled in the reviewed literature and mentioned in the general literature only in obscure newsletters or magazines.

The fact that a first book took so long appear is an interesting commentary on the people and institutions involved who are, as we have seen, at the margin of education and not very dependent on publishing in education. The nature of the work is often not perceived by the innovators as sufficiently deep or important for academic publication. To some extent this is a question of perception: to a scientist, the development of a new probe or interface, of new software, or of an innovative MBL curriculum is not a profound accomplishment and does not reach the level of originality of contribution that the scientific community would require for a reviewed article. At best, a new hardware development might rate a note in a column on lab techniques.

There is another force at work that helps account for the paucity of the literature: competition for resources. An innovative project never seems to have adequate resources; there is always another application, probe, unit, or field test that should be done. When exploring new terrain and developing new material, it is hard to stop. Furthermore, the time and costs of development are easy to underestimate. It is always difficult to allocate sufficient time to turn a functioning prototype into a stable, reliable, classroom-ready system with adequate documentation for students, teachers, and technologists. These pressures to produce conflict with the time and resources needed to reflect and publish. Given these pressures and the limited professional rewards for publication, the observed lack of publications is predictable.

The result has been the near-invisibility of MBL to many in education. Compared to education-generated innovations like collaborative student work, the MBL literature is vanishingly small. This book should begin to address this, but is only a first step. The eight papers in Part IV include, for instance, only a small sampling of the MBL hardware and software application and issues, and the seven papers in Part III just scratch the surface on ways MBL can be integrated into interesting curricula.

The Nature of Proof

Newcomers to an idea such as MBL are often skeptical and ask for proof of its value. This seems to be a reasonable request, but one that turns out, once you try to define what proof would be needed, to be meaningless. Most serious researchers understand this problem and have, therefore, avoided "racetrack" kinds of studies in favor of either case studies or well defined research of student learning in MBL-rich environments.

The first problem with a "proof" of the value of MBL is that the technology cannot, by itself, teach anything. MBL must be embedded in a curriculum, school, and social context, and any study of the successes of students in that context might be irrelevant in another context. More importantly, the MBL is only as good as its use in that context and it will only be one of many contributors to the success of a teaching experiment.

To make this point more clearly, consider two examples of MBL "failure." In one, a teacher had distributed MBL-based challenges to groups of students. As soon as any group came upon a problem that they did not understand, they raised their hands and stopped working. Within minutes, most groups in the classroom were stalled waiting for the harried teacher who could not possibly attend to all the problems at once. Somehow, this instructor had encouraged helplessness in his students and had not taught them to take risks, make mistakes, and forge ahead on their own. This was interpreted as an MBL failure because the MBL required a style of open-ended investigations for which neither the teacher nor his students were prepared.

The second "failure" involved a careful study of two classes that were to investigate the cooling curve of a substance as it went through a liquid-solid phase transition. The MBL class had a teacher who used half his class time to explain in excruciating detail every keystroke the students needed to collect MBL data for the cooling curve. In the remaining time, the students did the experiment as instructed. They proved to have learned no more than the students in the parallel class who lacked this technology but did the same experiment. Unfortunately, neither class really understood the difference between heat and temperature that this experiment was supposed to illustrate.

This second "failure" is an example of a "racetrack" study that pits an innovation against its conventional counterpart. Of course the MBL class learned no more; there was no difference in the underlying instructional design. Neither class had the opportunity to experiment, investigate, or to make mistakes; both represented terrible instructional designs.

There is an inherent contradiction in any racetrack study that uses the treatment/ control paradigm to evaluate a new technology or media like MBL that creates new opportunities to learn: either the new opportunities are avoided in order to make the two approaches similar, or the instruction in the treatment and control groups are incompatible. The second "failure" mentioned above was part of an attempt to make the treatment and control groups identical except for the presence of the MBL technology. This required throwing away most of the advantages that MBL had; the only difference between the two groups was that the data were recorded by hand in one group and by computer in the second. The researchers chose to keep everything constant except the presence of the technology. The problem with this design is that no reasonable person would expect that the mere presence of technology would have any value at all.

The alternative design would have been to create a learning situation where students used the MBL to its full advantage by exploring the cooling of several substances with and without phase changes, first making only general observations, and then following up with careful study of any aspect of cooling curves each group chooses. The MBL makes all this possible in the same time that the control students are laboriously doing one cooling curve by hand. The MBL class would certainly learn more, but would this be a "fair" comparison? Would it be fair to test the control students on the concepts behind cooling curves they had never observed? After all, the MBL curriculum was significantly different from that of the control group so that the differences observed would be due not to the MBL but to the curriculum it enabled.

Although proof of MBL's superiority is meaningless, there are a number of strategies for demonstrating the value of MBL. One is to develop instructional packages that use MBL that are superior to other approaches as Ron Thornton, Pricilla Laws, and David Sokoloff have as described in their papers. Ron has refined an approach to teaching the concepts of classical dynamics that uses MBL with an ultrasonic motion detector and force probe that is simply better than any other instructional strategy he has found. He demonstrates that the scores of students on a test he has developed improve much more when they go through his MBL-based instructional package than they improve using any other instructional strategy. Pricilla Laws has developed an alternative approach to dynamics designed to overcome the persistent problems students encounter when trying to understand and apply Newton's Third Law. She has developed a new sequence of experiments that rely heavily on measuring force and motion using MBL. Her results show dramatic reductions in student errors.

The work of Thornton and Laws may appear to be a proof of the value of MBL, but it is really a proof of the value of particular instructional strategies in classical mechanics, which happen to feature MBL, as measured by tests that may favor the MBL treatment. Their work does suggest that other superior instructional packages might be devised that rely heavily on MBL. This is what David Sokoloff has done in electricity and needs to be extended to other topics. Each new strategy needs to be carefully developed and tested against a measure of student progress that tests knowledge that is generally accepted as valuable as Ron has done in dynamics.

Another strategy for exploring the value of MBL is to develop excellent MBL-based instructional situations and carefully examine the resulting learning. This is the approach reported by Marcia Linn and Ricardo Nemirovsky in their papers. These and similar studies help us understand the role of MBL in learning without attempting to say whether MBL is essential or better in some sense. Those looking for a definitive answer to whether they should invest in MBL may find this approach wanting, but it is valuable because it helps us understand why MBL can facilitate deep learning in certain situations.

Contrary to the impression one might get from the excellent collection of educational research included in this book, the MBL area has suffered from a paucity of research that has some of the same roots as the paucity of literature noted above. The professional orientation of the developers and the pressure to complete underfunded projects have both contributed. In addition, there is another problem: many of the professionals who do educational research are divorced from evolving technologies. All the research reported in this book has required specialized software or hardware adapted to the needs of the researchers. Because few educational researchers have the capacity or orientation to undertake this technical work, far less research has been undertaken that in comparable fields. This would seem to indicate that a closer collaboration is required between traditional educational researcher and the developers of technology-rich fields like MBL.

Standards

When this conference was conceived, we hoped that it would provide an international forum where standards for hardware and software would be developed that would simplify the interconnection of probes, interfaces, computers, and software and thereby speed the development of the field and reduce costs to schools. The papers by Wilson and McFarlane address this important dream. McFarlane has had real success in gaining acceptance within the UK of MBL standards.

Except for her success, however, the attempt to establish international MBL standards seems doomed. The reality is that it takes a huge, dynamic market to drive the adoption of standards; a market that is largely lacking in education. The standards that are evolving for MBL are mostly those that are developing for other markets and then are adopted. A perfect example is provided by the RS-232 serial interface standard. While not a good standard for high-speed data, it is increasingly used for MBL applications because it is there, enforced by far larger market needs than MBL. For data transfer, MBL is increasingly maintaining compatibility with spreadsheets, for the same reasons. For MBL software modules, educators will not create their own optimized standards, they will use applets and the inter-process communications protocols that gain market acceptance.

Fusion of Educational Technologies

Many of the authors in this volume make the point that MBL does not exist in a vacuum. Educators willing to make a commitment to information technologies and willing to change instruction to accommodate them would be well advised to consider a suite of innovations that might include MBL as well as modeling, video, and multimedia instruction.

Modeling is frequently mentioned as a natural partner to MBL. The term "modeling" means quite different things to different authors. The general idea is to generate data based on theoretical models that can be compared to the observed data. This is important because students learn more when they commit themselves to a prediction prior to making an observation; if the prediction is right they are elated and if not, they are ready to try to understand why. It could well be that learning is directly tied to the richness of the interplay between student predictions and observations. To the extent that students understand models used to predict the results of an experiment, their learning can be enhanced by modeling.

For several authors, including de Beurs and Ellermeijer, the point of comparison between theory and data is in the form of a graph; the model and the experiment both generate graphs that can be compared. This is why student understanding of graphs is so essential and why we have included Roger's paper on this that has nothing explicitly to do with MBL. This is also why Nemirovsky's paper about student understanding of an experiment through graphs is important. Heuer reminds us, however, that graphs are not the only way to compare theory and experiment, and that other graphic constructs may be more valuable. Indeed, our imagination about how to make these comparisons is probably too constrained by our pre-computer experience.

Fuller makes a convincing argument for an alternate form of data input, using digitized video images. The video camera, when coupled with a digitizer and software that helps students extract data from the resulting images, is a very general and attractive way to collect data. This technology was impossibly expensive only a few years ago but is now built into many mass-marketed computers. Ellermeijer, Handheer, and Molenar make the point that as computers increasingly have video capacity, MBL should be integrated into multimedia instructional material.

The thrust of all these perspectives is that MBL is simply one valuable tool that needs to be part of the course designer's repertoire, and that its value in enhanced when used appropriately in conjunction with other tools. The future points toward a fusion of all these techniques in service of learning. The research problems we discussed in respect to demonstrating the value of MBL are even more severe when these complex mixtures are evaluated. Since the possible combinations of technologies are endless, we will have to rely increasingly on the intuition of experienced observers to select the best mixture of instructional strategies for any particular instructional task. Research can help guide in these choices and can evaluate whether the result is, in fact, effective learning.

The dream is that the coming fusion of technologies will involve a range of approaches that includes MBL and that are crafted by experienced educators into educational experiences that are extraordinary and far better able that current instruction to unleash the intellectual potential of all students. As computers become smaller, networking more ubiquitous, and lab interfaces more powerful, there will be far more possibilities for rich, experience-based learning taking place in a variety of contexts. The challenge is for educators to exploit these possibilities and use them to support fundamental changes in education that will vastly increase what all students learn.

Part I

Overview

1. From Separation to Partnership in Science Education: Students, Laboratories, and the Curriculum

Marcia C. Linn

University of California

Abstract. This paper examines the role of the science laboratory in science learning. By examining historical views of students, laboratories, and the curriculum, it describes growing understanding of the context in which science laboratories are likely to be effective.

Historically, those concerned with laboratories in science have gone from "separation" to "partnership." Separation characterizes early interest in science education because the various individuals concerned with science education worked separately. Initially, curriculum materials were developed primarily by natural scientists. For example, Millikan (1906) wrote a precollege physics textbook. At the same time, precollege educators who utilized those textbooks had little interaction with those who created the textbooks. Laboratories played many roles, ranging from vocational to motivational.

The period starting in the 1950s is characterized by the interaction between those concerned with science education, especially natural scientists and precollege professionals. Recently, a number of partnerships have been formed that involve experts in all areas concerned with science education. These partnerships are particularly apparent in efforts to incorporate Microcomputer Based Laboratories (MBLs) into the curriculum. Modern partnerships typically involve experts in natural science, experts in technology, expert precollege professionals, and leaders in pedagogy.

Initially, researchers compared laboratories to demonstrations and, for many goals, found no advantage for student-conducted investigations. In the 1960s natural scientists, inspired by Bruner and Piaget, designed laboratories to engage students in active learning. Students could conduct experiments like scientists. These experiences motivated students but did not necessarily contribute to understanding of science. Recently, MBLs have added the tools of scientists to the laboratory. Projects involving partnerships with science teachers, cognitive scientists, natural scientists, and technology experts have designed laboratories that engage students in knowledge integration. In partnership projects the goal of emulating the experiences of research scientists by providing a science laboratory is often realized. Students participate in a community of investigators, use powerful scientific tools, and investigate problems of their own choosing.

The trend from separation to partnership has been gradual. Naturally, there are examples of interactions and partnerships throughout the history of science education. Yet the predominant early theme was separation, and an emerging theme is partnership. The separation period continued to about 1950. The interaction period predominated from 1950 to 1975. The partnership period began to emerge in the late 1970s (for further discussion, see Linn, Songer, & Eylon, in press).

Although these trends are apparent in many science topic areas, this paper takes examples from physical science, since MBLs have been used most extensively there. Many of the comments and examples from the physical sciences apply to other sciences, although unique aspects of other sciences also deserve scrutiny.

1.1 Overview

To understand the role of the laboratory in science education, this paper will examine the historical ideas about (a) the nature of the student, (b) the role of the science laboratory, and (c) the focus of the science curriculum. The main themes are identified in this section and elaborated in subsequent sections.

1.1.1 The Science Student

Historically, more and more attention has been paid to the nature of the science student. Questions include: What motivates students to learn science? Which students are best suited to science understanding? How do students acquire science knowledge and what factors contribute to students changing their ideas about scientific phenomena?

Recently researchers have focused on how information about science is represented and linked by students. Rather than coordinating, linking, and integrating their ideas, these studies show that students tend to isolate, memorize, and mindlessly apply narrow scientific concepts.

1.1.2 The Science Laboratory

The science laboratory has taken a number of different roles in the history of science education. Its prominence and importance has varied historically. Science laboratories were instituted to provide vocational skills or demonstrate techniques. During the 1950s there was great interest in motivating students to participate in science courses. During this period, the laboratory was viewed as an opportunity to engage students in the activities of scientists. Laboratories were designed to provide students with firsthand evidence of the interest and excitement of science. More recently, those examining science education have questioned whether hands-on science laboratories do communicate the excitement of science. For example, some have complained that hands-on science is not always minds-on science: students may engage in laboratory procedures without thinking or reflecting about what they have done.

Under these circumstances, laboratories can be viewed as mundane rather than motivating. Students do not learn what makes science exciting for scientists by following cookbook laboratory procedures. MBLs can provide opportunities for more autonomous, independent, and exciting scientific investigation. Furthermore, MBLs provide students with one of the tools of expert scientists and allow students to engage in activities that resemble those in active scientific laboratories. Whether these opportunities are realized will be addressed in the section on the laboratory.

1.1.3 The Science Curriculum

What is the role of the science curriculum in science instruction? The primary purpose of the curriculum has been to set the goals for the science class and define the role of the teacher. Often textbooks list the goals for each section of the course as well as the overall objectives for science instruction. Commonly, the curriculum specifies both the subject matter to be learned and the role of the teacher in communicating that information.

Historically, the teacher has been seen as the authority or the source of scientific information in the classroom. More recently, many have called for a renegotiation of the authority structure in the science classroom. Frequently, innovative science curriculum materials emphasize that the teacher should move from the foreground of the classroom to the background, serving not as a source of information, but as a coach or tutor for students as they construct their own views of scientific phenomena (Collins & Brown, 1988; Collins, Brown, & Holum, 1991). The teacher guides students and arranges conditions to help students make sense of scientific phenomena. Teachers, in this view, do not dictate the structure, content, or organization of students' knowledge.

This trend reflects a general belief that students acquire scientific understanding as a result of a constructive process rather than as a result of absorbing or accumulating the information presented in a science classroom. More importantly, however, this change reflects a new view of the nature of scientific knowledge. Rather than viewing and communicating that scientific knowledge is primarily established and available in science textbooks, this approach emphasizes the dynamic nature of science knowledge. Students are invited to participate in the process of understanding scientific phenomena, not the process of learning the scientific ideas that have been developed by others. The MBL serves as a catalyst for change since many believe that this technological tool facilitates students' contributions to scientific understanding.

Thus, historically, the view of the student, the role of the science laboratory, and the goals of the science curriculum have changed. The impact of MBLs is best understood in this context. The next three sections expand on these themes and chart promising directions for MBLs. The final section provides some links and some connections among these related topics.

1.2 Developing Views of the Science Learner

During the three periods identified in this paper, a change in the view of the science learner can be identified. During the separation period, the pedagogical perspective of behaviorism reigned. During the interaction period, the information processing perspective was predominant, and in the most recent, partnership, period, a constructivist view of the learner has taken hold. The shift from behaviorism to constructivism has accompanied a growing realization that learning of science must be examined separately from, say, learning of Spanish.

1.2.1 Separation Period: General Principles

The separation period featured experts in learning who sought general principles that they believed would apply to all topics, including science. Details of science learning were left to those designing the science curriculum and the science laboratory.

Both general laws of learning and general human abilities were characterized during this period (see Linn et al., in press, for details). This search for fundamental principles coincided with the discovery of basic laws governing matter in the physical sciences. In the area of learning, psychologists including Watson (1913) and Thorndike (1910) identified laws to explain learning across a broad range of situations. These laws were thought to apply universally: to rats, pigeons, and humans (Calfee, 1981). Behaviorist theory asserted that learning was governed by the process of strengthening bonds between stimuli and responses. To investigate these bonds, researchers studied basic learning activities such as memorizing nonsense syllables or acquiring skill in navigating mazes. One implication of behaviorist theory was the emphasis on identifying goals for instruction in terms of "behavioral objectives." The behaviorist tradition focused educators on describing the outcomes of instruction.

Behavioral objectives identified desired outcomes but often the paths to achieving those objectives were not only unspecified, but in fact, unknown. The relationship between the mechanism of strengthening bonds between stimuli and responses and accomplishing more complex objectives such as "perform an experiment" was not clear.

At the same time, during the separation period, the factors that disposed learners to acquire scientific understanding were also conceptualized in general terms. General ability, interest in science, motivation to learn science, and mathematical and spatial abilities were predominant in the view of the learner. As a result, those who learned science were assumed to be those with the aptitude to respond to the existing curriculum.

During the separation period, student motivation to study science was generally assumed to be innate. The behaviorists viewed motivation as stemming from intrinsic interest in novelty that could be shaped by experience. Providing experiences in science courses was deemed sufficient to motivate students to participate in science. Furthermore, the view of the learner as governed by characteristics such as general ability led to the reasonable conclusion that those who would have interest in science would display it. The more complicated view that emerged later placed some responsibility for attracting students to science on those who designed the materials for teaching science.

Placing the responsibility for learning science primarily on the learner became more and more problematic as time went on. By the 1950s there was concern that insufficient numbers of students were entering scientific fields. Educators sought methods for attracting students to science. In addition, individuals who had been excluded from scientific pursuits, including women, became more vocal in their desires to be represented (Rossiter, 1982). Both of these trends spurred some rethinking of the view of the learner. This rethinking is reflected in the interaction period.

1.2.2 Interaction Period: Information Processing, Development, and Problem Solving

By the 1950s researchers were looking more microscopically at the process of learning. The *information processing perspective* alerted science educators to the complexity of many scientific ideas and helped clarify the difficulties that students face in learning advanced scientific material. The *developmental perspective* motivated those in science education to pay more attention to students' reasoning about the natural world. The *problem-solving perspective* led to a separation of general strategies and discipline-specific knowledge.

Information Processing. Those taking the information-processing perspective studied how students responded to information presented in small laboratory experiments. Information-processing studies of learning took place primarily in laboratories rather than in school settings. These laboratory studies alerted science educators to important considerations. In particular, well-documented constraints on the amount of information that students could process at once and on the processes of taking information from short-term to long-term memory raised awareness of some of the reasons that students were not making progress in science. Information processing theorists analyzed memory processes. Many conjectures concerning the nature of memory were investigated. Research on memory capacity revealed limitations on information processing capacity (Miller, 1956). Besides constraints on memory processes, researchers during this period also became interested in the work of a Swiss psychologist, Jean Piaget, who was describing general constraints on the reasoning processes of learners. Thus, both developmental psychologists and learning theorists were exploring constraints on the learner.

Curriculum materials were reviewed to reduce information-processing demands. Building on Thorndike's (1910) word lists, science educators sought to simplify vocabulary and provide opportunities to learn science concepts.

At the same time, researchers from the information-processing perspective were looking more microscopically at the behavior of students. Task analyses helped clarify difficulties in learning science and clarified the skills science students need. For example, the Science A Process Approach (SAPA) materials were designed with this hypothesis in mind. Serious problems arose when trying to make these materials accessible to the processing capabilities of students. The divide-and-conquer approach to teaching information-processing skills had the undesired consequence that students were often unable to put together the separate pieces that they acquired. Just as vocabulary is not a sufficient part of the science curriculum, isolated concepts are not sufficient for understanding science topics.

Thus, the response to the realization that the information-processing demands of science were great was to use a divide-and-conquer strategy. Students were helped to learn vocabulary separately from science concepts. It was assumed that if all the separate information that students needed to know was presented in an educationally accessible fashion, then somehow students would put the information together. As will be seen in the partnership period, this hope did not become a reality. Instead, science courses were often reduced to drill on vocabulary, and students gained isolated bits of science knowledge.

Reasoning Processes. During this period, in addition, more serious attention was paid to how students developed understanding of scientific material. Piaget's developmental theory argued that students simply could not reason abstractly until they had a large amount of concrete experience. Often, curriculum developers interpreted the need for concrete experience as the need for laboratories in science courses. Thus, the concrete experience provided to students was frequently that of the laboratory experience, as elaborated in the next section.

Piaget studied how students spontaneously developed understanding, not how they learned from instruction. American researchers expanded the theory and carried out a broad range of investigations to determine how development played out in science education courses (Lawson, 1985). There was widespread agreement in the science education community that science courses were too difficult because students were not ready to learn the material available. Thus, innovative materials were developed, more attuned to the intellectual characteristics of students. A prime example is the Science Curriculum Improvement Study. This curriculum featured both hands-on learning experiences and a learning cycle intended to build on the concrete experiences known to be beneficial to students and to help students consolidate those experiences into more abstract ideas about science (Karplus, 1975).

Interaction between science educators and developmental psychologists led to a much more careful analysis of how students learn scientific ideas. By attempting to understand developmental accomplishments, science educators spent time interviewing students and asking them to explain scientific phenomena. These interviews were enlightening. Often students' ideas about scientific events were quite different from those anticipated by educators. During this period, the roots of research on conceptual development in science emerged.

In particular, many researchers discovered that students developed ideas based far more on their observations and perceptions of the world than had first been anticipated. Students frequently developed ideas that reflected their experiences. This general observation became a primary tenet of the partnership period and also illustrated the limitations of developmental influences.

Problem Solving. During the interaction period, problem solving was studied primarily using tasks involving straightforward rules such as Tower of Hanoi. As a result, many described problem solving as the process of strategy acquisition. Furthermore, there was a tendency to distinguish between general problem-solving skills and specific strategies for individual tasks. It was argued that students needed some general problem-solving strategies, such as divide and conquer, and, in addition, some specific strategies to do Tower of Hanoi. This distinction between abstract problem-solving skills or critical-thinking skills and specific task-dependent strategies was consistent with Piagetian theory. On the other hand, this approach suggested that educators might primarily teach critical thinking or general problem solving and assume that students would be able to apply these skills to any task they encountered. Yet, students were often unable to apply general problem-solving strategies to specific problems. This observation had already been raised in Piagetian studies where students were able to perform a particular logical strategy such as conservation or separation of variables for one task, but failed to do it for another similar task. This and other indicators suggested that problem solving was more problem specific than anticipated. In science, then, students needed to learn to solve science problems, not general problems.

Just as information processing and developmental influences on the learner were being identified, much more complicated scientific ideas were being added to the curriculum. Modern physics, with its focus on atomic models and quantitative solutions to problems, replaced much of the descriptive material that had characterized physics instruction in the past. These two trends worked against each other during the interaction period.

Motivating students to participate in science was also viewed as somewhat separate from making science instruction effective. Thus, curriculum developers believed that motivation resulted from participating in science activities similar to those of scientists. In the end, these efforts were unsuccessful. The number of students participating in science increased slightly, and the percentage of participants decreased (Welch, 1979).

In summary, during this interaction period, the learner was examined more carefully. Researchers studied more complex tasks than had been studied in the past and looked more precisely at the actual behaviors of students. Techniques such as task analysis were used to analyze learning at a more precise level than had been the case in the past.

1.2.3 Partnership Period: Cognitive Theory

During the partnership period, those concerned with science education developed a "cognitive" perspective on learning, building on the information-processing, developmental, and problem-solving research. This view argues that learners make sense of the world by weighing alternatives, sorting out ideas, building on their observations, and generally engaging in a process of knowledge construction. The partnership projects studied the thinking processes engaged in by students and experts. New methods such as protocol analysis were developed to accomplish this task.

Groups jointly contributed ideas and drew on expertise from a broad range of disciplines to make sense of scientific understanding. In particular, discipline experts, technology experts, precollege professionals, and developmental psychologists were more and more likely to form collaborative groups. Unlike the interaction period, where scientists led the curriculum reform effort, during the partnership period leadership was more evenly distributed across groups concerned with science learning.

Thus, the partnership projects view learners as responsible for their own learning, rather than as absorbing information. They seek to help learners sort out experiences, reconcile alternative explanations, and integrate diverse observations. Researchers study more complex problems than had been investigated in the past. A major issue during the partnership period concerns whether logical reasoning strategies are independent of disciplinary knowledge or embedded in disciplinary problem solving. Spurred by evidence that students construct quite similar views of diverse scientific phenomena, researchers have become more and more convinced that reasoning is associated with disciplinary knowledge rather than independent (Eylon & Linn, 1988; Linn, 1986; West, Pines, & Sutton, 1984).

Evidence for this view comes from studies that reveal that students construct descriptive or phenomenological models of scientific phenomena that differ substantially from those of experts. At first, these views were labeled "misconceptions" because they departed from modern scientific views (Champagne, Klopfer, & Gunstone, 1982). Others saw evidence of reasoning that scientists typically engage in to make sense of observations even though the observations themselves differed from currently accepted scientific views (Linn & Songer, 1991). This distinction between what might be called "misconceptions" and what might be called "intuitive ideas" has persisted and fuels debate within the field. Careful examination of the processes that students use to achieve scientific understanding suggests that both views have some validity. In some cases, students adopt quite superficial strategies for making sense of scientific phenomena. In those situations, ideas are called "action" knowledge by Linn and Songer (1991) and might be labeled as misconceptions. In other cases, students struggle to integrate and synthesize their conjectures into what could be called intuitive theories.

Partnership period research suggests that the cognitive processes that govern student knowledge construction take the form of reflection, integration, and organization. Instead of stating the outcome of these processes, those investigating student understanding are examining the process leading up to the views that students develop. These processes differ from student to student depending on their dispositions to learn science. Some students indicate that their process for learning scientific information consists of memorization, and others describe activities leading to knowledge integration and reconciliation of diverse experiences. Researchers continue to investigate this cognitive perspective on student understanding, extending the debate and discussion to examination of the social construction of scientific knowledge (Eckert, 1990; Lave & Wenger, 1991; Linn & Songer, 1991).

During the interaction period, the focus on processes of memory motivated researchers to view the learner as governed by external forces and constrained by internal processes. In contrast, during the partnership period, learners' abilities to monitor and control their learning are emphasized. Processes involved in self-monitoring and reasoning about one's own reasoning are seen as controlling the application of information-processing capabilities. Thus, learners are viewed as actively involved in the process of knowledge construction rather than as primarily constrained by memory capacity. The learner is capable of recognizing overload and able to engage in activities that reduce the memory demands of a particular situation. In addition, the instructor has the responsibility to guide this process. Thus, during the partnership period, instruction helping students develop self-monitoring processes is central (Collins et al., 1991).

In summary, the partnership projects focus on understanding within a disciplinary context. Processes of self-monitoring and meta-reasoning are studied within a discipline, rather than across disciplines. The boundaries between disciplines remain unclear. To contextualize this perspective, an example from the Computer as Learning Partner project is described.

1.3 An Example: The Computer as Learning Partner

The Computer as Learning Partner (CLP) collaborative group developed during the partnership period to take advantage of MBLs. Collaborators include experts in pedagogy, experts in classroom teaching, experts in technology, experts in physical science, and experts in cognition. They seek new approaches for teaching science and for characterizing how students come to understand science. The group is jointly constructing a cognitive view of effective science learning by building on experiences from the past, analyzing experimental teaching, examining student learning, and drawing on experiences of other research projects.

A major focus of the CLP project is to help students integrate the science topics taught in science courses with common, everyday experiences. Students growing up in America have a large number of similar experiences that they draw on to make sense of scientific phenomena. These similar experiences shape descriptive or phenomenological models of science that are commonly held by students (e.g., diSessa, 1983; diSessa, 1988; Eylon & Linn, in press).

The CLP project both develops instructional materials and examines how students draw on these materials to make sense of thermodynamics. The group investigates how students make sense of thermal events, especially those that they encounter in their lives. Problems such as how to keep beverages cold for lunch, how to dress to stay warm on a cold, winter day, how to interpret experiences gained from touching wood and metal objects in the same environment, and other common, everyday, thermal events characterize the instruction.

The project develops materials to teach these skills to students using a process of trial and refinement. Initially, a semester-long science curriculum was designed and investigated. This curriculum has been reformulated and administered to new groups of students over a five-year period. A number of publications describe this trial and refinement process (Linn, 1992; Linn, Songer, Lewis, & Stern, 1993; Songer & Linn, 1991). The materials developed by the project draw heavily on computer technology. Initially, the investigators sought to take advantage of tools used by scientists for real-time data collection and dynamic display of information. These are the basic elements of MBLs. Subsequently, more powerful scientific tools such as simulation and modeling environments have also been incorporated into the MBLs. The materials support cognitive activities such as (a) predicting outcomes of scientific experiments, (b) reconciling experimental results with their predictions, (c) reflecting on the results of groups of experiments, and (d) integrating everyday and laboratory investigations (Linn & Songer, 1991).

Early in the process of developing instructional materials and investigating how they help students integrate scientific knowledge, the CLP group noted that the goals for science courses are often inappropriately matched to the knowledge, constructive processes, and experiences of students. In particular, in thermodynamics, it is common for students in the middle school years to encounter models involving molecular kinetic theory and to compute changes in temperature and calories. Computing change in calories or temperature seems to place a veil of numbers over thermal events and send the wrong message about scientific problem solving (Songer & Linn, 1991). Molecular kinetic models are simply not understood by many of the students engaged in sense making in middle school and, as a result, are frequently viewed as separate from everyday scientific phenomena. The scientific ideas in the typical course are not linked to complex everyday problems or to scientific ways of thinking. The result is that students adopt memorization as the mode for learning scientific ideas in school, think of scientists as unconcerned about practical problems, and decide that school science is not at all relevant to everyday science. To overcome these serious impediments to fostering student knowledge construction, CLP has sought to identify descriptive or *pragmatic principles* to help students understand practical scientific events as well as the nature of scientific inquiry. The pragmatic principles identified by the CLP project are shown in Figure 1.

1.3.1 Scaffolded Knowledge Integration

To help students construct understanding, culminating in these pragmatic principles, the CLP project has developed an approach called *scaffolded knowledge integration* (SKI). Scaffolded knowledge integration involves four key components. First, scaffolded knowledge integration makes the process of reasoning about science visible. Knowledge construction is made visible by providing students with models of thermal events and insight into processes of linking related ideas, reflecting on similar experiences, and identifying generalizations that explain multiple events. The process of making this reasoning visible is supported in software. For example, in Figure 2, the process of students constructing a *principle* to explain the results of an experiment is illustrated. In addition, in Figure 3, the process of linking the results of a classroom experiment to what is referred to as a *prototype* is illustrated. Prototypes

Flow Principles:

Heat Flow Principle: Heat energy flows only when there is a temperature difference.

Direction of Heat Flow Principle: Heat energy flows only from objects at higher temperature to objects at lower temperature.

Rate Principles:

Surface Area Principle: When only surface area differs, heat energy will flow faster through the larger surface area.

Mass Rate Principle: When only mass differs, the temperature of the larger mass will change more slowly. Initially, heat energy will flow at the same rate from both objects to a cooler surround.

Temperature Difference Principle: The greater the temperature difference between objects and their surround, the faster heat energy flows.

Conductivity Principle: A good conductor allows heat energy to flow faster than a poor conductor.

Material and Rate Principle: When only material differs, the temperatures of the two objects will change at different rates. Initially, heat energy will flow at the same rate from both.

Total Heat Flow Principles:

Thermal Equilibrium Principle: Eventually all objects in the same surround become the same temperature unless an object produces its own heat energy.

Mass and Total Heat Flow Principle: When only mass differs, more heat energy flows from the larger mass to a cooler surround.

Material and Total Heat Flow Principle: When only material differs, more heat energy flows from one material than the other to a cooler surround.

Integration Principle:

Heat Energy and Temperature Principle: When each part of an object has the same temperature, more heat energy will flow to a cooler surround from the whole than from each part.

Figure 1   Examples of pragmatic principles

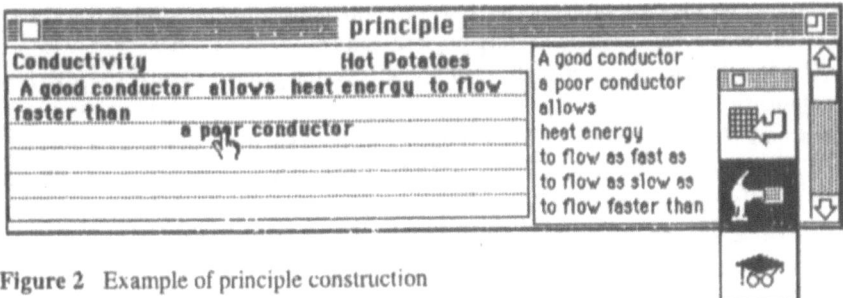

Figure 2   Example of principle construction

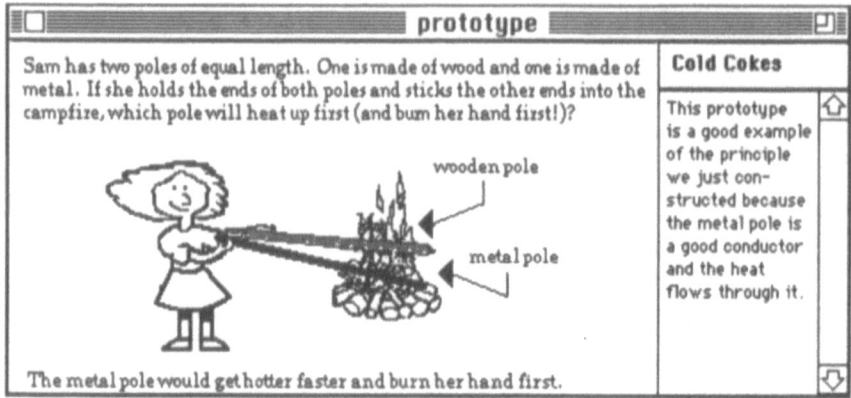

Figure 3 Example of linking ideas using a prototype

in the CLP project are situations where students have accurate predictions about the outcome of the event but lack a mechanism for explaining why the outcome occurs. These examples of activities that make the thinking process visible are described in greater detail in a number of CLP publications (Linn, 1992; Linn et al., 1993; Songer & Linn, 1991).

A second component of scaffolded knowledge integration concerns encouraging autonomous learning. This is accomplished in the CLP project in several ways. First, students critique their own and others' laboratory write-ups. Students jointly construct criteria for critiquing activities and then provide feedback to other students much as expert scientists might referee scientific papers. Second, students respond to prompts to reflect on their ideas such as "I predict this because..." or, "My results explain this everyday event because...." These and other activities balance the making thinking visible goal with opportunities for student autonomy (see also Linn, in press; Linn, diSessa, Pea, & Songer, 1994).

The third component of scaffolded knowledge integration concerns identification of appropriate goals for science courses. These goals need to be aligned with the knowledge that students have and consistent with the processing capabilities that students bring to their courses. As a result, a component of scaffolded knowledge integration for the CLP project is the focus on pragmatic principles rather than models of thermal events used by leading-edge researchers.

The fourth component of scaffolded knowledge integration involves supporting the students' learning process as much as is necessary, drawing on the social context of learning (Linn & Burbules, 1993). Students work in small groups and appropriate each others' ideas (Newman, Griffin, & Cole, 1989). They become responsible for their own learning because they try their own ideas first and seek help in making sense of phenomena, rather than seeking answers from authority figures. This component requires careful design to avoid reinforcing stereotypes (Agogino & Linn, 1992 May–June).

1.3.2 A Longitudinal Study

To investigate how students in the CLP project make sense of thermal events, a longitudinal study was carried out by Lewis (1991). Lewis identified three groups of students: converging, progressing, and oscillating students. In a replication study (Lewis & Linn, 1994), this initial study was extended. In the second investigation, a total of 30 students, randomly selected from those in the CLP course, were interviewed on five separate occasions. These eighth-grade students were asked a series of questions about four topics in thermodynamics at each interview. The topics were (a) insulation and conduction, (b) thermal equilibrium, (c) heat and temperature distinction, and (d) heat flow. The first interview focused on students' elaborations of their responses to the pretests that all students took. Example pretest questions are provided in Figure 4. More details on the interviews and the questions used in the interviews can be found in Lewis and Linn (1994).

Students' views of the four main thermal concepts were rated at each interview. Students were identified as holding one of six levels of explanations at each point in the longitudinal study. The levels and their characteristics are provided in Figure 5.

To illustrate the knowledge construction process, the characteristics of the three types of students are described: Converging, Progressing, and Oscillating. For each category of students, a single example is given with some excerpts from the actual interviews. These excerpts show how each group of students approaches the problem of constructing a view of thermal events.

1.3.3 Converging Students

Converging students comprise 15% of the total population of students who participate in the CLP project. These students expect to make sense of their prior experiences with thermal phenomena and expect to integrate those experiences with their classroom activities. They work hard to compare and contrast their observations and experiences in the natural world with the observations and experiences that occur in science instruction. These students warrant their conclusions with evidence. The students utilize evidence from experiments as well as from observations. They expect that their experiments and their observations can be reconciled. Even seemingly contradictory information is incorporated as these students attempt to form a coherent perspective. Finally, these students seek abstract explanations for the phenomena they observe. As a result, they are eager to adopt the pragmatic model for thermal events that characterizes the CLP materials.

The converging student described here is asked to discuss a variety of familiar, complex situations in the second interview. During this interview, there is already considerable evidence that she is attempting to make sense of everyday observations and classroom experiences. For example, when asked whether all the objects in a cold cabin in the snow would be the same temperature, she responds, "Because I think that everything would be room temperature, it's just that they feel different."

1. a) You want to keep a soda cold for your school lunch. What is the best thing to wrap it in?

    b) What is the main reason for your answer?

2. a) In general, are heat energy and temperature the same or different?

        (circle one)   same   different

    b) What is the main reason for their similarity or difference?

        Give an example that explains your answer.

3. a) Give an example of a good conductor. _____

    b) What makes the substance you picked a good conductor?

4. a) Give an example of a good insulator. _____

    b) What makes the substance you picked a good insulator?

5. a) A metal spoon and a wooden spoon were put into a 65°C oven for 2 hours. What do you predict their temperatures will be after two hours in the oven?

                temperature of the metal spoon _____

                temperature of the wooden spoon _____

    b) What is the main reason for your answer?

6.   If a metal plate and a Styrofoam plate are in the same room, what will happen?

    a) (check one)   _____ the metal plate will be warmer.

                           _____ the Styrofoam plate will be warmer.

                           _____ both plates will be equally warm.

    b) Give the main reason for your answer.

7.   You are shoveling snow without wearing gloves.

    a) Which shovel handle would you rather use? (circle one)

        metal-handled shovel   wooden-handled shovel   either shovel handle would be okay

    b) Fill in the blanks to make a principle that applies to these shovels:

        If two objects that differ only in material are in the same surrounding, heat energy will flow into and out of the object that is the better _____ faster than it will flow into and out
                                            insulator/conductor

        of the object that is the poorer _____.
                                            insulator/conductor

---

Figure 4 Sample questions from Fall 1989 Semester Pretest/Posttest. Print and size have been significantly reduced due to space limitations.

Level of Explanations

1 — **Intuitive Conceptions:** responses to questions and explanations consist primarily of intuitive conceptions, e.g., choice of aluminum foil to wrap a cold soda because foil holds the cold in; wool wrapped around a cold object would cause it to warm up faster than an unwrapped object.

2 — **Encoding New Facts without Explanations:** students' responses display the encoding of new factual data without an understanding of that information or the ability to apply it, e.g., after a class experiment, students become convinced that objects in the same environment are the same temperature. However, they cannot explain why they are the same temperature or why they feel different to the touch.

3 — **Mixed Predictions, Idiosyncratic Explanations:** students at this level give some correct and some incorrect predictions which were typically cued by the surface features of a problem. Their explanations appear to result from an attempt to preserve intuitive conceptions but explain experimental outcomes, e.g., a student might state that metals feel cold in a cold environment because they don't really absorb the cold, it just sits on the surface of the metal. Insulators, on the other hand, were said to absorb cold into them so that they do not feel cold to the touch.

4 — **Mixed Predictions, Explanations:** students make both correct and incorrect predictions and give explanations that are a mixture of both correct and incorrect conceptions. Incorrect explanations are usually tied to initial intuitive conceptions. These mixed explanations exclude any idiosyncratic explanations.

5 — **Good Predictions, Mixed Explanations:** students make excellent predictions and give explanations that are a mixture of target conceptions, intuitive conceptions, and intermediate conceptions.

6 — **Target Conceptions:** predictions and explanations are consistent with target conceptions (pragmatic principles).

Figure 5   Level of explanations used by Lewis (1991) to analyze students' interview responses

She is grappling with her experience that metals feel cold, but that the temperature of metals and woods in the same environment is identical.

When asked a more precise question accompanied by concrete material, she elaborates her answer: The student is asked to examine a piece of wood and a piece of metal. She predicts that the two will feel differently and confirms the prediction by holding the piece of metal and the piece of wood in her hand. She responds, "It's a conduct—it conducts heat really well, and, so, like when you touch it, the heat goes out of your hand and into the block making your hand feel cold. But the wood isn't, so it doesn't feel cold." This student has an explanation for observations and recites it when asked to explain an everyday phenomenon.

Later in the interview, this student is asked to explain an everyday situation where both metal and wooden strips have been in the trunk of a car for several hours on a hot summer day. Here she is extending her ideas about thermal equilibrium to a hot environment. She says, "Well, I think that the metal would be a hotter temperature, but, then, I was thinking that one thing I was saying about the room temperature, how they're all the same, but I think it's different than the room temperature." So, she is not sure whether the thermal equilibrium idea should be applied to the trunk. The experimenter asks her why the trunk would be different from the situation where objects are all at room temperature, and she responds, "Because. Because it's like in metal, and metal, I mean, well, like the sun's shining outside, and in a room, there wouldn't be, you know. Even if it's hot, it's not like shining right on it, and I think that would make it hot." So, basically, this student is wondering about what would happen if the sun shines on a metal trunk. The interviewer helps her to distinguish these two situations by saying, "So, the sun shines on something, that's different than just being in a room." The student responds, "Because I think, I think, like the sun's shining on it could almost be like it's in the oven, and that wouldn't be room temperature." The interviewer explores this idea with a few more questions, and the student starts to apply her thermal equilibrium idea. She recognizes that everything in the trunk would probably be at the same temperature. First, she says, "First of all, because like, okay, the sun's like, if it's like shining on the trunk, that's metal. It would be hot like conducting through the metal. Then, it would go, maybe, I don't know, into the trunk making that hot. But even if, I think, well, I think it would just get really hot from the sun." Then the interviewer asks whether the metal and the wood in the trunk would be the same temperature or a different temperature. After a few exchanges, the student replies, "Oh, the heat or maybe they're both the same temperature, but when you touch the metal, it's a good conductor, so the heat goes from the metal into your hand, but it doesn't happen with the wood. So, you can't feel it as much."

Clearly, the student is grappling with thermal equilibrium as studied in science class and extending the idea to a hot environment. She has difficulty identifying the salient features in each situation. However, once she has settled on thermal equilibrium applying to the trunk, she integrates the new situation with the one that she studied. Clearly, this student's ideas are converging on the heat flow model.

1.3.4 Progressing Students

Compared to converging students, progressing students are less inclined to seek global coherence of ideas. Often these students are satisfied with local coherence and these students are prone to construct school science and non-school science views of thermal events. Nevertheless, progressing students seek local coherence, and often go on to construct more global views of thermal situations. One has the feeling in talking with progressing students that if they were given sufficient time and encouragement, they would achieve the same understanding as converging students. Various factors stand in the way of this accomplishment. In particular, progressing students have less confidence in their own observations and experiments than do converging students. These students often rely on authority from the teacher instead of on their own experiments and observations. As a result, they are less likely to reflect on the information and more likely to accept it without making it coherent with their other ideas. In addition, progressing students seem disposed to reflect but less energetic about the reflective process than converging students. This may stem from their reliance on authority and it may also stem from their belief that their efforts at making sense of classroom experiences and everyday observations have not been successful in the past. In some isolated cases, progressing students even told interviewers about situations where they felt that classroom experiences were imposed on them without appropriate explanation.

Thus, there is a sense in which progressing students seem to be negatively impacted by previous science experiences. These students sometimes explained that when they tried to understand scientific ideas and reconcile them with observations, they were less successful in their science courses than when they memorized the material presented in class. It is likely that these students had difficulty reconciling the science model taught in class with their observations. Successful students may, in their efforts to seek coherence, actually abandon some of the precepts of molecular kinetic theory in favor of models of thermal events that are less sophisticated but more useful.

In summary, progressing students are similar to converging students in that they are disposed to integrate ideas and seek coherence. However, they differ from converging students in that they primarily focus their integration activities on local rather than global events, that they rely on authority figures for evidence as well as on their own observations and experiments, and that they often accept pronouncements from authority and textbooks without reflection. These students sometimes contort the information in textbooks to align it with their own observations or evidence.

It is these progressing students whose efforts at knowledge construction may be underappreciated in science courses. These students often construct views of scientific phenomena that differ from those held by experts. They may interpret criticism of their ideas as being aimed at the process they use to construct the ideas rather than as being aimed at the outcome of their constructive process. Thus, feedback provided to these students needs to carefully differentiate between the methods that they use to gain scientific ideas and the actual ideas that they construct.

A progressing student is chosen to illustrate this process. This student is responding to questions about thermal equilibrium similar to the questions that were asked of the converging student described previously. The student is asked to describe how a metal stove filled with wood would feel before the stove was lit. The student is not sure whether the metal stove and the wood would be the same temperature, although he is sure that they would feel differently. The student's first explanation is to admit that he really does not know what the temperatures would be. After some probing, this student says, "Because like Mr. K said, all of it. It's just like room temperature. It's been in the same place, it's been in the same place for the same time." Thus, the student is stating the thermal equilibrium principle but rather than basing his confidence in this principle on his own observations and experiences, he is basing his confidence in the principle on the authority of the teacher (Mr. K). The student is seeking local coherence about thermal equilibrium and also local coherence about how metals feel and how wood feels. Reconciling these two coherent views is difficult and the student relies on the teacher's authority rather than on a reflective integration of these two ideas.

This process of partial integration continues. At the fourth interview, the student is asked to explain the relationship between surface area and temperature for mashed potatoes. He is contrasting the situation where the mashed potatoes are in a heap and where they are spread out on a plate. The student reports, "Well, the more surface area, the more heat energy." Here, he is clarifying the relationship between surface area and heat energy transmission. The student justifies this generalization by citing evidence from a laboratory where surface area was investigated. However, this information is not completely integrated. The interviewer asks him what principle explains the situation in the mashed potatoes experiment, and he says, "Heat energy. The higher temperature always flows to the lower temperature." Thus instead of persisting to link the surface area principle back to the mashed potatoes situation, the student invokes another principle having to do with difference in temperature and applies it to the situation. This is a useful principle as well since soon after the mashed potatoes have started to cool, one heap will already have a different temperature from the other and the temperature-difference principle will apply. Nevertheless, this student has not integrated the two principles in applying them to the mashed potatoes situation but rather has achieved local coherence for each principle separately.

In summary, progressing students are satisfied with local rather than global coherence. In addition, they frequently rely on authority rather than on their own observations and experiments. Furthermore, they state information about situations in a serial fashion rather than recognizing that the information would be most effective if it were integrated. As can be seen from the example described here, progressing students continue the process of knowledge integration throughout the semester, but they do not converge on all of the target heat-flow concepts identified for the course. It is likely that these students will continue to construct more powerful views of thermal events, even after they leave the course. They have appropriate strategies for doing so. Nevertheless, they also have the potential for being confused or mislead by authority figures or sources that do not provide sufficient context for science information.

1.3.5 Oscillating Students

A third pattern of knowledge construction observed in these longitudinal studies is called oscillating. Students with an oscillating perspective treat each situation they encounter separately rather than linking one situation to another. They may, when asked, attempt to create ad hoc explanations for phenomena but they do not seek to relate these ad hoc explanations to ideas that they have generated previously. Many times oscillating students indicate that they cannot explain phenomena because they do not see any links between the phenomena they are observing and previous information. Oscillating students, like progressing students, rely on authority to explain situations when they can remember pronouncements made by the teacher or other authoritative materials. Frequently, these students cite authoritative information even when it contradicts their observations, and often these students fail to see the contradictions apparent to others.

For example, one oscillating student was asked to explain an experiment where a piece of wood and a piece of metal are in the same environment. The student is asked whether these two objects would have the same temperature or not. He says, "Well, I think it's probably the same thing [as the experiments we did in class] except one of them would be ... Well, one of them would be, well, see, the—the leg of the chair would be colder than the metal that was in the trunk, but it's almost the same thing because I don't know." Then, when asked to clarify, he says, "I don't ... well, it's kind of the same thing, but different temperatures that's all." Then, when asked to explain again, he says, "Well, it seemed like they were colder, but Mr. K said it was room temperature." This student put forth three different explanations and lacked criteria for selecting one over another.

In a later interview, he is asked to explain about the difference in temperatures between a wooden board and metal bowl placed in the same oven for a long time. First, he says, "They would all get hotter." Then, when asked whether the metal bowl would be hotter than the oven, he says, "No." However, he says that the metal bowl would be hotter than the board. Then, when asked again, he changes his mind, and says, "The metal bowl would be hotter than the oven because probably the oven is at the same temperature at the same time." Since this does not seem to make any sense, the experimenter asks for further explanation, and he says, "Since the bowl has been in there for a long time, and it just keeps on getting more heat and more heat and it gets hotter and hotter." Again, the student generates a series of isolated ideas and oscillates among them.

Overall, oscillating students apply contradictory ideas to the same events and lack criteria for distinguishing their ideas. They rely on authority figures and observations to generate these ideas. They seem genuinely unable to integrate information from one situation to another.

Thus, in this longitudinal investigation during a semester-long middle-school course, three patterns of knowledge integration emerge. One pattern is characterized by a propensity toward coherence seeking. Converging students actively integrate information and attempt to find over-arching explanations. In contrast, progressing students are satisfied with local rather than global coherence. They rely on authority rather than observation or experimentation to explain many events. They are inclined toward reflection, but rely on this technique only some of the time. Oscillating students apply contradictory ideas to the same events and rarely reflect on their ideas.

Converging students comprised 10 to 15% of those in the CLP program, progressing students comprised approximately 70% of the students, and oscillating students comprised between 10 and 15% of the students. Overall, therefore, 85% of the students are making progress in understanding thermal events. Some are seeking global coherence and others are seeking local coherence. The curriculum is helping students to integrate their understanding of thermal events, and, in addition, this process of knowledge integration is clearly difficult and time consuming.

Examination of the knowledge construction process illustrates the value of scaffolded knowledge integration training in the CLP curriculum. Students need support in integrating ideas and reflecting on them. Asking students to explain why two situations with similar outcomes are the same does seem to encourage them to identify abstractions or principles. The heat-flow model and its accompanying principles are accessible to both progressing and converging students. Even oscillating students occasionally mention these principles, although their ability to use them for knowledge integration is not apparent.

The 10 to 15% of the students classified as oscillating clearly need additional training in order to benefit from the CLP curriculum. In particular, these students do not seem disposed toward integrating their ideas and, therefore, are not benefiting from the curriculum materials. It may be that the best approach for these students is to simultaneously address knowledge integration and the view of themselves as integrating ideas. Once this view is embraced, then oscillating students might be more successful in differentiating among the various conjectures that they have about particular situations.

These examples of students' knowledge construction are elaborated in Lewis (1991) and Lewis and Linn (1994). These longer reports characterize more completely the alternative processes of knowledge construction followed by students. Taken together, these analyses provide strong evidence for the importance of considering the processes that students use to make sense of science in designing instructional materials.

Careful observations of the constructive process in students supports the conjecture that instruction must help students make sense of their existing ideas and relate their existing knowledge to new ideas. Instruction that attempts to eliminate existing ideas is likely to inadvertently encourage students to adopt the oscillating perspective that was observed for a small percentage of CLP students. In the CLP project, most students seem empowered to integrate their ideas and willing to take some risks in trying to link observations and experiences with more abstract principles. In order for science materials to more generally take advantage of these investigations and to help students construct understanding, the scaffolded knowledge integration framework needs to be further elaborated.

Evidence from this longitudinal study reinforces the notion that the knowledge construction process is strongly rooted in disciplinary knowledge. A general disposition toward knowledge integration seems to distinguish converging students from the others. However, logical reasoning skills seem less important than specific evidence gained from observation and methods for making sense of evidence.

These excerpts from the longitudinal study show that students using MBLs need appropriate loci for their knowledge-construction activities. MBLs cannot succeed unless they are part of a curriculum that meets student needs. Identifying appropriate loci is a complicated and difficult task. The CLP Project is also investigating other techniques besides the pragmatic principles for helping students engage in knowledge construction. One technique, for example, is to use models of thermal events to help students organize their knowledge (Lewis, 1991; Thomas, 1993).

Next, the place of the laboratory and the role of the curriculum in fostering students' knowledge construction will be discussed. A primary question concerns how laboratories in science curricula can help progressing and oscillating students. Research suggests that converging students are likely to make sense of scientific phenomena and build robust and integrated understanding almost independent of the curriculum materials that are provided for them. Indeed, often those designing new materials are lulled into the belief that materials are effective when students typical of the converging students in these investigations succeed. Curriculum materials that are only effective for 10 to 15% of the students are not going to help students who most need effective instruction. Thus, the best way to test materials utilizing MBLs is to determine their effectiveness for students in the middle of the distribution.

1.4 The Role of the Science Laboratory

How has the role of the science laboratory changed throughout the history of science education? By examining the history of the science laboratory, it will be easier to place the MBL in perspective.

Laboratories have been used in science courses since at least the late 1800s when science became a sporadic part of the curriculum. Starting in the early 1900s, laboratories were not always included. However, one of the objectives of early science instruction was to prepare students for vocations in science. It was believed that the manual skills required in science ought to be provided in science courses. Therefore, one of the earliest goals for the science laboratory was to provide vocational skills (Raizen, 1991). This role became less important as the view of science learning changed. It was eventually relegated to a small portion of the science laboratory experiences available in schools. One can think of science teachers and material developers as constructing a view of the role of the science laboratory much as one thinks of students as constructing understanding of scientific phenomena.

1.4.1 Separation Period: Laboratory or Demonstration

During the period prior to 1950, research was conducted to determine the benefits of the science laboratory. Many students were encouraged to take science so that they could be prepared for specific careers. The laboratory was seen as a component of the science course in providing that preparation.

The science laboratory was sometimes intended to illustrate facts and principles and at other times to provide career training in science methods. For example, Cureton (Cureton, 1927; Curtis, 1931) reviewed all available textbooks, teachers' guides, and pamphlets on teaching junior-high science and found widespread support for the view that the purpose of the science laboratory was to transmit to students specific skills and methods useful either at present or in the future real life.

Researchers compared the student-conducted laboratory to the teacher-demonstration of the laboratory. A review article in 1946 reported little progress in understanding the benefits of the science laboratory (Cunningham, 1946). These comparison studies were inconclusive for a number of reasons. First, the treatments were described very generally as laboratories or demonstrations. Laboratories that required students to follow a protocol were grouped with those that allowed for some opportunity to design investigations. Teacher demonstrations that emphasized safety and technique were grouped with demonstrations that described historical trends in experimentation. Second, outcome measures consisted almost exclusively of paper and pencil achievement tests that were often poorly linked to the laboratory activities. For example, tests often measured recall of the technique and outcome, information likely to be communicated by a demonstration just as well as a student-conducted laboratory. Third, although statistical analyses available at the time lacked the power needed to infer subtle effects, they were accorded great status and often led to acceptance of reliable differences without sufficient reflection on the validity of the study.

Results of these comparison studies were contradictory. This prompted some to complain that "no very great benefit can be gained by more group studies" (Cunningham, 1924; Curtis, 1926, p. 103). Others, however, relied on the study that supported their viewpoint.

Pella (1976) summarized the history of the journal *Science Education* to reinforce this view. He reported that science educators described promising practices, science teachers described their successes, and researchers described their findings but results were not integrated. Pella complained that researchers took a disorganized and incoherent view towards the accumulation of knowledge concerning science instruction. He commented, "Ignorance, vagueness, and mere good intention are of no more virtue in science education than in any other human enterprise" (Pella, 1976, p. 439).

1.4.2 Interaction Period: Hands-on Science

Starting around 1950, there was a growing demand for more students trained in science. An effort was undertaken to increase the number of students who participated in science. This effort was spearheaded by natural scientists who, partly as a result of the Russian superiority in the race for space, partly as a result of the establishment of the National Science Foundation (NSF), and partly as a result of a general belief that this problem was best solved by natural scientists, stepped in to improve the instruction in science. The laboratory served three purposes in the reformed curriculum: (a) to motivate students, (b) to permit active learning, and (c) to illustrate the scientific method.

Primarily, the natural scientists who reformed the science curriculum in the 1950s wanted to motivate students to participate in science. Natural scientists believed students who had access to some of the tools and materials that scientists used in their own investigations would be excited. For example, Zacharias at MIT was enamored of the "air track" and the "ripple tank" as experimental apparatuses that bring science to life. Zacharias wanted students to use a ripple tank and an air track to get firsthand experience with waves and with frictionless motion. As a result, he developed science materials featuring these apparatuses. This effort was funded by the NSF and resulted in the Physical Science Study Curriculum (PSSC).

Besides using laboratory experiences to motivate students to pursue science, reformers also endorsed the developmental views of Piaget (1952) and Bruner (1966; 1968) concerning the advantages of active over passive learning.

During the interaction period, natural scientists enlisted precollege professionals to pilot test new curricular materials. Many reports on the interactions during this period, however, suggest that natural scientists were reluctant to modify materials in light of trial data. Often the reports of precollege professionals were discounted or ignored by the natural scientists. Welch explains, "Scientists were usually hesitant to accept the criticism of their 'science' from school teachers unless very convincing substantiation data were provided. More often than not, decisions on revisions were based on debates and arguments among the project staffs ... [this] explains the high difficulty level of most if not all of the newer curricula" (Welch, 1979, p. 288). The status differences between these groups precluded a natural exchange of ideas at this time.

Thus, the process of determining the role of the laboratory in science learning was fueled by a belief in the importance of active learning and as well by a need to recruit more students to science. However, pilot data concerning how students respond to science curriculum materials were not incorporated by the natural scientists who were designing the materials. The resulting materials were too difficult for most students.

An additional objective for including laboratory experiences in the science curriculum had to do with the methods of science. The natural scientists who designed new curriculum materials during the interaction period often emphasized the scientific method. One way for students to gain insight into the scientific method was to conduct experiments of their own. Of course, it is difficult to conduct experiments. In particular, mechanics, often the first topic addressed in the physics curriculum, is most straightforward in a frictionless universe. Thus air tracks were popular. Students explored idealized scientific experiences rather than experiences that students had in everyday life. This separation of laboratory experiments from students' experiences and observations of the natural world simplified experimentation but made knowledge integration more difficult (Physical Sciences Study Committee, 1960).

In addition, the goal of providing students with insight into the scientific method as a result of working in the science laboratory proved difficult to achieve. Often experiments did not go as expected or students could not apply the scientific method to more interesting and challenging problems than those arranged in the laboratory. Frequently, students gathered data that were not compatible with the theories they were investigating and had difficulty explaining the differences (Welch, 1979).

Furthermore, the scientific method as it was portrayed in the textbooks of the interaction period differed from the scientific method as it was practiced in the laboratories of scientists. Scientists frequently capitalized on serendipitous events, conducted uncontrolled experiments, and drew on observations of the natural world. In contrast, the laboratories offered in school settings neglected these modes of reasoning. As a result, many complained that the scientific laboratory presented an inaccurate view of the scientific method (Burbules & Linn, 1991).

Thus, the scientific laboratory served two possibly competing goals. One was to provide an opportunity for learners to actively engage in scientific processes and perhaps discover ideas on their own. The other was to use and learn the scientific method. For many experiments, students who were engaged in active investigation abandoned the requirements of the scientific method as stated in the curriculum, because the steps were too rigid. In addition, the logistics of laboratories in classrooms were a major problem for classroom teachers. Successful curricula provided all the necessary materials from beakers to living organisms (Karplus & Thier, 1967).

In summary, during the interaction period, laboratories were recognized as a potential motivator for students to participate in science courses. The primary justification for this participation was that students could actively engage in the processes of science. Ideally, students would actively engage in the processes of science, use the scientific method, work in well-equipped labs, have ample time for experiments, and have teachers who could supervise these laboratory experiences. Realistically, many of these conditions were not met and, furthermore, the costs were considerable.

The NSF provided funds for the development of curriculum materials with strong laboratory components, publishers participated by developing laboratory manuals and distributing apparatus, and the NSF also funded a large number of teacher training workshops to prepare teachers to use lab materials in the classroom. In spite of all these effects as noted above, the primary goal of increasing the number of students entering science was not achieved.

1.4.3 Partnership Period: MBLs

During the partnership period, some of the drawbacks to laboratory investigation characteristic of the interaction period were removed. A primary force in modifying the science laboratory was the MBL. These laboratories provided students with tools more appropriate to the goal of conducting experiments following the practices of expert scientists. Students could collect data in real time, watch the data appear on their computer screens, print out the results of their experiments, and analyze the findings. These tools reduced some of the logistic problems faced by science teachers, although they also introduced others, specifically those having to do with computers in science classrooms.

The purpose of the laboratory during the partnership period was broadened and refined by experts in technology, pedagogy, natural science, and precollege teaching. Following Piagetian theory, some believed that hands-on learning experiences were necessary to foster the development of logical reasoning strategies. Yet, Piaget had specified that this development took considerable instructional time, and research investigations were often unsuccessful when they set out to show that experience in the laboratory helped learners understand scientific phenomena.

Another view of the science laboratory was put forth by Papert, who argued that students needed to learn to construct knowledge rather than learn about science. He argued that the teacher did the knowledge construction in the past. Now it was necessary to provide students with the tools and resources appropriate for constructing their own knowledge. Papert believed that the computer language Logo would be a sufficient inducement to help students learn the process of construction. A large amount of research on the impact of Logo suggested that Logo in and of itself was certainly not sufficient, although Logo and an effective curriculum helped students construct more scientific understanding (Papert, 1980).

Both of these approaches to the science laboratory were unsuccessful in part because they tried to separate reasoning strategies from the context in which they were applied. Knowledge construction is not so much general as specific (Brown, Collins, & Duguid, 1989; Vygotsky, 1962). Students might employ a scientific investigative technique in one context but neglect it in another. General logical reasoning strategies might be available to students but not employed as predicted.

Consider research on students investigating displaced volume and trying to understand whether the weight of the object influences how much of the volume is displaced (Burbules & Linn, 1988). In one experiment (Linn, 1983), students were engaged in experimentation with metal cylinders and plastic cylinders being placed in beakers of water. All the cylinders sank and were completely covered by the water. Many students predicted that the metal cylinder would displace more liquid than the plastic cylinder. This was even though they were well aware that the amount of water in the glasses would not change and that the volume of the cylinders would not change. Thus students had the logical reasoning skills necessary to solve this problem, but they had a belief about the nature of weight that stood in the way of their reaching an appropriate logical conclusion. Instead, these students asserted that the heavier object would displace more liquid.

When allowed to conduct experiments to investigate their hypotheses, students frequently stuck with their beliefs. In one particularly poignant example, a young man observed two cylinders, which differed only in weight, displace the same amount of liquid. He accused the experimenter of bringing "magic water" and argued that the experiment would have worked correctly if the conditions were appropriate. Clearly, the reasoning that students perform in that context involves their understanding of the situation. Separating logical strategies from scientific understanding neglects these connections.

This view raises complicated questions concerning how students generalize and expand their knowledge. The conjecture that students generalize their knowledge by using logical reasoning is incomplete. Hands-on learning experiences also address only part of the problem. Disciplinary knowledge and logical reasoning must be integrated.

1.4.4 MBLs in the CLP Curriculum

CLP illustrates the potential role of MBLs in supporting science knowledge integration. CLP, as mentioned, has been exploring the benefits of MBLs for more than five years and refining the scaffolded knowledge integration framework. To illustrate the power of these refinements, performance of different cohorts of students is shown in Figure 6. The figure shows the percent of students in each cohort who could explain how heat and temperature differ and give several concrete examples to illustrate their point of view. As can be seen in Figure 6, uninstructed students rarely provide adequate answers to the question. Thus, the challenge to the CLP group was substantial. Performance on this item increased as the curriculum was refined.

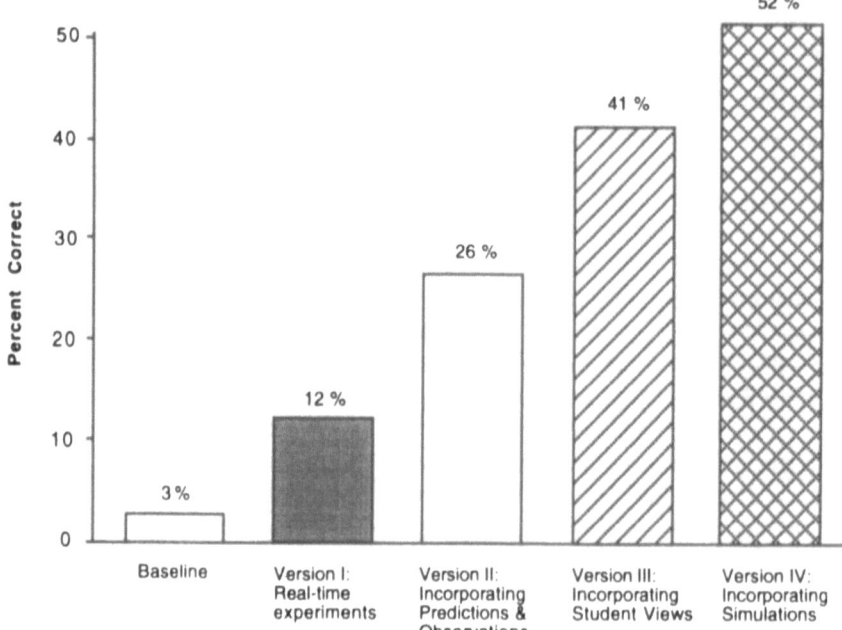

Figure 6 Improvement in ability to distinguish heat energy and temperature as the result of curriculum refinement

The reformulations of the CLP curriculum are briefly described. More details on the process of reformulation and the modifications to the curriculum that led to improvements in student understanding can be found in Linn (1992).

The first version of the CLP curriculum implemented MBLs for standard laboratories. MBLs collected information during scientific experiments. Students conducted such experiments as heating and cooling liquids and displayed these data graphically. This curriculum helped about 12% of the students to adequately answer the criterion question.

During these early trials using MBLs, students often did not even observe what was happening in their experiment while the graph was appearing on their computer screen. Frequently they were unable to label the lines on their graphs after the graphs had been completed because they were not sure which probe was associated with which experimental condition. Students were not actively engaged in constructing understanding in this situation, although they were using MBLs.

Based on these observations, the curriculum was modified to version 2. Friedler, Nachmias, and Linn (1990) instituted two conditions: one featuring predictions of graphical results, and the other featuring observations of experimental changes. Both conditions were considerably more successful than the first version. Students paid more attention to the laboratory results. In addition, the students were rarely surprised by the outcomes of their experiments. Instead, they were intent on trying to reconcile the experimental outcomes with their predictions or observations. In version 2, both predictions and observations were required (see Figure 7).

The experimenters were still concerned that students' progress was limited. Only about 30% of the students were gaining a fully integrated understanding of thermodynamic principles. The experimenters observed closely in the classroom to determine why students were still confused. They found that students were still having considerable difficulty reconciling the experiments in the classroom with their everyday experiences. The experiments focused on activities that could be conducted using laboratory apparatus. Thus students heated different volumes of water, they examined cooling in dishes with different surface areas, and they measured thermal equilibrium by detecting the temperature of objects throughout the classroom. Yet, students often separated their everyday experiences from their science-laboratory experiences (Songer, 1989). The curriculum was modified in version 3 to pay more attention to the views of students.

In version 4, everyday science experiences were added to the computer curriculum in the form of simulations. Due to the logistic difficulties of conducting these experiments using apparatus in the classroom, simulations were designed and included in the curriculum. Now students conducted some experiments with real-time data collection and others with simulations. As can be seen, this modification resulted in better performance on the criterion question (Figure 6).

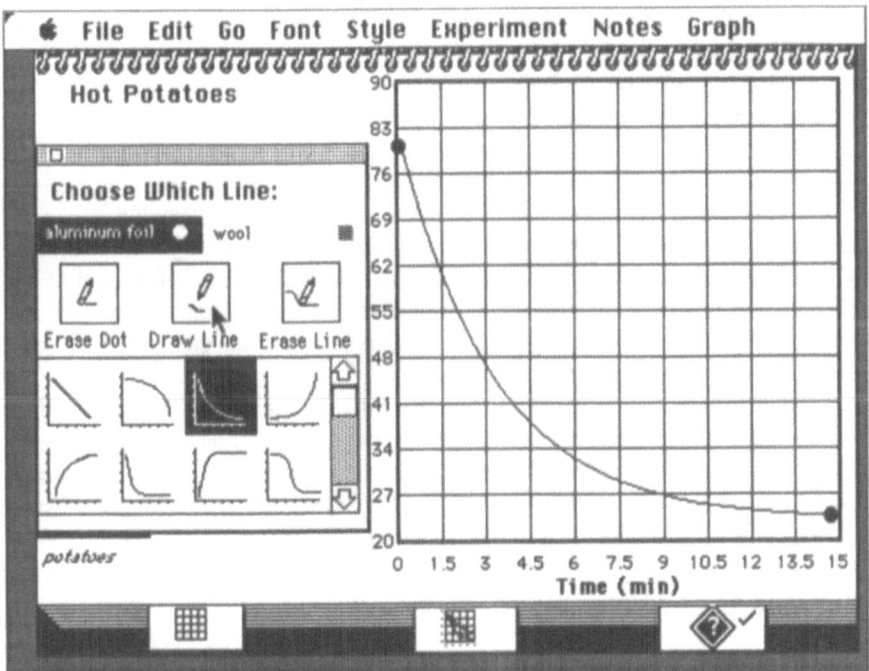

Figure 7 Prediction interface. Students indicate the outcomes they expect and then gather data to confirm or refute their predictions.

Approximately half of the students gained a fully robust and integrated understanding of the difference between heat energy and temperature. Comparing these results to the results reported earlier for progressing, oscillating, and converging students, it can be seen that looking just at the understanding of heat energy and temperature, about half the students develop a fully integrated perspective. In contrast, about 85% of the students develop some progressing ideas about thermodynamic concepts and only about 15% develop a fully integrated view linking insulation and conduction with thermal equilibrium and relative rates of heat flow as well as specific heat.

The CLP trial and refinement process illustrates that MBLs are not a panacea in and of themselves. In order to be effective, they must be embedded in a powerful curriculum. In implementing laboratory experiences for students, instructors must diligently pay attention to how learners integrate understanding, and design experiences that allow the learner to construct powerful and effective views.

1.4.5 Implications

The development of effective science laboratories requires trial and refinement. The idea that science laboratories are motivating has guided the use of the science laboratory throughout its history. Laboratories are intrinsically motivating because they release students from their desks. They do not necessarily motivate students to become scientists or even to gain understanding of scientific ideas. It is by taking the science laboratory as an opportunity to foster knowledge integration that the motivational characteristics of the laboratory can be capitalized on, and students can gain a more powerful understanding of scientific phenomena.

In CLP the scaffolded knowledge integration framework guided refinement of the curriculum to achieve benefits. It is neither sufficient nor appropriate to give students the tools of scientists without creating a culture around the tools. Scientists use realtime data collection to answer more complicated questions than those facing students. Many scientists delegate the actual experimental work to others and spend most of their time looking at the results of experiments and trying to make sense of groups of experiments. In examining the purposes and advantages of MBLs, these ideas need to be taken into consideration.

This is a complicated process. For example, frequently MBLs are used in conjunction with tools that allow students to compare experimental outcomes with ideal models for those experimental outcomes and to analyze how close the experimental data comes to the idealized model. Such activities are appropriate for a lesson on modeling. If students are still grappling with understanding the basic science concepts, then modeling might be confusing and actually stand in the way of knowledge integration.

Overall, the CLP experience suggests the importance of carefully examining the outcomes and student understandings that result from MBLs. It also suggests the importance of making sure that the goals of the course that employ the MBL are well understood so that student progress can be measured against those goals.

One important goal for science laboratory courses is to help students understand how to explain naturally occurring and everyday phenomena. Yet these phenomena are often far more ambiguous and complex than experiments conducted using air tracks or ripple tanks. MBLs broaden the range of problems that can be studied experimentally but require decisions about what complexities to introduce. The CLP project added simulations to the curriculum in order to help students broaden their understanding and gain a more robust view of everyday events. This lesson might be applied to MBLs for other science topics.

Of course the science laboratory cannot be examined independent of the science curriculum. Indeed, throughout its history, the science laboratory has complemented the science curriculum. Sometimes the relationship between the laboratory and the curriculum has been fairly close and at other times it has been somewhat distinct. The next section examines curricular goals in more detail.

1.5 The Goals of the Science Curriculum: How Have They Changed?

Two key issues have guided goal selection in science education. First, natural scientists have continuously sought to provide a more modern version of the science information presented to students. Second, science instructors have continuously demanded that the curriculum be more teachable. These two processes work in opposition to each other. That is, as science knowledge has expanded, the curriculum has become less teachable. Modern ideas are more abstract and less descriptive, and as more and more topics have been crowded into the curriculum, less time is available to cover each one. In addition, the tension between science for future scientists and science for citizens has led to a search for goals that bridge these audiences. Such goals should provide science information that helps citizens make sense of phenomena in their everyday lives and lays a firm foundation for future scientists. This can be accomplished by emphasizing the repertoire of models of scientific phenomena that can be applied to the same problem. Models from natural science, from observation, or from engineering can be contrasted. Furthermore, this approach emphasizes that science is comprised of multiple models and that distinguishing and selecting models is part of science.

1.5.1 Separation Period: The Curriculum as a Source of Knowledge

Prior to 1950, the primary goal of the science curriculum in physics was to impart information about the knowledge held by scientists. At first, the information was provided by scientists and communicated to students (e.g., Millikan & Gale, 1906).

Starting in about 1930 there were also science textbooks put together by publishers and teachers. This tradition arose when teachers complained that the curriculum materials were unteachable and asked publishers to provide books that met their needs. Publishers cooperated by determining what it was that teachers wanted to have in the textbooks and coordinating teams of teachers to write textbooks. For example, the well-known textbook initially edited by Dull, Metcalfe, and Williams (1960) has gone through revisions, each revision reflecting surveys and suggestions from teachers. Both the textbooks written by the well-known scientists and the textbooks put together by the publishers have persisted and are found in each of the periods discussed in this paper.

Early textbooks communicated what was known about science at the time. Typically, making the text more teachable meant that certain topics were left out of the curriculum because they had proven to be too difficult for students to understand.

Both teachers and students complained that the physics course was packed with too much material. As early as 1920, there were requests by those concerned with physics education that the materials be more carefully selected (Curtis, 1926). By 1950, this problem had become acute because modern physics concepts had been added to the classical theories without any reduction in topics.

During the separation period when textbooks did indicate what the goals of the course were, they were frequently stated in behavioral terms. This was particularly true of the publishers' texts.

1.5.2 Interaction Period: Modern Science

The push to modernize the curriculum was one of the factors leading to massive curriculum reform efforts of the 1950s and '60s, led by natural scientists. There seemed to be little or no appreciation that the modern view of mechanics, thermodynamics, electricity, magnetism, and other topics would turn out to be much more difficult for students to understand than had been the case with the classical view.

Scientists may not have appreciated this problem because they added the modern view to their well-established understanding of classical physics. To scientists, modern views were parsimonious and elegant. Students, on the other hand, jumped immediately into modern physics and often found it confusing. Modern physics often featured atoms and electrons that were not observable features of the physical world. Students had difficulty mapping their understanding and knowledge of the physical world onto these models. In addition, the beauty of modern physics was its computational elegance. Maxwell's equations, Fourier transforms, and other mathematical formalisms made it possible to elegantly describe physical phenomena. These elegant formulations were of great value to research scientists but often less useful for explaining everyday problems. These reforms made knowledge integration more difficult.

The funding provided by the NSF led to innovative and exciting curricula for physics. The PSSC (1960) course discussed earlier is one example. Another is the Harvard Project Physics (HPP) course, developed by Rutherford, Holton, and Watson (1970). This innovative approach to physics instruction placed physics concepts in a broad perspective. Just as the PSSC course seemed to reflect the goals that an experimental physicist might have had for a high-school course, the HPP course appeared to be focused on the sort of course that a historian of science would have wanted.

For example, in discussing heat and temperature, the HPP curriculum describes the steam engine developed by James Watt. In discussing the Watt steam engine, the book assumes that the reader will understand the effect of the metal cylinder walls on the efficiency of the steam engine. Concerns about phase change and other aspects of thermal phenomena must be inferred from the historical discussion. The chapter goes on to introduce the concept of horsepower, to discuss the effect of the steam engine on the industrial revolution, and to provide modern examples of steam engines and their effectiveness. The next topic in the chapter is placed in the context of the experiments of Joule. Students are expected to understand the concepts of heat, temperature, specific heat, and thermal conductivity to understand this discussion. Students next apply these concepts to energy in biological systems. The textbook brings together a broad array of related scientific ideas and models the kind of integrated understanding that one would like students to acquire. However, the scaffolding required to acquire this kind of integrated understanding from the textbook is left to the teacher and the laboratory.

Later textbooks reflected the growing realization that students had difficulty with concepts from modern physics. The most famous of these is *Conceptual Physics* developed by Hewitt (1987), an instructor at the city college in San Francisco, who initially developed a college text and then collaborated with high-school teachers to create a high-school text. The Hewitt book features both quantitative and qualitative models of thermal events and follows the same perspective in other topic areas. In addition, the book draws attention to naturally occurring problems. For example, the chapter on heat opens with a discussion of the conundrum presented when one walks on the grass and on the pavement on a hot day. The Hewitt book links everyday experiences to science concepts. For example, specific heat is introduced by pointing out that some foods remain hotter much longer than others: one might burn one's tongue on the filling but not on the crust of a pie. The Hewitt book links everyday experiences to molecular kinetic theory. In contrast, the CLP curriculum links these experiences to a heat-flow model.

In spite of these changes in the curriculum, examination of the exercises in textbooks, from Dull, Metcalfe, and Williams (1960) or Stollberg and Hill (1965), to HPP and PSSC, reveals considerable uniformity. Virtually every textbook asks students to compute changes in calories or temperature. The problem of mixing water at different temperatures and predicting the final temperature is common to all of the textbooks. Computing how much heat is required to heat a particular substance by a particular number of degrees is found in every textbook. Texts vary in whether they emphasize computation, theories of experts like molecular-kinetic theory, pragmatic ideas like heat flow, or links to everyday problems, but they all require students to develop a computational model that they can apply to problems found at the end of the chapter. In the partnership period, models were contrasted more explicitly.

1.5.3 Partnership Period: Multiple Models of Scientific Phenomena

In the partnership period, MBLs became a catalyst for new curricular goals in science based on alternative models of science concepts. MBLs enable students to contrast models of phenomena and to emulate scientists. Scaffolding students as they compare and contrast models and helping students identify abstractions to link their ideas is common in partnership projects. For example, Barbara White and John Frederiksen (1990) have investigated this approach for teaching concepts in electricity, and Andy diSessa (1992) has used the approach for teaching concepts in mechanics. Research shows that students are learning much less physics than had been anticipated (National Assessment of Educational Progress [NAEP], 1978; 1988), and studies suggest that these trends can be reversed (Laws, 1986, June; Thornton & Sokoloff, 1990).

1.5.4 CLP: An Example

In the CLP project, MBLs have motivated students to develop and refine their models of thermal events. As our longitudinal study shows, students need ideas that they can apply to their own experiences to begin the knowledge integration process. Heat flow is a descriptive model, it is accessible for students, it is easy to apply to relevant everyday problems, and it is useful for interpreting MBL graphs. Students use their model to explain complex events; teachers scaffold students as they gain understanding.

Supporting knowledge integration requires the renegotiation of the authority structure in the classroom discussed above. Rather than viewing the teacher as the source of knowledge and the absolute authority, students view the teacher as someone who can help and support them. Teachers encourage this change by asking students to critique their own and each other's investigations, much like scientists. Instructors help students view themselves as monitoring their own understanding and identifying what they know and do not know by encouraging self-criticism.

In renegotiating the authority structure in the classroom, the relationships between the teachers and the students are modified, as are the relationships among the students themselves. Students are most effective when there is mutual respect for ideas and mutual support in gaining understanding. In contrast, the norms for student interaction in classrooms are quite different. Typically, normative behavior involves pointing out the weaknesses of students, reinforcing stereotypes concerning who can learn science, and engaging in other practices that involve only some students in knowledge integration (Linn, diSessa, Pea, & Songer, 1994).

1.6 Conclusions

Research investigations of science laboratories have proceeded from separation to partnership. Small laboratory studies guided early thinking about the science learner. During the interaction period, curricular innovations were evaluated by those interested in science learners. During the partnership period, classroom-based investigations such as those of CLP have become more and more common.

In spite of the complexities of classroom-based research, considerable insights can be gained from these investigations. For example, comparing one version of the CLP curriculum to another can pinpoint the effects of innovations. Laboratory studies, evaluation studies, and classroom investigations each contribute distinct information. Rather than selecting one over another, what is needed is the right mix of these different types of studies.

Although MBLs often work well when they are implemented by those who develop them, they are frequently difficult for others to use. In particular, teachers starting to use MBLs for the first time need to become familiar with (a) conducting laboratories, (b) managing computers and their associated breakdowns and technical difficulties, (c) managing software, (d) understanding interfaces, (e) dealing with students' unequal prior experience with technology, and (f) creating a curriculum. These logistic issues are nontrivial and need to be considered as MBLs became more available.

In studying the science learner, scientists have shifted from a search for general laws to a focus on describing the process of knowledge acquisition. Designing science courses that meet the needs of citizens as well as future scientists increases the complexity of science teaching. Current courses meet the needs of future scientists but often bore or frustrate students who wish to be educated citizens.

Curriculum policy has also proceeded from separation to partnership. Constructing a responsible curriculum policy requires substantial cooperation from a broad range of individuals. Partnerships with strong input from all participants are essential for curriculum reform. Questions facing policy-makers include: How should MBLs be incorporated into the curriculum? Who should pay for these innovations? Are these innovations cost effective? If the desire is to change the science curriculum, how can that be accomplished? Who should be in the vanguard? Is a trickle-down approach appropriate, where new teachers learn about these technologies at universities and carry the message out to existing school systems? Is an in-service approach more appropriate? Policy questions are often best answered within the context of an individual school or school district since local conditions determine whether an innovation will be successfully incorporated into a classroom or not. Without partnerships and techniques for tailoring innovations to learning contexts, progress is unlikely.

From classical to modern to everyday physics, the goals of science courses have occupied the discussions of those concerned with science education continuously. Even stating the goals of science courses has proven considerably more difficult than anticipated. Behavioral objectives, lists of topics, and cognitive goals each have their strengths and limitations. Often in discussions these separate kinds of goals are not even distinguished, further inhibiting consensus in the community.

As the goals of the science curriculum are considered, new criteria have emerged. When the goal was to prepare students to become scientists, it seemed reasonable to get students as close as possible to the understanding of experts. With the new focus on providing basic technical competence to a broad range of students, understanding everyday events seems far more important. Identifying effective ways to help students gain insights into everyday phenomena is a major challenge for the partnerships that are forming to improve scientific understanding. Given the current lack of success of the curriculum, it seems useful and reasonable to consider radical alternatives to the current view and to conduct experiments investigating a broad range of alternatives and approaches to the curriculum. The CLP curriculum, for example, spends twelve weeks on concepts of thermodynamics, far more than would occur in a typical course. Investigating other broad departures from the usual way of imparting physics knowledge seem justified given the current dismal success of the existing curriculum.

Historically, science laboratories have been viewed as vocational training centers, motivational tools, and inquiry environments. Each of these reformulations of the goals of the science laboratory has built on the previous view. Vocational labs were motivating to students and their motivational character was then capitalized on during the interaction period when scientists wished to attract more students to science studies. Careful examination of science laboratories revealed that they also provided opportunities to construct scientific ideas. Determining just how the constructive process can be guided, as well as providing the right amount of scaffolding and encouragement, is a research topic that will demand the attention of the community for some time to come.

MBLs have provided a new impetus for resolving the discovery-learning/constructivist dilemma in the laboratory. These new tools make learning in the science laboratory far more powerful and potentially abstract. By judicious use of MBLs, a dramatic improvement in the science curriculum can be made.

Technological tools can facilitate the renegotiation of the authority structure in the classroom. For example, it is possible to free the teacher to tutor rather than pass on authoritative information by placing straightforward information in databases and help facilities. A constructivist approach to the science laboratory underscores the importance of mutual respect. Teachers and students can be viewed as a community of scholars with a common goal. Yet, such respect is difficult to achieve and our current methods of laboratory learning do not lend themselves to this approach. An important question for the future will be how to capitalize on powerful, new laboratory tools to enhance the social construction of science knowledge.

References

Agogino, A.M. & Linn, M.C. (1992 May-June). Retaining female engineering students: Will early design experiences help? [Viewpoint Editorial]. In M. Wilson (ed.), *NSF Directions, 5*(2), pp. 8–9.

Brown, J.S., A. Collins, et al. (1989). Situated cognition and the culture of learning. *Educational Researcher 18*(1): 32–41.

Bruner, J.S. (1966). *Toward a theory of instruction.* New York: Norton.

Bruner, J.S. (1968). *Processes of cognitive growth: Infancy.* Worcester, MA: Clark University Press.

Burbules, N.C. & Linn, M.C. (1988). Response to contradiction: Scientific reasoning during adolescence. *Journal of Educational Psychology, 80*(1), 67–75.

Burbules, N.C. & Linn, M.C. (1991). Science education and the philosophy of science: Congruence or contradiction? *International Journal of Science Education, 13*(3), 227–241.

Calfee, R. (1981). Cognitive psychology and educational practice. In D.C. Berliner (ed.), *Review of research in education* (pp. 3–73) (Vol. 9). Washington, D.C.: American Educational Research Association.

Champagne, A.B., Klopfer, L.E., & Gunstone, R.F. (1982). Cognitive research and the design of science instruction. *Educational Psychologist, 17*(1), 31–53.

Collins, A. & Brown, J.S. (1988). The computer as a tool for learning through reflection. In H. Mandl & A.M. Lesgold (eds.), *Learning issues for intelligent tutoring systems* (pp. 1–18). New York: Springer-Verlag.

Collins, A., Brown, J.S., & Holum, A. (1991). Cognitive apprenticeship: Making thinking visible. *American Educator, 15*(3), 6–11, 38–39.

Cunningham, H.A. (1924). Laboratory methods in natural science teaching. *School Science and Mathematics, 24,* 709–715, 848–851.

Cunningham, H.A. (1946). Lecture method versus individual laboratory method in science teaching: A summary. *Science Education, 30,* 70–82.

Cureton, E.E. (1927). Junior high school science. *The School Review, 35,* 767–775.

Curtis, F.D. (1926). *A digest of investigations in the teaching of science in the elementary and secondary schools.* York, PA: Maple Press Co.

Curtis, F.D. (1931). *Second digest of investigations in the teaching of science.* York, PA: Maple Press Co.

diSessa, A. (1983). Phenomenology and the evolution of intuition. In D. Gentner & A.L. Stevens (eds.), *Mental models.* Hillsdale, NJ: Lawrence Erlbaum Associates.

diSessa, A. (1988). Knowledge in pieces. In G. Forman & P. Pufall (eds.), *Constructivism in the computer age* (pp. 49–70). Hillsdale, NJ: Lawrence Erlbaum Associates.

diSessa, A. (1992). Images of learning. In E. De Corte, M.C. Linn, H. Mandl, & L. Verschaffel (eds.), *Computer-based learning environments and problem solving.* NATO ASI Series F, Vol. 84. Berlin: Springer-Verlag.

Dull, C.E., Metcalfe, H.C., & Williams, J.E. (1960). *Modern physics.* New York: Holt, Rinehart and Winston, Inc.

Eckert, P. (1990). Adolescent social categories: Information and science learning. In M. Gardner, J.G. Greeno, F. Reif, A.H. Schoenfeld, A. diSessa, & E. Stage (eds.), *Toward a scientific practice of science education* (pp. 203–218). Hillsdale, NJ: Lawrence Erlbaum Associates.

Eylon, B. & Linn, M.C. (1988). Learning and instruction: An examination of four research perspectives in science education. *Review of Educational Research, 58*(3), 251–301.

Eylon, B. & Linn, M.C. (in press). Models and integration activities in science education. In E. Bar–On, Z. Scherz, & B. Eylon (eds.), *Designing intelligent learning environments.* Norwood, NJ: Ablex Publishing Corporation.

Friedler, Y., Nachmias, R., & Linn, M.C. (1990). Learning scientific reasoning skills in microcomputer-based laboratories. *Journal of Research in Science Teaching, 27*(2), 173–191.

Hewitt, P.G. (1987). *Conceptual physics: A high school physics program* (Teacher's ed.). Menlo Park, CA: Addison-Wesley Publishing Company.

Karplus, R. (1975). Strategies in curriculum development: The SCIS Project. In J. Schaffarzick & D.H. Hampson (eds.), *Strategies for curriculum development.* Berkeley, CA: McCutchan Publishing Corp.

Karplus, R. & Thier, H.D. (1967). *A new look at elementary school science: Science Curriculum Improvement Study.* Chicago: Rand McNally & Co.

Lave, J. & Wenger, E. (1991). *Situated learning: Legitimate peripheral participation.* Cambridge, MA: Cambridge University Press.

Laws, P.W. (1986, June). *Workshop physics.* [Proposal to the Fund for the Improvement of Post Secondary Education, U.S. Department of Education], Carlisle, PA: Dickinson College,

Lawson, A.E. (1985). A review of research on formal reasoning and science teaching. *Journal of Research in Science Teaching, 22,* 569–617.

Lewis, E.L. (1991). *The process of scientific knowledge acquisition among middle school students learning thermodynamics.* Unpublished doctoral dissertation, University of California, Berkeley, CA.

Lewis, E.L. & Linn, M.C. (1994). Heat energy and temperature concepts of adolescents, adults, and experts: Implications for curricular improvements. *Journal of Research in Science Teaching, 31*(6), 657–677.

Linn, M.C. (1983). Content, context, and process in adolescent reasoning. *Journal of Early Adolescence, 3,* 63–82.

Linn, M.C. (1986). Science. In R. Dillon & R.J. Sternberg (eds.), *Cognition and instruction* (pp. 155–204). New York: Academic Press.

Linn, M.C. (1992). The computer as learning partner: Can computer tools teach science? In K. Sheingold, L.G. Roberts, & S.M. Malcolm (eds.), *This year in school science 1991: Technology for teaching and learning.* Washington, DC: American Association for the Advancement of Science.

Linn, M.C. & Burbules, N.C. (1993). Construction of knowledge and group learning. In K. Tobin (ed.), *The practice of constructivism in science education* (pp. 91–119). Washington, DC: American Association for the Advancement of Science (AAAS).

Linn, M.C., diSessa, A., Pea, R.D., & Songer, N.B. (1994). Can research on science learning and instruction inform standards for science education? *Journal of Science Education and Technology, 3*(1), 7–15.

Linn, M.C. & Songer, N.B. (1991). Cognitive and conceptual change in adolescence. *American Journal of Education, 99*(4), 379–417.

Linn, M.C., Songer, N.B., & Eylon, B. (in press). Shifts and convergences in science learning and instruction. In R. Calfee & D. Berliner (eds.), *Handbook of educational psychology.* Riverside, NJ: Macmillan.

Linn, M.C., Songer, N.B., Lewis, E.L., & Stern, J. (1993). Using technology to teach thermodynamics: Achieving integrated understanding. In D.L. Ferguson (ed.), *Advanced educational technologies for mathematics and science* (pp. 5–60) (Vol. 107). Berlin: Springer-Verlag.

Miller, G.A. (1956). The magical number seven, plus or minus two: Some limits on our capacity for processing information. *Psychological Review, 63,* 81–97.

Millikan, R.A. & Gale, H.G. (1906). *A laboratory course in physics for secondary schools.* Boston: Ginn and Company.

National Assessment of Educational Progress (NAEP) (1978). *Three national assessments of science: Changes in achievement 1969–77.* Denver: Education Commission of the States.

National Assessment of Educational Progress (NAEP) (1988). *The science report card: Elements of risk and recovery: Trends and achievement based on the 1986 national assessment.* Princeton, NJ: Educational Testing Service.

Newman, D., Griffin, P., & Cole, M. (1989). *The construction zone: Working for cognitive change in school.* London: Cambridge University Press.

Papert, S.A. (1980). *Mindstorms: Children, computers, and powerful ideas.* New York: Basic Books.

Pella, M.O. (1976). Guest editorial: Sixty years of science education. *Science Education, 60*(4), 433–439.

Physical Sciences Study Committee (1960). *Physics.* Boston: D.C. Heath.

Piaget, J. (1952). *The origins of intelligence in children.* New York: National Universities Press.

Raizen, S.A. (1991). The state of science education. In S.K. Majumdar, L.M. Rosenfeld, P.A. Rubba, E.W. Miller, & R.F. Schmalz (eds.), *Science education in the United States: Issues, crises and priorities* (pp. 25–45). Philadelphia: The Pennsylvania Academy of Science.

Rossiter, M.W. (1982). *Women scientists in America: Struggles and strategies to 1940.* Baltimore: Johns Hopkins University Press.

Rutherford, F.J., Holton, G., & Watson, F.G. (1970). *The Project Physics course handbook.* New York: Holt, Rinehart & Winston.

Songer, N.B. (1989). *Promoting integration of instructed and natural world knowledge in thermodynamics.* Unpublished doctoral dissertation, University of California, Berkeley, CA.

Songer, N.B. & Linn, M.C. (1991). How do students' views of science influence knowledge integration? *Journal of Research in Science Teaching, 28*(9), 761–784.

Stollberg, R. & Hill, F.F. (1965). *Physics: Fundamentals and frontiers.* Boston: Houghton Mifflin Company.

Thomas, M. (1993). *Thermal Model Kit* [software]. Berkeley, CA: University of California.

Thorndike, E.L. (1910). The contribution of psychology to education. *Journal of Educational Psychology, 1*, 1–14.

Thornton, R.K. & Sokoloff, D.S. (1990). Learning motion concepts using real-time microcomputer-based laboratory tools. *American Journal of Physics, 58*, 858–867 (Appendix B).

Vygotsky, L.S. (1962). *Thought and language.* Cambridge, MA: MIT Press.

Watson, J.B. (1913). Psychology as the behaviorist views it. *Psychological Review, 20*, 158–177.

Welch, W.W. (1979). Twenty years of science curriculum development: A look back. In D.C. Berliner (ed.), *Review of research in education* (pp. 282–308). Washington, DC: American Educational Research Association.

West, L.H.T., Pines, A.L., & Sutton, C.R. (1984). In-depth investigations of learners' understandings of scientific concepts and theories. In A.L. Pines & L.H.T. West (eds.), *Cognitive structure and conceptual change.* New York: Academic Press.

White, B.Y. & Frederiksen, J.R. (1990). Causal model progressions as a foundation for intelligent learning environments. *Artificial Intelligence, 24*(1), 99–157.

2. Trends and Techniques in Computer-Based Educational Simulations: Applications to MBL Design

Betty Collis and Ivan Stanchev

University of Twente

Abstract. Computer-based educational simulations are seen as a subset of the larger set of instructional approaches whose goal is to help learners come to a better understanding of real, complex systems. In this paper, a conceptualization is developed for the domain "understanding complex systems" and within this, a scheme is offered whereby computer simulations are considered relative to interrelated cognitive and instructional aspects. Techniques and trends in simulations relative to visualization, interactivity, and intelligence are discussed within the framework of the scheme. The relationship between microcomputer-based laboratory environments and computer-based simulations in science education is considered, as well as the emergence of MMLs—multimedia laboratories. MBLs and MMLs are compared with respect to the trends of visualization, interactivity, and intelligence as a way of identifying common aspects with simulations in science education. Promising directions for improving the effectiveness of both MBLs and simulations are suggested.

2.1 Introduction: Computer-Based Simulations and MBLs in Science Education

This paper presents an analysis of certain concepts, strategies, and techniques for understanding complex systems. In particular we limit our focus to computer-related learning environments for science education. We first discuss simulation software, and then argue that this type of software overlaps in many ways the category related to Microcomputer-Based Laboratories (MBLs). We particularly focus on certain techniques and trends in the design of these learning environments, in particular, visualization, interactivity, and intelligence. We also introduce MMLs—Multimedia Laboratories—and consider these as well in terms of the three major trends discussed in terms of simulations and MBLs.

Computer-based simulations (for convenience, referred to only as simulations during the rest of this paper) have been defined and categorized in many ways, for example, sometimes related to the degree to which the variables within the system being simulated are well defined (Collis, 1988), to the degree of learner control of events within the simulation (Gredler, 1986), or if the system being simulated is natural, man-made, or imaginary (Schaick Zillesen, 1990). Regardless of the perspective, a simulation can be very simply defined as computer software that takes as input some values of certain scientifically interesting variables, processes them in some way, and then presents them in processed form to the learner, who may or may not have the opportunity to further manipulate them. The goal of the experience is to better understand the complex system represented by the variables. In this broad view, microcomputer-based laboratories (MBLs)[1] are generically similar to simulations, although of course the origins and types of data and variables and the ways in which data are input into the computer differ. Although it is not the case that MBLs are a subset of simulations, there is enough functional and didactic overlap between the two categories of electronic learning environments that key observations relative to the instrumentation of simulations can also be useful to the design of instrumentation for MBLs. We will focus primarily on simulations in the first part of this analysis, but in the second part make the extension to MBLs.

2.2 Trends and Techniques

Our focus in this analysis is also on the instrumentation of simulations (and of the software component of MBLs). Instrumentation aspects include screen design, design of output display, instructional design variables in the software itself, choices available as design options in the software, and other issues controlling the designer of the software. In particular, we consider trends and techniques relative to visualization, interactivity, and intelligence as important aspects of instrumentation design.

2.2.1 Visualization

The rapid evolution of the technology related to the display and manipulation of visualization in computer environments is of course well known. The increasing use of interactive video in schools, the ability to store huge amounts of visual material on a single CD-ROM, the capacity to digitize photographs and even moving video so that it can be manipulated within software environments via video windows or by using DV-I technology to compress and decompress moving video so that it can be manipulated within a simulation environment have led to a corresponding increase in the quality and quantity of visualizations in educational software and particularly in simulation environments. We see the quality, the speed of appearance, the "look" of graphics in simulation software improving enormously even by the year. And we see a strong interest in interactive video as a component of science simulations. A comparison of computer simulation programs on the commercial market over the past ten years makes this line of development abundantly clear.

---

[1] Microcomputers interfaced with traditional laboratory apparatus to carry out functions of data collection, handling, and display. The term "MBL" was introduced by Tinker at TERC.

But we are not convinced that this visual explosion always (or even very much) is being driven by cognitive and instructional theory, or by instructional needs identified by a prior examination of the domain "understanding complex systems." We suspect a technology push motivates some aspects of interest in increased visualization. Thus one purpose of this paper is to map various aspects of visualization onto a cognitive-instructional framework for simulation but also more generally to contribute to the larger question of the relationship of visualization to learning given the emerging possibilities of multimedia (Moonen & Stanchev, 1992).

2.2.2 Interactivity

Aside from visualization, we are also very much aware of two other trends that are now of considerable interest not only with respect to computer simulations but to educational software in general and even more broadly, to many aspects of education. These are interactivity and what we might call metacognitive support, or "intelligence." Interactivity is easier to discuss. Vygotskyan theory, where learning is seen to occur as a result of social interaction, has contributed to a broader view of interactivity as part of computer-augmented learning experiences (Forman & McPhail, 1989). In addition, ideas about computer-supported cooperative learning are becoming increasingly influential in the design of learning activities involving educational software (see, for example, Scardamalia, Bereiter, McLean, Swallow, & Woodruff, 1989). Interactivity has always been seen as an important feature of educational software, but this was generally understood to be learner-software interaction. Now we see stress on the importance of embedding learner-software interaction within human interaction in order to motivate and produce learning (Tennyson & Thurlow, 1987). But even within the learner-computer interaction framework, many new options are becoming available with simulation software, such as those related to model exploration (i.e., implementing a hypertext-like exploration of a model, allowing a change of view and zooming in/out of concept domains; see Hoog, Jong, & Vries, 1991, p. 376). Thus it is appropriate to also consider interactivity in our analysis of trends and techniques in simulation environments and to attempt to relate these interactivity options to a cognitive-instructional framework. In addition, science telecommunications networks, such as the National Geographic Society's Kids Network Program, bring interpersonal interactivity during the collection and analysis of scientific data far outside the boundaries of the classroom (Songer, 1989).

2.2.3 Intelligence

The trend we call "intelligence" is harder to define but it involves a deliberate focus on metacognition and the stimulation or support of metacognitive processes in the learner. Thus "cognitive tools" are becoming popular, and software making use of hypertext and hypermedia organization of learning materials are fueling this interest (Kommers, Jonassen, & Mayes, 1991). We see "idea processors," "semantic mapping," and other kinds of "mind tools" assumed to be valuable and appearing more and more as embedded tools in educational software (Jonassen, 1988). The assumption is, among other things, that such tools stimulate metacognitive processes such as reflection and epistemological analysis, processes considered important for learners using computer simulations (Jong & Njoo, 1990).

We also see as another direction of this trend toward more "intelligence" in simulation software the increasing interest in including diagnostic or tailored (i.e., "intelligent") tutoring as functions to better steer or support the learner during interaction with the simulation (Hollan, Hutchins, & Weitzman, 1987; Towne, Munro, Pizzini, Surmon, Coller, & Wogulis, 1990). The extent to which this intelligence is electronically steered by the software or is available (either as part of the electronic environment or apart from it) as an optional resource for the teacher or student is also an emerging area of interest with respect to simulation environment design (Thomas & Hooper, 1991).

2.3 Purposes of the Analysis

As specialists in the design of instrumentation for new technologies we interact with colleagues who are engrossed in the implementation of these trends relative to visualization, interactivity, and intelligence in educational software and in particular in simulations. We see their, and their students', enthusiasms, and we hear similar enthusiasms at conferences and vendor exhibitions. However, we feel the need for a systematic way to consider these trends and techniques in an instructional/cognitive setting. We want to have a better sense of the critical problems students encounter in understanding complex systems with the help of simulations, and from this we want to identify the types of techniques of most help in dealing with those problems. Simply stated, we want to confront the urge to be fascinated by new developments in technique and technology. In a time of mushrooming new technical possibilities we want to reconsider the domains in which those possibilities can be applied so we can better identify the best fit and most promising cutting edge for their application in learning.

Thus the purposes of this discussion are:

- At the most concrete level, to suggest design guidelines based on instructional/ cognitive principles for the implementation of various techniques related to visualization, interactivity, and intelligence in educational computer simulations and by extension, to MBL software design.

- At a more general level, to stimulate reflection on "critical attention areas" from a cognitive/instructional perspective in the design of computer-based instrumentation whose aim is to help students better understand complex systems.

- At the most general level, to suggest a new view of the domain, "understanding complex systems," in which computer simulations, MBLs, and also the emerging "multimedia laboratories" (MMLs) are educational tools.

2.4 Conceptualizing the Domain, "Understanding Complex Systems"

We begin by suggesting a framework for the development of students' capabilities to understand complex systems. First we clarify various aspects of this domain and the cognitive-instructional dimensions we will suggest to represent increasing capability within the domain. Following this, we relate the domain to simulation software (and later to MBL software) and use the framework to focus on the three trends of visualization, interaction, and intelligence.

2.4.1 Systems

A first step is to define "complex systems." There are many formal definitions, of course, varying in their components and terminology according to the discipline (and author) making the definition. We will define a system as a set of interrelated variables. For our purposes, a complex system is a system where the interrelationships between the variables influences the current state of the system, so one cannot say that the output or current state of the system can be fully understood by knowing the value of any one variable within it or even by knowing about the values of the variables in isolation. The relationship among them is part of the system.

Systems have component parts and also have state-transition relationships. Sometimes our interest with respect to "understanding" a system stops at knowing the component parts and their hierarchical or taxonometric relationship (example, the organizational chart of a company). Other times our interest focuses on the input-output aspect of a system. This can be addressed using a "black box" approach, where what goes in and what comes out matter, but how it gets transformed within the box is of little or no direct importance. Most often, we are interested in both aspects.

Sometimes systems are well defined. A machine, for example, has a finite set of components and well-defined relationships. The output of the system has a functional relationship to the current state of the components (although any system can malfunction or unexpected variables can influence its performance). In theory at least we can develop an algorithm to describe the output of the system based on knowledge of the values of a specified set of variables. Often this algorithm is expressed mathematically. In reality 100% certainty is not possible for any system; however, we will define a well-defined system as one where a functional relationship can explain the "usual" output of the system.

In contrast, there are many systems which are not well defined. These systems operate at best in a probabilistic manner; we can fashion an algorithm, but the probability that the algorithm can predict the output of the system given input on the variables called for by the algorithm is middle to low. This can be because we cannot in a practical sense specify all the variables that influence a system, those that we can identify we can only imperfectly measure, and we cannot state more than general tendencies in the relationship between the variables. Most complex systems fall in this category, particularly those involving human behaviour (i.e., politics, performance of organizations) but also systems relating to the environment, and to settings such as agriculture, fisheries, or human/animal/plant physiology. With the latter type of systems the best we can do is deal probabilistically within them. We can, of course, "pretend" that a system is well defined, and that a certain set of variables and a certain algorithm will give a reasonable resultant state of the system, but our result must be understood to be an estimation if we are to understand the complex system as it "really" is.

2.4.2 Understanding

A second basic clarification with respect to the domain "understanding complex systems" relates to the word "understanding." We have already noted that "understanding" can mean to know the component parts of a system and their structural relationship, to know the most likely output of a system given a certain set of inputs, or a combination of these. Then there are intellectually different levels of understanding as well. These can range from having only a general overview of a system, through knowing key concepts and variables and how they are interrelated in the system, through being able to solve problems in the system, and also to being able to "stand above" the system and interrelate it with other systems or see how to change the system itself in order to change the likelihood of a certain output state. These levels relate to the learner's cognitive activity.

Another aspect of understanding relates to the actions and intentions of the teacher or instructional designer to help bring about understanding. These can be categorized in terms of instructional techniques—in particular, providing appropriate help, providing clarification, giving feedback, providing task guidance, providing appropriate tools, and providing a rich learning environment. They can also be categorized in terms of the overall intention of the teacher's instructional activity. Such intentions can include providing a motivating orientation, developing specific concepts, guiding the application of concepts to new settings, or stimulating reflection and analysis.

The meaning of "understanding complex systems" is thus related both to the cognitive characteristics of the learner as she is engaged in the act of "understanding" and the strategies for facilitating that understanding employed by the teacher or by the designer of learning materials. These two dimensions are interrelated, with "recalling"/"providing motivation" as likely to be associated with a minimally sophisticated level of system understanding, and "creating new perspectives/stimulating higher order thinking about the system" as a combination approaching rich understanding of the system. Figure 1 shows this relationship. Also illustrated in Figure 1 are four so-called critical transitions in the development of understanding of complex systems. These are illustrated by the numbered dots and the resultant-like vectors emerging from one dot and pointing to the next.

It is clear we have made simplifications here. The cognitive-activity axis, for example, should be seen as a continuum, where "noticing" precedes "recalling/connecting" with previous experience, which leads into an assimilation/accommodation loop (to use terminology of Piaget), leading to a gradually enlarged and/or strengthened knowledge base,

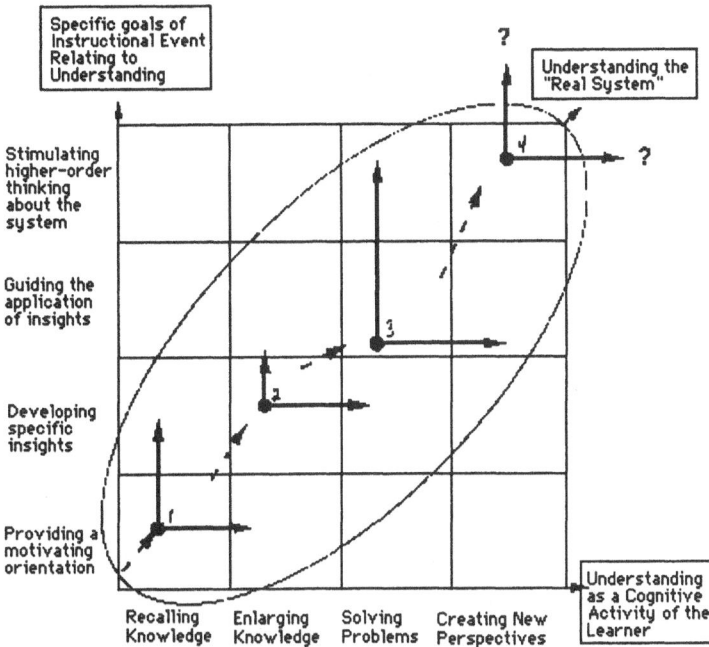

Figure 1   Critical transitions in increasingly sophisticated understanding of complex systems

which is then applied to solving problems of various degrees of complexity, which in turn can lead to synthesizing and creating new perspectives. However, this level of cognitive sophistication can also loop back to reflective strengthening of one's knowledge base, so that, also in turn, new things about the problem space can be noticed and the whole cognitive progression entered again from the origin.

The dimensions of the cognitive-instructional framework shown in Figure 1 are relatively familiar. What might be new is our conceptualization of four critical transitions relative to the framework that we believe are of particular importance to the design and employment of computer simulations. We are defining as critical transitions in the cognitive/instructional matrix those periods when a learner is not "ready" on his own to adequately perform a level of cognitive activity but needs or will particularly benefit from instructional support. The large dots on the grid symbolize points of learner "self-sufficiency" as she progressively comes to better understand a complex system, the vertical arrows leading from the dots represent the categories of instructional goals which seem most appropriate to lead/support the learner as he progresses from one to the next level of cognitive "self-sufficiency." The movement from the origin to the top northeast corner of our cognitive/instructional matrix represents the progression toward maturity with regard to understanding a complex system. We also include an elliptical area on Figure 1 to indicate an hypothesis of the boundaries of variation of instructional strategies most appropriate to different cognitive-activity levels. Clearly, relative to the learner and the definition of "understanding" in a given situation, the endpoint of the understanding process may stop at or between any of the dots within this elliptical path.

With these critical points identified relative to cognitive/instructional understanding axes, we are ready to consider visualization, interactivity, and intelligence as design aspects of computer simulations.

2.5 Visualization and the Critical Transitions

2.5.1 Representing Complex Systems through Visualizations

When we approach the task of "understanding complex systems," it is not often that we can directly deal with the one and only one exemplar of a system without any need for representation of the system. We must represent the system in some way, either by providing data that exemplify the system for direct examination (as with MBLs), or by providing a model to represent the system, as with simulations. Here we have another major issue in the domain of "understanding complex systems"— how well does a set of data or a representation reflect the "real" system? The fidelity of the representation of a system relative to the "real" system is a major issue in simulations (Hoog, Jong, and Vries, 1991, for example, discuss input, output, and time fidelity). Of course, the level of fidelity by definition must be weak in non-well-defined systems. Fidelity is both a quantitative and a qualitative concern, dependent on the level of understanding of the learner and focal point of understanding (structure or behavioural) that is important to given learners. Reigeluth and Schwartz (1989) for example, argue that "full" fidelity might be overwhelming for novice learners, and Towne and Munro (1989) argue that multiple levels of representation complexity should be available within a simulation, to be chosen according to the knowledge level of the learner. Thus while system representative could be expressed through abstract symbols such as formulas or through numeric variable values, it is likely that visualization will be important to both input and output fidelity.

2.5.2 Visualization Complexity

Visualization in simulation software can vary on several dimensions: for example, from symbolic to realistic, or from still to moving. Gradations of detail also occur as another dimension. A graph is an example of a symbolic visualization, often still, but it could be shown in motion, relative to changes in the values which it is symbolizing. In simulations, we can identify five major types of visualizations: still and moving graphs, sketches and drawings, digitized photographs, animations, and moving video. The category sketches and drawings is most ambiguous, in that sketches may include abstract iconic representations or minimalistic realistic representations and thus the boundary between sketches and drawings and other types of visualizations is one of gradation rather than demarcation. Currently, the use of windows for overlapping graphics in simulation is also of interest (Schaick Zillesen, 1990).

2.5.3 Relating Visualization to the Critical Transitions

Over the past decade we have noticed a definite trend with respect to visualization in simulations, especially those for science education. In the early 1980s, if simulations had visualization aspects, these were typically simple line drawings or graphs. Simple animation, such as a fish responding to another member of its pertinent food chain, also were popular. However, in the course of time, the quality of high-resolution graphics available on school computers consistently improved, as did the attempting-to-be-realistic visualizations found in simulation software. New trends are, of course, still occurring in hardware. CD-ROM storage allows the capture and display of large numbers of digitized visualizations. Interactive video systems are becoming more affordable and popular. "Multimedia," often involving an integration of graphic visualizations, animation, and even moving video (perhaps only yet in a window and of short duration and grainy quality, but nonetheless, moving video) now dominate software catalogues and educational software exhibitions. Thus it would seem that the trend relative to visualizations in simulations is simple to describe—toward more visualizations, better quality visualizations, and moving video.

In reference to our cognitive-instructional grid in Figure 1, it would seem that one of two assumptions about visualizations in simulations may be justified. One of these is that "more and moving" is better in terms of visualization, so that no matter where one is on the cognitive-understanding grid, making available quality video in digitized environments is generally a good thing. Another assumption may be that simple visualizations may be best for "simple" cognitive-instructional locations on the grid, but the more complex the cognitive and instructional task becomes, the more complex visualizations are desirable. Thus if we use a different meaning for an overlay sketch of an ellipse, where we interpret narrowness of the ellipse as simplicity (i.e., line drawing, simple graphic) of visualization and greater width of the ellipse as complexity of visualization (complex animations, moving video), the first assumption would involve an overlay such as a rectangular prism over the roughly diagonal path from no to mature understanding shown in Figure 1. In contrast, the second assumption would place a wedge-shaped figure over the diagonal path, narrow at the base, but continually widening as instructional-cognitive complexity increases.

We, however, do not endorse either of these approaches. Relative to the first critical transition area (the lower-left dot in Figure 1), we think rich and realistic moving video may well be best for providing a motivating overview and for triggering the maximal number of recollections for the learner. However, we think the tasks associated with the next two critical transition areas (the middle two dots in Figure 1) are more likely to benefit from simple, representational visualizations, where the learner's attention is guided as cleanly as possible to a focus on relevant aspects of a concept or problem situation. Rich and detailed visualization may in fact distract the student rather than help him focus at these stages of understanding. However, as the learner approaches Critical Transition 4, an overall look at the full system in its complexity, but with the ability to zoom in and out, to digitize the visualization of the system so as to be able to experiment with the manipulation of its components in ways outside the model of the simulation designer may be highly effective techniques.

2.5.4 Visualization Guidelines: The Figure 8

Thus, as a generalization, we suggest that in software for understanding complex systems, interactive video may be particularly useful for orientation; graphs, minimalistic drawings, and simple animations may be best for enlarging knowledge and solving problems; but that advances in digitized video within the simulation environment may be best recommended for learners approaching the "Creating new perspectives/Stimulating higher-order thinking" Critical Transition. We visualize our hypotheses about visualization in Figure 2. The area repre-

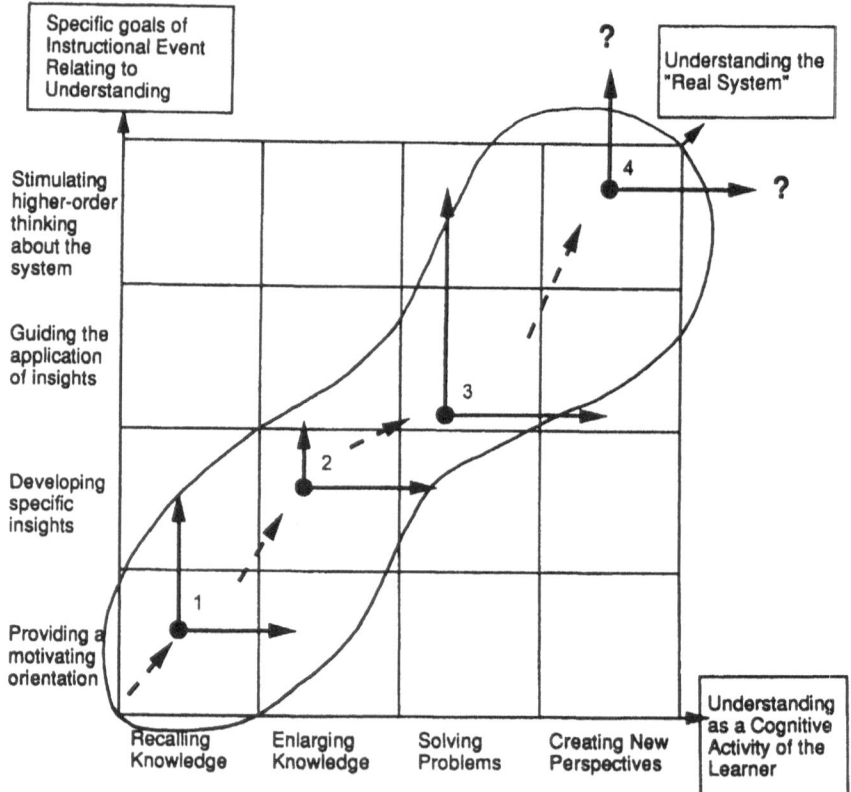

Figure 2 Projection of visualization trends in educational computer simulations. The wider the ellipse, the more appropriate a complex type of visualization such as moving video. The narrower the ellipse, the more appropriate a "simple" visualization such as a sketch or graph. Thus complex visualization is hypothesized as an appropriate design option in the areas of Critical Transitions 1 and 4, and simple visualizations in the areas of Critical Transitions 2 and 3.

senting our hypothesis about visualization guidelines has a resemblance to a figure 8, thus our choice of terminology.

2.6 Interactivity and the Critical Transitions

Earlier we noted trends in interactivity in educational simulations, in particular, trends toward more interpersonal interactivity as part of the simulation-use experience, and trends toward the provision of more options in the software for learner-choice of where and what he will do, browse through, link with, zoom to, experiment with, manipulate, and hypernavigate. Again, the two assumptions discussed with graphics could also be argued as reasonable guidelines for interactivity in simulations. Either give everyone as many options as possible, with as much social interaction as possible, or provide a steadily increasing gradation, so that more mature learners have a wider range of tools, options, and possibilities for collaborative social interaction (even with interaction partners in other countries, through telecommunications-facilitated interaction).

2.6.1 Guidelines for Interactivity: The Ellipse

As before, we do not support either of these as guidelines for interactivity in simulation software. Instead, we see interactivity as premature, perhaps overwhelming or counterproductive near Critical Transition 1, but also perhaps less desirable near Critical Transition 4, where the stage of development of deep understanding in an individual may be distracted by inequities in the comprehensive level of partner interactors, by limitations on time to reflect and speculate, or by constraints on one's imagination imposed by the ideas of the designer of various tool options. Thus, we see the best place for interactivity, either with other students or with a variety of options and tools, as near the "middle" of the cognitive-instructional diagonal, that is, in the area of Critical Transitions 2 and 3. Figure 3 shows our elliptical visualization of guidelines for interactivity in simulations.

2.7 Intelligence and the Critical Transitions

Finally, with regard to embedded intelligence, we also have an hypothesis that can be visualized on the cognitive-instructional grid. This hypothesis, however, corresponds with one of the assumptions we considered and rejected in the cases of visualization and interaction as trends. Our hypothesis with respect to embedded cognitive tools is the "wedge-like" situation, where one's productive use of tools increases with one's cognitive-instructional maturity, relative to the complex system under consideration. Thus Critical Transition 1 may be least appropriate for self-choice and self-use of embedded tools, Critical Transitions 2 and 3 can involve respectively more use of such tools, but Critical Transition 4 offers the greatest possibility for productive use of embedded cognitive tools.

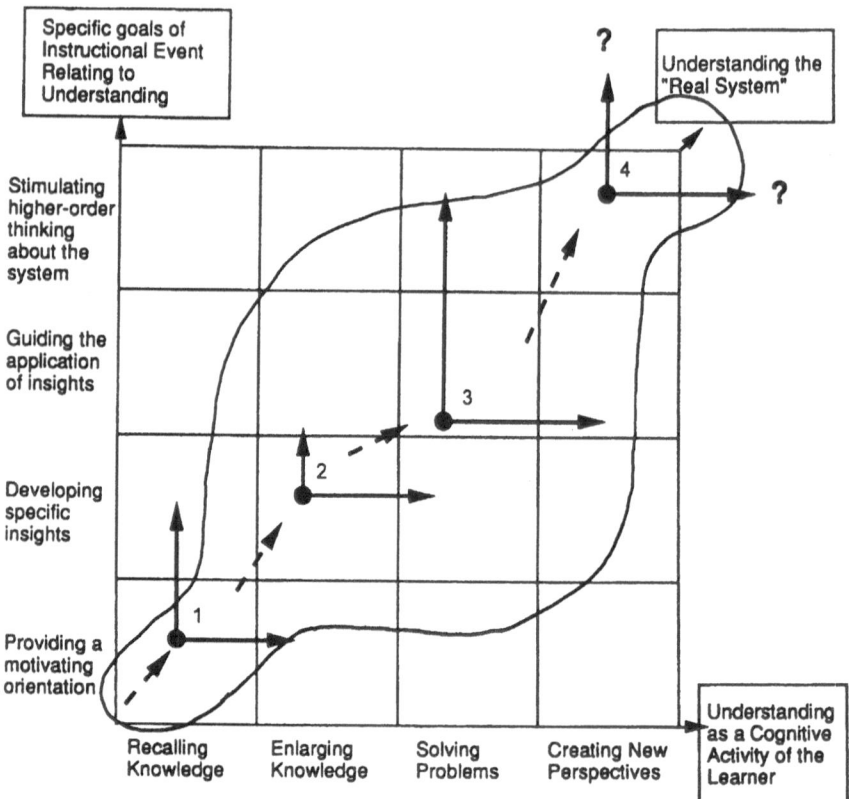

Figure 3   Projection of interactivity trends in educational computer simulations. The wider
the ellipse, the more appropriate the instructional integration of rich and complex interactions. It is thus hypothesized that complex interactions may even be
counterproductive in the areas of Critical Transitions 1 and 4 but are valuable
design options in the areas of Critical Transitions 2 and 3.

2.7.1 Guidelines for Intelligence: The Wedge

Thus we see a wedge as the most appropriate guideline for embedded intelligence in
simulation. Figure 4 illustrates this hypothesis. We are less comfortable with this
guideline, however, relative to our other guidelines for visualization and interactivity,
in that the embedding of more intelligent diagnostic coaching or tutoring probably is
best represented by the Ellipse hypothesis associated with interactivity in Figure 3
than it is with the wedge guideline shown in Figure 4. The wedge guideline, however, does appear to relate to intelligent tools such as tools for cognitive mapping or
for modelling (see Miller et al., 1993).

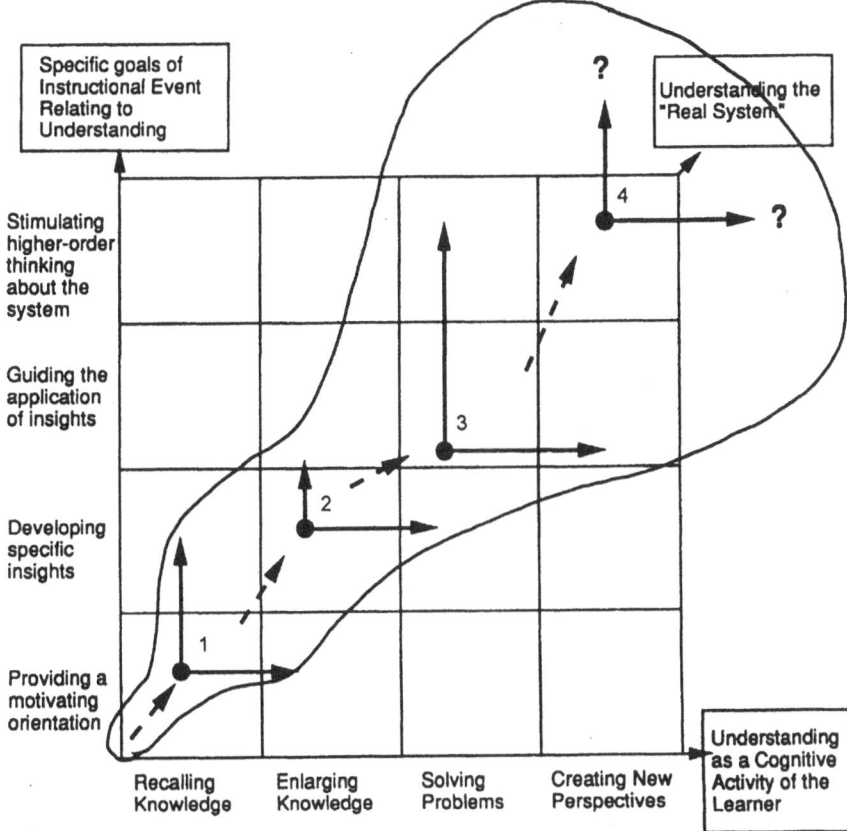

Figure 4   Projection of intelligence trends in educational computer simulations. The guideline is that tools to support metacognitive functioning (i.e., modelling tools and
concept-mapping tools) become increasingly more appropriate as cognitive/in-
structional maturity increases.

2.8 Applying the Trend Analysis to MBLs

So far we have focused on simulations as electronic learning environments with the
aim of helping the student better understand complex systems and relationships within
those systems. However, we feel that MBLs share relevant aspects of simulations
and thus can also be considered relative to the cognitive-instructional grid and our
visualization, interactivity, intelligence trends and guidelines. We feel this application to be especially valuable because of the limited attention that has so far been
given to the design of the software component of MBL environments. Where do
MBLs stand now with respect to visualization, interactivity, and intelligence? What
are guidelines for their future development? To address these speculations, we briefly
look first at the evolution of MBLs.

2.8.1 Evolution of MBLs

During the beginning of the 1980s, when microcomputers were first being introduced into schools, one of the ideas that challenged teachers' imaginations was the possibility of using the computer for collecting, processing, and plotting data as graphs. This was connected with science and math education and later became a separate direction of research and development with respect to the implementation of computers in educational environments. In the middle of the 1980s many projects involving the use of the computer to capture, process, and display experimental data as graphs were started in the US and also the UK. The main goals of these projects were, from one side, to improve the use of the microcomputer in schools for the above-mentioned goals and activities, but also, from the other side, to explore the educational results and cognitive effects of using computers for science teaching in nontraditional ways, including the idea that computers could augment some of the standard activities in the school science lab.

A landmark project in this evolution was the "Computer as Lab Partner (CLP) Project" (Stein, Nachmias, & Friedler, 1988). This project was designed to examine the cognitive consequences of MBL for various aspects of eighth-grade students' science learning. Each MBL system, used by a pair of students, included an Apple IIe, temperature probes hooked to the computers, and MBL software. Students spent more than half their time performing laboratory activities which investigated thermal phenomena. It was found, among various results, that students improved their ability to identify graph trends and to extract the meaning of the information presented (Linn, Layman, & Nachmias, 1987); and that students' graphing misconceptions were replaced with more accurate conceptions (Mokros & Tinker, 1987).

2.8.2 Visualization and MBLs

Together with the invention of the term, "Microcomputer Based Lab," the educational goals of using MBLs came to be defined more clearly, such as various aspects of the development of graphing skills. For example, students during one of the CLP experiments were to create graphs showing the interdependence among different variables and to see the dynamic relationships between those variables (Linn, Layman, & Nachmias, 1987). This main goal of MBLs was thus connected with some specific and measurable cognitive gains for the learners. Cognitive learning goals came to be further defined, relative to comprehension and manipulation of graph features and graph templates, to graph-design skills and graph problem-solving skills (p. 245–247). It became possible also, together with the visualization of existing physical laws, to make the next step and to interpret already-given graphs, being able to articulate some of the interdependencies between variables, and thus to increase the graph-related problem-solving skills of the students. An example of this is the "Back to the Future" graph approach (Mokros & Tinker, 1987), in which students had to interpret a graph that goes backward in time. This continual refinement of cognitive expectations for interpretation of MBL displays also led to an enhancement of the requirements for better visualization inside the MBL environment, which in turn led to even more complex cognitive goals, such as evaluation of different kinds of complex information using MBLs. Another study from the Berkeley group involved with the CLP Project, which was devoted to evaluations of science lab data and the role of computer-presented information (Nachmias, & Linn, 1987), illustrates this evolution of visualization in MBLs. The study was directed toward three purposes: "to assess the extent to which students critically evaluate computer-presented graphs, to examine the effect of an extensive use of MBL on students' critical evaluation skills, and to assess the effect of enhanced explicit instruction on the development of these skills" (p. 493). In this study, the instruments for critical evaluation of graphs opened another direction of evolution of MBLs—interactivity.

2.8.3 Interactivity and MBLs

The instrument in Nachmias and Linn's 1987 study was devoted to the analysis of the causes behind five cases of invalid or unreliable graphs: errors in graph scaling, in probe setup, probe calibration, probe sensitivity, and errors occurring through experimental variation. This analysis needed interactivity for its investigation. More work began to occur to increase the interactivity aspects of MBL use, for example through the provision of computer-generated feedback as a design feature within the MBL (Friedler, Nachmias, & Linn, 1988). Not only for increasing the level of interactivity but also for fostering students' thinking skills, the on-line feedback provided by the computer in conjunction with an appropriate instructional environment was expected to foster students' thinking skills. This indeed was also an implicit investigation relating to increasing the "intelligence" of the MBL.

2.8.4 Intelligence and MBLs

The goals in this aspect of Friedler, Nachmias, and Linn's 1988 study were to examine the use of MBLs to foster true scientific reasoning skills, observation, and prediction. There were three activities in this study, all carried out within an MBL environment. These were: "(a) off-line activities and games that introduced the concepts of observation and prediction and their role in the process of problem solving, (b) domain general computer games, and (c) a series of experiments investigating the temperature flow of liquids during heating and cooling." This study also showed that, except for collecting, recording, and manipulating data, the MBLs existing at that time did not in themselves serve other activities connected with knowledge acquisition and processing. Only the operational part of the problem-solving process was supported and improved through the use of the MBLs, but not the higher-order thinking-skills aspect. It will have to depend on what the teacher and student do with the MBL environment that will bring its use into the metacognitive domain.

In the last few years, however, initiatives to enhance the intelligence of MBLs, such as building into the MBLs models for simulation and for system modelling in order to increase the understanding of the system, have begun to appear. New concepts were developed to express trends in this direction—such as The Computer as a Lab Partner Curriculum (Linn & Songer, 1990). As an example of these concepts, an MBL was integrated with a model of thermodynamics in order to provide a coherent explanation for a class of interrelated problems, rather than risk MBL use being associated with isolated understanding of individual problems. This conjunction of MBL and intelligent support tools can come to help the science educator improve science education by focusing not only on fundamentally important knowledge domains, but also on the strategies necessary for problem solving in those domains.

Another good example is the IP-COACH system (see Hartsuijker, Bart, & Zandbergen, 1992), a modular MBL program developed by a team at the Department of Physics Education of the University of Amsterdam. Their original MBL was extended with two modules, CALCSHEET and MODEL-ENVIRONMENT, where the last module is comparable with some existing programs for dynamic system modelling. The MODEL-ENVIRONMENT module allows the student to put in parameters and starting values of variables, which are interpreted in a mathematical model in the form of a differential equation, and later to compare the results received from the simulation model and from the original MBL.

2.8.5 MBLs and the Three Trends: Where Are MBLs Now?

Referring to the three main trends—visualization, interactivity, and intelligence—and projecting the evolutionary process of MBLs, as described above, on the cognitive-instructional grid used earlier for the consideration of educational computer simulations (Figure 1), we offer the conclusion that MBLs at this moment are "in the middle of the road." As a field, we are trying, often successfully, to solve problems and to guide the application of insights using MBLs, but we still have limited experience with moving students to Critical Transition 4—"Create new perspectives/Stimulate higher-order thinking"—within the traditional design of MBL software. That is why, reflecting on the visualization, interactivity, and intelligence trends, and looking for new areas of enlargement and enrichment of the MBL concept, we suggest it is time to rethink some design aspects of MBLs. For example, MBLs frequently assume various limitations (amount of data that can be captured and analyzed, number of dimensions available for graphing interdependencies [MBLs now typically use only two dimensions], etc.). Using the contemporary techniques that are now enriching simulations, we may be able to overcome some of these design limitations in MBL software and thus reach more effectively the desired educational goals. In terms of our construct of the idea of Critical Transitions in educational simulations (Figure 1), the above-described evolutionary process of MBL software appears to be a bottom-up procedure, starting at the origin of the graph and moving "up," toward the middle of the grid, now generally located somewhere around "Enlarging Knowledge" as a cognitive activity and "Developing Specific Insights" as an instructional strategy.

Looking at the same grid but in the framework of the guidelines for visualization, interactivity, and intelligence that we discussed for simulations (see Figures 2, 3, and 4), we interpret this as implying that the current state of MBL software is typified by relatively simple visualizations (thus most appropriate for Critical Transition areas 2 and 3), not a very high level of interactivity (thus most appropriate for Critical Transition areas 1 or 4), and only a limited intelligence (thus most appropriate for Critical Transition areas 1 and 2).

However, although there is currently an insufficiency or mismatch—of visualization, interactivity, or intelligence in MBL software—work is already underway to try to find ways to balance these insufficiencies. Thus we next turn to what we see as top-down evolutionary processes using multimedia techniques in science teaching. We believe these media-driven techniques may be able to augment some of the inadequacies in MBL software environments from a cognitive/instructional perspective. We call this area of development "Multimedia Labs."

2.8.6 Evolution of MML (Multimedia Labs)

Looking into product descriptions in the latest catalogues of multimedia educational software, we can see technically sophisticated products using video and computer graphics for explanation of basic principles in biology, physics, and chemistry. We call these packages MMLs—multimedia labs. For the accompaniment of experiments, fixed simulation models are recorded on the same videodisc as that where extensive collections of graphical images and video sequences are available. For example, the IBM Biology Series[2] is described as being developed around a tutorial-oriented instructional design, but it includes colour graphics, simulations and animations, as well as support tools such as on-line help and glossary, so that the student can better understand difficult concepts. The package, "Chemical Life Processes Explained," of the IBM Biology Series, is a particular example. It works with the same parameters as many MBLs—pH, temperature, and substrate. The package "Discover by Exploring" (also distributed by IBM) shows how an interactive videodisc can be integrated with such multimedia learning environments. Here the videodisc works with software that guides its use in 30 sequenced lessons. Students can, via the videodisc, observe chemical reactions, plan and carry out kinetics experiments, and be guided as they work through simulated experiments. Students can see the results of their mistakes, watching on video and observing feedback delivered via graphics.

A weakness in these multimedia collections is that the models underlying the simulations used within them cannot be changed; thus the system is described once through a mathematical model, and the student is constrained in his exploration and capacity to simulate and change the system under investigation, to try different alternatives. Instead, the simulation is only of one fixed process. Thus the cost of increased visualization via storage on videodisc or CD-ROM is a limitation on what can be explored in a simulation environment.

There are also examples in the "IBM Physics Discovery Series" which are very similar to the starting point of MBLs, but in a highly visualized version. The package "Applying the Laws of Motion" is an example. (Early MBLs often worked with pH

---

[2] IBM, Educational Fulfillment Center, P.O. Box 666, Dayton, NJ 08810-9988

or velocity as measured by an ultrasonic sensor, used for the "discovery" of Newton's Laws.) In MML packages such as the discovery series, combining characteristics of both simulations and MBLs, students can build hypotheses as to how changing certain characteristics of a moving object (i.e., a car) affects its performance, but mainly the students can change only one variable (in the case of the car, the time variable).

Following the trends of evolution of MBLs, the MMLs (multimedia labs) are trying also to increase the level of visualization available to the student (see, for example, the use of "Quicktime movies" in the package "Operation: Frog"[3]), or conversely are trying to increase interactivity at the cost of decreasing visualization. The product "Interactive Physics II"[4] is a physics simulation laboratory that allows students to build and model simple and complex experiments by facilitating users to draw and build objects on the computer screen, define physical properties such as mass and velocity for each object, set up and run experimental environments, then save the animations they create as "Quicktime" movies. The need for systems-oriented thinking and explanation in science is of considerable importance during the educational process, and is leading to the development of very complex MMLs such as "Rediscover Science"[5] and "Science 2000"[6]. The former includes a series of lessons organized in separate modules connected with an increasing number of suggested lab activities; ideas for science projects; and reading, writing, and thinking activities. These products are all available stored on a single CD-ROM disk, but the producer suggests supplementing the disk with videotapes available from the vendor, Encyclopedia Britannica. The "Science 2000" package includes two videodiscs and "hands-on manipulative kits" as well as software and a teacher's guide. This kind of bundling is typical for this stage of the evolution process of the MMLs, in that many producers are now trying to include hands-on manipulative kits as part of their multimedia packages, and to break somehow the limitations of the simulation models recorded on fixed storage media (i.e., videodisc or CD-ROM). Indicative of this direction is a research project now in progress at the University of Amsterdam, which connects hands-on MBLs with multimedia using interactive video. This represents an attempt to connect the stronger interactivity elements of the MBLs with the stronger visualization features of MMLs so that the best elements of each can complement the relative weaknesses of the other. From our guidelines related to Critical Transition areas, however, this may not be the best direction of design development, in that we hypothesized rich, moving-video visualizations as best for Critical Transition areas 1 and 4 and interactivity complexity as best for Critical Transition areas 2 and 3.

[3] Scholastic Software, Jefferson City, MO 65102
[4] Knowledge Revolution Inc., 15 Brush Place, San Francisco, CA 94103
[5] Edunetics Corp., Arlington, VA
[6] Decision Development Corp., San Ramon, CA

2.9 MBLs, MMLs, and Simulations: Mutual Enrichment

We see that the evolutionary processes associated with MBLs and MMLs can fit very well with our theoretical model, not only because of the content of the process of how we build simulation models, but because of the fact that all these types of instrumentation are striving toward the same goal—the better understanding of complex systems and processes in science. In this sense, sometimes it is very difficult to say, for some flexible and open-ended tools used in secondary math and science education, such as "The Explorer Series: Physics and Biology Explorer,"[7] where the border is among simulations, MBLs and MMLs, as they are more and more being integrated together. Especially good examples of these sorts of integrated products with the goals "system-thinking improvement" and "complex-systems understanding" are software packages for environmental education, such as "Biology Explorer: Population Ecology,"[8] "A Field Trip to the Rain Forest,"[9] and "Interactive Nova—Race to Save the Planet."[10] In the area of environmental studies, often it is very difficult to illustrate the complex relationships among the different species that live in an ecosystem. That is why for this type of content area, a product like "Field Trip to the Rain Forest" that includes illustrations showing each species in its natural habitat, but also accompanying books, sets of disks, on-line guides, and data cards providing information about organisms' homes, food, enemies, and friends, can be educationally appropriate. Using all these resources, students can simulate and graph (as with MBLs) food-chain activities and identify relationships between the different organisms. The package "Population Ecology, Discovering Ozone Module"[11] has software for graphing data as well as manipulating simulation models. Many of the environmental packages have video components.

There is still a long way to go from our current levels of simulations, MBLs, and MMLs to the "perfect" science lab. We need the balanced use of all available technological resources to present to the student the richness and dynamic behaviour of the real world in its full complexity. But we also need to use such resources judiciously, as more is not necessarily better in terms of visualization, interactivity, and embedded intelligence in computer simulations, MBLs, or MMLs. Our simple hypotheses of a so-called Figure 8 guideline for visualizations, an Ellipse guideline for interactivity, and a Wedge guideline for embedded intelligence are offered as a contribution to this design problem for simulations, MBLs, and MMLs.

---

[7] Wings for Learning, 1600 Green Hills Road, P.O. Box 660002, Scotts Valley, CA 95067-0002
[8] Wings for Learning
[9] Wings for Learning
[10] Scholastic Software, P.O. Box 7502, Jefferson City, MO 65102
[11] Wings for Learning

References

Collis, B. (1988) *Computers, curriculum, and whole-class instruction*. Belmont, CA: Wadsworth.

Forman, E., & McPhail, J. (1989, March) *What have we learned about the cognitive benefits of peer interaction? A Vygotskian critique*. Paper presented at the Annual Meeting of the American Educational Research Association, San Francisco.

Friedler, Y., Nachmias, R., & Linn, M.C. (1988) *Using microcomputer-based laboratories to foster scientific reasoning skills*. Internal paper, Graduate School of Education, University of California, Berkeley, CA.

Gredler, M.B. (1986) A taxonomy of computer simulations. *Educational Technology*, 26(4), 7-12.

Hartsuijker, A.P., Bart, C. van, & Zandbergen, P. van. (1992) *Courseware development and educational research in the Dutch approach to MBL in chemistry education*. Paper prepared for the NATO Advanced Research Workshop, Microcomputer Based Labs: Educational Research and Standards, Amsterdam, 9–13 November, 1992.

Hollan, J.D., Hutchins, E.L., & Weitzman, L.M. (1987) STEAMER: An interactive, inspectable, simulation-based training system. In G. Kearsley (ed.), *Artificial intelligence and instruction: Applications and methods* (pp. 111–134). Reading, MA: Addison-Wesley Publishing Company.

Hoog, R. de, Jong, T. de, & Vries, F. de. (1991) Interfaces for instructional use of simulations. *Education & Computing*, 6, 359–385.

Jonassen, D.H. (1988) Integrating learning strategies into courseware to facilitate deeper processing. In D.H. Jonassen (ed.), *Instructional design for microcomputer courseware* (pp. 151–181). Hillsdale, NJ: Lawrence Erlbaum Associates.

Jong, T. de., & Njoo, M. (1990, September) *Learning and instruction with computer simulations: Learning processes involved*. Paper presented at the NATO Advanced Research Workshop on Computer-Based Learning and Problem Solving, Leuven, Belgium.

Kommers, P.A.M., Jonassen, D.H., & Mayes, J.T. (1991) *Cognitive tools for learning*. NATO ASI Series F, Vol. 81. Berlin: Springer-Verlag.

Linn, M.C., Layman, J., & Nachmias, R. (1987) The cognitive consequences of microcomputer-based laboratories: Graphing skills development. *Journal of Contemporary Educational Psychology*, 12(3), 244–253.

Linn, M.C., & Songer, N.B. (1990) *Teaching thermodynamics to middle school students: What are appropriate cognitive demands?* Internal paper, Graduate School of Education, University of California, Berkeley, CA.

Miller, R., Ogborn, J., Briggs, J., Brough, D., Bliss, J., Boohan, R., Brosnan, T., Mellar, H., & Sakondis, B. (1993): Educational tools for computational modelling. *Computers and Education*, 21(3), 205–261.

Mokros, J.R., & Tinker, R.F. (1987) The impact of microcomputer-based labs on children's ability to interpret graphs. *Journal of Research in Science Teaching*, 24, 369–384.

Moonen, J., & Stanchev, I. (1992) *Theory and instrumentation for visual communication in teaching and training*. Internal document, Faculty of Educational Science and Technology, University of Twente, Netherlands.

Nachmias, R., & Linn, M.C. (1987) Evaluations of science laboratory data: The role of computer-presented information. *Journal of Research in Science Teaching*, 24(5), 491–506.

Reigeluth, C.M., & Schwartz, E. (1989) An instructional theory for the design of computer-based simulation. *Journal of Computer-Based Instruction*, 16(1), 1–10.

Scardamalia, M., Bereiter, C., McLean, R.S., Swallow, J., & Woodruff, E. (1989) Computer-supported intentional learning environments. *Journal of Educational Computing Research*, 5(1), 51–68.

Schaick Zillesen, P.G. van. (1990) *Methods and techniques for the design of educational computer simulation programs and their validation by means of empirical research.* Doctoral dissertation, Faculty of Educational Science and Technology, University of Twente, Netherlands.

Songer, N.B. (1989) Technological tools for scientific thinking and discovery. *Reading, Writing, and Learning Disabilities*, 5, 23–41.

Stein, J.S., Nachmias, R., & Friedler, Y. (1988) *An experimental comparison of two science laboratory environments: Traditional and microcomputer-based.* Internal paper, Computer as Lab Partner Project, University of California, Berkeley, CA.

Tennyson, R.D., & Thurlow, R. (1987) Problem-oriented simulations to develop and improve higher-order thinking strategies. *Computers in Human Behavior*, 3, 151–165.

Towne, D.M., & Munro, A. (1989) Artificial intelligence in training diagnostic skills. In D. Bierman, J. Breuker, & J. Sandberg (eds.), *Artificial intelligence and education: Synthesis and reflection* (pp. 291–297). Amsterdam: IOS.

Towne, D.M., Munro, A., Pizzini, Q.A., Surmon, D.S., Coller, L.D., & Wogulis, J.L. (1990) Model-building tools for simulation-based training. *Interactive Learning Environments*, 1, 33–50.

3. MBL, MML and the Science Curriculum— Are We Ready for Implementation?

Ard Hartsuijker[1], Yael Friedler[2] and Frits Gravenberch[1]

[1] National Institute for Curriculum Development (SLO)

[2] Hebrew University of Jerusalem

Abstract. In this paper dilemmas are raised concerning the development and implementation of microcomputer-based laboratories (MBL) and multimedia laboratories (MML) in science education for students in the age range 12–18 years old. Our purpose in doing this is to initiate a discussion on the advantages of MBL and MML in science education. Recent trends in science curricula show various areas where MBL and MML could contribute to improve science education. Any evidence of advantage should be a basis for subsequent curriculum development and educational research.

3.1 Introduction

The aim of this paper is to raise some major dilemmas in curriculum development and educational research concerning the development and use of MBL and MML in science education. With these dilemmas a discussion on the contributions and advantages of implementing MBL and MML in science education will be encouraged. Firstly, in order to do so, some concepts to be referred to in this paper are defined.

1. "Science education" refers to the traditional education programs for the sciences— biology, chemistry and physics, as well as to interdisciplinary topics, such as environmental studies and the relevant topics in technology education.

2. In the paper the focus is on general education for students in the age range from 12 to 18 years. Depending on the national educational system, this involves lower and higher secondary education, or junior high and high school. In some countries all the issues raised can also be applied to the first and second years of college.

3. The concept curriculum, as used in the paper, refers to the broad term as appears for example in Walker (1990), which he abbreviates as follows:

   A curriculum refers to the 'content' and 'purpose' of an educational program together with their 'organization'.

A science curriculum plan informs about: aims, syllabus, and curriculum materials. In functionality the plan should act as an educational framework, indicating how a program and teaching materials for science education can be effected.

- The "aims" include a definition of cognitive goals as well as affective goals.

- The syllabus includes a general hierarchy of concepts and skills development, as well as a general time sequence, including some more detailed specified targets related to topics and themes.

- The "curriculum materials" include the description of the materials (i.e., teacher guides, students' books, lab experiments, field work) by which the intended curriculum can be realized.

- Often the curriculum plan is completed with information about pedagogical methods and student evaluation.

4.   The following definitions for MBL- and MML-applications in science education are used:

- "Microcomputer-Based Labs" (MBL) are programmed microcomputers interfaced with laboratory sensors and actuators, which allow students to carry out automatized laboratory experiments and enable them to collect, process and represent data.

- "MultiMedia Laboratories" (MML) are programmed microcomputers, which are used in science education to present, explain and explore scientific principles and processes by means of audio, video, animations and simulations. MML could be accompanied by sensors.

The paper considers MBL as well as MML. This is based upon two facts related to students learning activities in science education:

- MBL can not only be realized with traditional classroom laboratory experiments but also with real-life situations which are stored in multimedia databases, as is explained in the paper of Ellermeijer, Landheer and Molenaar (1992).

- In science education educational computer simulations, MBL and MML are striving towards the same goal—to accomplish a better understanding of complex systems in science, as is noticed and explained in the paper of Collis and Stanchev (1992).

3.2 Trends in Science Curriculum Development

The first section of this paper considers some general trends in the history of curriculum development and educational research (section 3.2.1). This is followed with a discussion on recent trends in the aims of science education (section 3.2.2). Finally, curriculum development decisions, which are needed to combine old and new trends to new curricula, are discussed (section 3.2.3).

3.2.1 Historical Trends

The development of science curricula in most of the western countries can be roughly divided into three periods:

1. During the first period, the late 19th and the first quarter of the 20th century, the number of students in high schools was relatively low and the population was very homogeneous in terms of socioeconomic background and career expectation. The aim of high school teaching was to prepare students for university studies. As a consequence of this policy the topics learned in high school reflected the topics offered to the students in university. The "high school curriculum" was written by scientists from the university and its main goal was to teach the "up-to-date scientific knowledge."

2. The second period, up to the late '50s, is characterized by a growth in both the relative number of students and heterogeneity in their abilities. As a result, some particular scientific topics, which were suspected to be difficult to teach to less able students, were excluded from the curriculum. Besides, more effort was made to incorporate practical consequences of educational theories in the science curriculum. In this period more teachers and educators became involved in the development of the science curricula.

3. The third period, from the late '50s onward, is characterized by efforts to combine the two trends mentioned earlier: to teach "up-to-date scientific knowledge," an aspect which was neglected during the second period, and at the same time to take into consideration the large and heterogeneous population of high school students. This was the period when "new science curricula" were developed (such as: the BSCS and the PSSC in the USA, Chem Study and Nuffield in the UK, CMLS and PLON in the Netherlands). The emphasis in these curricula was on teaching scientific concepts and principles as well as inquiry skills.

Since the late '70s a major effort of educational researchers has been directed to the understanding of the students' cognitive structure, as well as the learning processes that occur with students from various levels and ages who develop scientific concepts. Many articles deal with students' misconceptions, naive conceptions and alternative frameworks. Some of these articles have a particular section on "implications for teaching." In this way research findings "found their way" into the science curricula (e.g., concept maps, V-maps, advanced organizers which make analogies to known everyday phenomena).

Another major development which affects science curricula is the growing availability of relatively inexpensive microcomputers. Since the early '80s many schools have been equipped with computers, and programs for courseware development were launched. As a result an increased number of groups started to develop courseware, computerized environments and microworlds, and investigated how these could be incorporated into the science curricula.

3.2.2 Recent Trends

More recent curricula still reflect some of the goals that guided the "new science curricula" developments in the late '50s and the '60s. However, some other goals result from an emphasis on new trends in society, development of new technologies, requests for better qualification by universities, and results of cognitive research as were mentioned above.

General goals of present science curricula can be divided into several domains. These include:

1. concept and skill development;
2. understanding the interaction between science, technology and society;
3. development of problem-solving skills;
4. development of inquiry and motor skills in scientific laboratory and outdoor field-work investigations;
5. preparation for further studies and future careers.

The next four sub-sections elaborate some of the new trends in science curricula.

Change towards "Realistic" Education. Initiated by the demands of society, there is a tendency to include experiences from everyday life in both science curricula and cognitive development of concepts and skills. In many countries the STS trend (science, technology and society) is enhancing the connections between science in schools and problem solving in real-life contexts.

With respect to this trend, the following goals are emphasized:

* to understand new concepts, processes and theories, which relate to the STS domain;
* to acquire knowledge of historical details and philosophical aspects of major turning points and inventions;
* to understand applications of science in technology and in other disciplines, such as medicine and agriculture, in society and everyday life;
* to develop positive attitudes concerning scientific and technological impact on society, and at the same time consider moral, sociological and economic issues;
* to develop awareness of the limitations of science and technology when it comes to solving problems that relate to mankind, environment and society;
* to prepare for responsible citizenship;
* and to develop awareness in consumer behaviour.

Emphasis on Scientific Approaches. With respect to the growing emphasis on teaching and learning in laboratory experiments and in outdoor fieldwork, the goals of the curricula are to:

* understand the epistemology of science and the way scientists work;

- develop inquiry skills—the ability to define a problem, to state hypotheses, to plan experiments, to report data and to conclude on the results of the experiments;

- develop critical thinking skills—the ability to criticise scientific data, to change ideas in relation to new data, and to discriminate between data, conclusions and interpretations;

- develop data manipulation skills—the ability to understand scientific reports, to search for data, to discriminate between major and marginal information;

- develop motor skills needed for learning in laboratory experiments and outdoor fieldwork.

Better Preparation for College and University. One of the major goals in the educational systems is to narrow "the gap" between high school and college/university. For example, from recent studies in the Netherlands (LICOR/HBO, 1992) it was concluded that students:

- lack the capability to transfer problem-solving skills;

- lack general skills of studying and learning.

With respect to this trend, science curricula goals which aim at strengthening the high school–college/university link are to:

- acquire general abilities and skills, such as metacognitive skills, which would help students in higher education;

- contribute to a better selection of favourable university studies and future career.

Emphasis on Individual Learning. Based on cognitive research, curriculum development groups consider different learning styles of different students and emphasize their individual needs. In trends on individual learning they foster the affective domain. The goals which reflect these trends are to:

- develop positive attitudes towards science;

- enhance motivation to learn science;

- obtain confidence to apply science knowledge and skills;

- develop social skills in cooperative science learning.

3.2.3 Curriculum Development Decisions

A science curriculum plan results from cooperation amongst experts involved in the process of combining "old" and "new" trends in science education. On the basis of research and development they design: a syllabus, teaching materials, classroom laboratory experiments, multimedia laboratories, outdoor fieldwork, and so on. Their expertise includes curriculum design, educational instrumentation, science and educational research, educational evaluation and educational writing. Walker (1990) characterises the development process as a "deliberate approach:"

> In broad outline the process of curriculum development is one of formulating a platform of ideas, using these ideas to conceptualise the problem and to generate promising preliminary versions of materials, assessing the merits of promising early versions, and revising them until they cannot be improved further.

Although science curricula differ within countries, generally speaking one might say that the curriculum document includes many implicit and explicit decisions regarding the elements: "content," "purpose" and "organization."

Examples of general decisions which should be taken in curriculum development are:

- Who is entitled to make demands and who decides what has to be included in or excluded from the national examination program and the curriculum?
- What total amount of science matter should be taught, in correspondence with the national examination program?
- What are the minimum and preferable number of hours available in the timetable for science education?
- What are preferable pedagogical approaches and taxonomies applicable in the process of teaching and learning?
- What is the preferable balance between the types and number of "theoretical" and "experimental" teaching and learning environments?
- What are bottlenecks with respect to concept and skill development?
- How can students vary regarding an individual learning route and strategy?
- What are good examples of teaching materials (textbooks, laboratory equipment, multimedia)?

  Answers to these questions are given on different levels of curriculum development:

- Some of these questions are decided at a national level and can be answered in the "national" science curriculum.
- Other questions are to be decided at school level or at the level of the individual teacher. The answers should become a part of the school- or classroom-curriculum or in the students' teaching materials.

3.3 MBL, MML and the Science Curriculum

How do MBL and MML applications match with the new goals of science curricula? What kind of curriculum decisions concerning MBL and MML should be taken? In order to answer these questions, the present achievements of MBL and MML technological developments are described (section 3.3.1), and a review of some of the major articles which look at MBL and MML from various points of view (section 3.3.2) are reported.

3.3.1 MBL and MML, Technological Developments

MBL was one of the early technologies which found their way into the science class-room after microcomputers entered the schools (Tinker, 1984). The MBL-era started with simple sensors interfaced to the computer. The sensors enabled students to collect physical data in a laboratory experiment (e.g., temperature, light intensity, heart-beat rate) and to observe real-time graphical representations of the phenomena on the computer screen.

The following directions in MBL technological developments are observed:

a. Improvement of hardware:

   - improving the quality of the input/output electronics;

   - adding different kinds of sensors and actuators, and improving the quality of the existing ones.

b. Improvement of software by adding different sub-programs for better:

   - data collection (i.e., time- and event-triggering, standard calibration data);

   - data processing (i.e., smoothing and filtering operations, complex manipulation by means of spread sheet operations);

   - data representation (i.e., various graphical representations);

   - data analysis (i.e., mathematical operations).

c. Improvement of user interface—faster speed of the computer processor, higher resolution of the computer screen, colours, and more friendly interaction with the student.

d. Extension of software by adding sub-programs for:

   - controlling laboratory experiments (i.e., steering actuators);

   - system dynamic modelling (to compare the outcomes of simulations with the real data gathered by the sensors);

   - data capturing from video representations of physical, chemical and biological phenomena in real-life situations.

MML technologies with audio, video, animations and simulations that aim to explain and present scientific principles and processes with data from science and the world outside school are still in their infancy. However, recent developments on DV-I and CD-I show that the technological limitations are declining.

3.3.2 MBL and MML, Curriculum Developments and Educational Research

The studies and research reports presented in this section have been categorized into four sections according to the major points discussed in each of the articles:

1. development of graph interpretation skills;

2. development of inquiry skills and critical thinking;

3. acquisition of subject-matter knowledge;

4. enhancing school implementation.

Development of Graph Interpretation Skills. Mokros and Tinker (1987) discussed the potential of MBL in developing graphic skills. They examined the intensive involvement of students using the three senses while operating the sensors and performing lab experiments. They evaluated the ability of junior high school students to understand graphs and found that students have two major difficulties in understanding graphs:

a. students look at graphs as pictures of the phenomena being studied;

b. students have difficulties understanding the meaning of the slopes of the graphs, while studying phenomena.

Mokros and Tinker illustrated that in using MBL, students improve in the understanding of graphs—even though graph-understanding was not directly part of the curriculum (which was aimed at learning scientific concepts).

Linn, Layman and Nachmias (1987) also found that MBL helps junior high school students to analyze information presented in graphs, as well as to build graphs and draw conclusions based on the information presented in the graph.

Brasell (1987) found that a teaching/learning period as short as a single class period with a MBL-motion unit is sufficient for high school physics students to improve their understanding of distance and velocity graphs when compared with a pencil and paper graph construction control treatment. She found that most of the improvement can be attributed to the real-time graphing feature of the MBL.

Adams and Shrum (1990) did not find any improvement in graph interpretation abilities among 10th grade biology students using MBL. They also found that traditional laboratory exercises, which allow students to practice graph construction skills, result in higher student achievement on graph-construction tasks.

Stein, Nachmias and Friedler (1990) found that junior high school students who use MBL draw equally valid conclusions from their graphed data when compared to students who learn in the traditional way. They also found that students need larger setup times and more off-task behaviour in the traditional mode.

DeBeurs and Ellermeijer (1992) found that knowledge of MBL-hardware (on a system level) helps students to interpret graphs and to overcome measurement problems.

Hartsuijker, Van Bart and Zandbergen (1992) confirmed the results of deBeurs and Ellermeijer. The researchers reported that using MBL within the chemistry domain results in an improvement of students' understanding of chemical processes. The students improved in acquiring critical thinking abilities and they also showed an increase of involvement in the chemistry subject matter.

Development of Inquiry Skills and Critical Thinking. Friedler, Nachmias and Songer (1989) and Friedler, Nachmias and Linn (1990) designed a teaching module for junior high school students to foster scientific reasoning skills and to assess its effect within the microcomputer-based laboratory environment. The authors report about successful instruction of inquiry skills in the MBL environment. They also stress the essential role of the teacher in guiding and coaching the students while they use the MBL to collect and analyze computerized data.

Nachmias and Linn (1987) assessed the extent to which students evaluate computer presented information and the effect of extensive use of MBL on the development of such a critical evaluation skill among the students. Following an instruction period, the students improved their ability to analyze the causes behind invalid graphs which were presented by the computer.

Acquisition of Subject-Matter Knowledge. Songer and Linn (1991) characterized middle school students' beliefs about science, the learning of science, and the relationship between these beliefs and the integration of scientific knowledge in the domain of thermodynamics. The study focused on students involved in the Computer-as-Lab-Partner curriculum (CLP) using MBL for real-time data collection. The same group of researchers (Linn and Songer, 1991) also designed a thermodynamics curriculum regarding understanding thermodynamic concepts by middle school students. These studies demonstrate substantial improvement in student understanding when the intellectual demands are put on the students, and not on the software. They also stress the advantages of using research on learning and instruction to reformulate the CLP curriculum.

DeBeurs and Ellermeijer (1992) report about the developments of MBL curriculum materials for physics, connected to an extension of certain aspects of computer science and information technology in physics. They postulate that if the aim is to prepare students for independent lab work in the physics classroom, more time and attention should be allocated to numerical approach and system dynamic model-building, and less time should be spent on the mathematics part of the curriculum. In a new syllabus they tried to teach physics concepts intertwined with computer science.

Ellermeijer, Landheer and Molenaar (1992) found that interactive video MBL (with real-life situations) helps to improve students' understanding of basic mechanics concepts. They also notice that students are not aware of the fact they learn more and differently, because they seem to feel they have not really done any physics.

Enhancing School Implementation. Hartsuijker, Bart and Zandbergen (1992) noticed severe problems while teachers try to implement MBL in chemistry education. The teachers experienced technological problems that resulted from the complexities/possibilities of the software. They also had difficulties in cooperating with their new role—coaching the MBL classes.

Voogt (1992) reported that clear student textbooks do promote the implementation of the curriculum. She also noticed that technical support personnel and easy-to-use software are important necessary conditions for implementation of a complex innovation. She concludes that students' motivation to use MBL supports the implementation of MBL, reducing some costs for the teachers.

3.4 Dilemmas on Implementing MBL and MML Science Learning Environments

One of the major findings which emerge from curriculum development and educational research (section 3.3.2) is the importance of curriculum materials which accompany the MBL and MML learning environments. In other words "no curriculum, no knowledge." This section of the paper raises some dilemmas connected with curriculum development (section 3.4.1) and implementation (section 3.4.2). With these dilemmas a discussion can be raised in order to find out what subsequent curriculum development and educational research will be necessary to influence and to contribute to the future implementation of MBL and MML learning environments in science education (section 3.4.3).

3.4.1 Curriculum Dilemmas

1. Should MBL and MML support science curricula in force (goals, contents and methodologies)

    – or –

    should MBL and MML evoke radical changes in the domain of the sciences in middle and high school?

    This dilemma applies to several levels:

    * trends in science education as in section 3.2.2;
    * structure of the science subjects in schools;
    * interrelation among scientific disciplines;
    * topics and the sequence thereof;
    * necessity to develop new teaching materials to replace existing ones.

2. Should science teaching be approached according to constructivism

    – or –

    instructionism? (Tinker, 1991)

    This dilemma applies to several levels:

    * classroom organization;
    * individual learning versus group learning;
    * memorised learning versus in-depth learning;
    * teachers.

3.4.2 Implementation Dilemmas

1. Should the implementation decisions be taken at the national (state, district) levels,

   – or –

   at school level?

   This dilemma applies to several levels:

   • finances;

   • teacher training, the number of trained teachers, and who's trained;

   • curriculum and materials at school level.

2. Should the emphasis be on new cycles of development and educational research

   – or –

   on implementation (teacher training, facilities)?

3. Should implementation be achieved by focusing on pre-service and in-service training

   – or –

   teachers/trainers?

4. What is the best balance between investment in hardware and software

   – or –

   investment in human resources?

3.4.3 Conclusions

Our conclusions may be summarized as follows:

• Decision making on integration of MBL and MML in classroom practice should be part of decision making on changing the curriculum as a whole.

• Crucial problems are both the pedagogical design of MBL and MML science learning environments and teaching materials and the long-term implementation in education.

We believe that the technological problems, which were not meant as the subject of this paper, will be solved in future by means of continuous research and scientific exchange on a national and international level. In order to improve science education by implementing MBL and MML learning environments, it is strongly recommended that the design and development of MBL and MML science learning environments and teaching materials should be preceded, as well as accompanied, by thorough curriculum development, educational research and application of adequate innovation strategies.

References

Adams, D.D. and Shrum, (1990) The effects of microcomputer-based laboratory exercises on the acquisition of line graph construction and interpretation skills by high school biology students. *Journal of Research in Science Teaching*, 27(8), 777–787.

Beurs, C. de and Ellermeijer, A.L. (1992) *Computer applications in physics, the integration of information technology in the physics curriculum.* Paper submitted to the NATO Advanced Research Workshop, Microcomputer Based Labs: Educational Research and Standards, Amsterdam, 9–13 November. Chapter 12 in this book.

Brasell, H. (1987) The effect of real-time laboratory graphing on learning graphing representations of distance and velocity. *Journal of Research in Science Teaching*, 24(4), 385–395.

Collis, B. and Stanchev, I. (1992) *Trends and techniques in computer based educational simulations: applications to MML design.* Paper submitted to the NATO Advanced Research Workshop, Microcomputer Based Labs: Educational Research and Standards, Amsterdam, 9–13 November. Chapter 2 in this book.

Ellermeijer, A.L., Landheer, B., and Molenaar, P.P.M. (1992) *Teaching mechanics through interactive video and a microcomputer based lab (IV/MBL).* Paper submitted to the NATO Advanced Research Workshop, Microcomputer Based Labs: Educational Research and Standards, Amsterdam, 9–13 November. Chapter 16 in this book.

Friedler, Y., Nachmias, R., and Songer, N.B. (1989) Teaching scientific reasoning skills: A case study of a microcomputer-based curriculum. *School Science and Mathematics*, 89(1), 58–67.

Friedler, Y., Nachmias, R., and Linn, M.C. (1991) Learning scientific reasoning skills in microcomputer-based laboratories. *Journal of Research in Science Teaching*, 27(2), 173–191.

Hartsuijker, A., Bart, C. Van, and Zandbergen, P. (1992) *Courseware development and educational research in the Dutch approach to microcomputer based laboratory (MBL) experiments in chemistry education.* Paper submitted to the NATO Advanced Research Workshop, Microcomputer Based Labs: Educational Research and Standards, Amsterdam, 9–13 November.

LICOR/HBO (1992) Internal report, in Dutch.

Linn, M.C., Layman, J., and Nachmias, R. (1987) The cognitive consequences of microcomputer-based laboratories: Graphing skills development. *Journal of Contemporary Educational Psychology*, 12(3), 244–253.

Linn, M.C. and Songer, N.B. (1991) Teaching thermodynamics to middle school students: What are appropriate cognitive demands? *Journal of Research in Science Teaching*, 28(10), 885–918.

Mokros, J.R. and Tinker, R.F. (1987) The impact of microcomputer-based labs on children's ability to interpret graphs. *Journal of Research in Science Teaching*, 24(5), 369–383.

Nachmias, R. and Linn, M.C. (1987) Evaluations of science laboratory data: The role of computer-presented information. *Journal of Research in Science Teaching*, 24(5), 491–506.

Nachmias, R., Stavy, R., and Avrams, R. (1990) A microcomputer-based diagnostic system for identifying students' conception of heat and temperature. *International Journal of Science Education*, 12(2), 123–132.

Songer, N.B. and Linn, M.C. (1991) How do students' views of science influence knowledge integration? *Journal of Research in Science Teaching*, 28(9), 761–784.

Stein, S.S., Nachmias, R., and Friedler, Y. (1990) An experimental comparison of two science laboratory environments: traditional and microcomputer-based. *Journal of Educational Computing Research*, 6(2), 183–202.

Tinker, R.F. (1991) *Thinking about science*. Unpublished paper.

Tinker, R.F. (1984) *Microcomputer-based laboratory*. Resource Guide, Apple Educational Affairs, Cupertino, CA, USA.

Voogt, J.M. (1992) *Microcomputer based laboratories in inquiry-based science education—an implementation perspective*. Paper submitted to the NATO Advanced Research Workshop, Microcomputer Based Labs: Educational Research and Standards, Amsterdam, 9–13 November. Chapter 10 in this book.

Walker, D. (1990) *Fundamentals of curriculum*. San Diego, CA: Harcourt Brace Jovanovich Publishers, pp. 5 and 472.

Part II

Research

4. Using Large-Scale Classroom Research to Study Student Conceptual Learning in Mechanics and to Develop New Approaches to Learning[1]

Ronald K. Thornton

Tufts University

Abstract. Microcomputer-based laboratory (MBL) tools and guided discovery curricula have been developed as an aid to all students, including the underprepared and underserved, in learning physical concepts. To guide this development, extensive work has been done to find useful measures of students' conceptual understanding that can be used in widely varying contexts. This paper focuses primarily on the evaluation of student conceptual understanding of mechanics (kinematics and dynamics) with an emphasis on Newton's 1st and 2nd laws in introductory courses in the university. Student understanding of mechanics is looked at before and after traditional instruction. It is examined before and after MBL curricula that are consciously designed to promote active and collaborative learning by students. The results show that the majority of students have difficulty learning essential physical concepts in the best of our traditional courses where students read textbooks, solve textbook problems, listen to well-prepared lectures, and do traditional laboratory activities. Students can, however, learn these fundamental concepts using MBL curricula and Interactive Lecture Demonstrations which have been based on extensive classroom research. Substantial evidence is given that student answers to the short answer questions in the Tools for Scientific Thinking Force and Motion Conceptual Evaluation provide a useful statistical means of evaluating student beliefs and understandings about mechanics. Evidence for the hierarchical learning of velocity, acceleration, and force concepts is presented.

4.1 Introduction

For about six years we have been studying how students in universities, high schools, and middle schools learn to understand the physical world. We have used

---

[1] This work was supported in part by the National Science Foundation under the Student Oriented Science Project and the MBL for Teaching Teachers project, by the Fund for Improvement of Post-secondary Education (FIPSE) of the U.S. Department of Education under the "Tools for Scientific Thinking" project and "Interactive Physics" at Tufts University, the US Department of Education, the Board of Regents of Massachusetts, and Apple Computer, Inc.

our understandings of student learning to design environments where students have been able to learn fundamental physical concepts that were seldom learned in more traditional environments. Our intention has been to create an environment where closer to 90% of the students learn fundamental concepts rather than one where only the top 20% succeed ([4] and see discussion below). In addition, we have intended to develop materials and methods that will be successful in widely varying contexts for students of different ages, cultural heritages, and preparation. Results from research in cognitive science and education substantiate the importance of basing development of scientific concepts and skills on concrete experience [1,8]. To these ends the "Tools for Scientific Thinking" project [12–16] at the Center for Science and Mathematics Teaching at Tufts University has developed microcomputer-based laboratory (MBL) tools and curricula that can help students make connections between the physical world and the principles which constitute scientific knowledge. The computer tools (which are being used in middle school, high school, and colleges, including college teacher preparation and enhancement programs) provide a convenient and effective means for students to collect and display physical data in a form that they can remember, manipulate, discuss and think about. The tools have enabled the development of curricula, based on research in science learning, that allow students to take an active role in their own learning.

To guide our work we needed to develop a practical means of assessing student conceptual knowledge in physics that would serve our many goals. This paper will identify the goals, discuss methods of assessing student learning in force and motion (kinematics and dynamics), use actual classroom research results, motivate the form of assessment chosen in light of the goals, discuss problems raised about the assessment methods, and provide evidence for validation of the assessment.

Although the evidence that students have not been learning fundamental physics concepts seems convincing to most physics education researchers, practicing physics teachers seem to require either overwhelming statistics or even measures of their own students to be convinced. It is useful to teachers involved in course design and modification to be able to evaluate student understanding of concepts that form a foundation for further learning in the subject. To satisfy these needs, we designed a short answer conceptual evaluation that can easily be used in many different contexts and that provides reliable information about student beliefs about motion and how those beliefs are being changed (or not) by the instruction.

One of the purposes of this paper is to show that the conceptual evaluation we have been using provides a useful measure of student understanding of mechanics concepts. We have been conducting studies of the effectiveness of traditional methods of teaching motion concepts in addition to those curricula that use MBL tools. As a result, we have much information to bring to bear on the question of evaluation. We have explored student understandings of physics concepts in middle schools, high schools, colleges and universities in calculus and algebra-based introductory physics courses, and in teacher preparation and enhancement programs. Substantial work on curriculum development and evaluation was done at the college and university campuses that are part of the "Tools for Scientific Thinking" project[2]. Professors Sassi of the University of Napoli and Professor Borghi of Pavia have done research (some of which is reported in this publication) involving the use of MBL curricula with Italian university students (including future teachers) and high school teachers and their students. The Workshop Physics Program [5, 6], under the direction of Priscilla Laws at Dickinson College, has provided a more ideal college learning environment into which the MBL tools have been adopted and some of the curricular pieces have been adapted. In more usual environments, we have used pre- and post-testing and other forms of evaluation to examine the understandings of thousands of college and university physics students under the Student-Oriented Science project (funded by the National Science Foundation). We have also collected data for a large sample of secondary and middle school students. All of these contexts have provided strong evidence for significantly improved learning and retention by students who used the MBL materials, compared to those taught in lecture [11–16] and these data also provide evidence to support the usefulness of our method of evaluation.

One learning environment, however, has provided the kind of information about student learning of mechanics that is of particular value in the context of this paper. David Sokoloff of the University of Oregon, a co-author of the curricular materials, has provided opportunities for extensive research on student learning [15] and the opportunity to experiment with new methods of teaching large classes (see Section 4.5 of this paper). A six-year collaboration with David Sokoloff has provided data on thousands of students. The university offers an introductory non-calculus physics class with two large lecture sections of between 150 and 200 students. The lectures are high quality traditional instruction. The laboratory is offered as a separate course, taught by graduate teaching assistants. Approximately one half of the students take the laboratory. We have had the opportunity to do very detailed evaluations of the students in this course, some of whom experienced only traditional instruction and some of whom have experienced the Tools for Scientific Thinking Motion and Force Laboratory Curriculum. The conclusions about student learning we have come to by working with Oregon students have been consistent with research results from other similar—and even very different—learning environments. However, the research at Oregon has been distinguished by the opportunity to take repeated measures on large numbers of students over many years. For this reason, the discussion that follows will focus upon student learning of motion and force at Oregon. To illustrate the generality of evaluation methods, however, the following section on kinematics concepts will use data from many universities.

---

[2] The original participating colleges and universities are Tufts University, University of Oregon, California Polytechnic State University, San Luis Obispo, Dickinson College, Massachusetts Institute of Technology, Muskingum College and Xavier University. Additional schools have been involved in developmental work including Arizona State University, Lee College, Ohio State, Joliet Community College, University of Texas, Austin.

4.2 Evaluating Student Learning of Motion (Kinematics) Concepts in Traditional and MBL Environments

One of the foundations for understanding force and motion (dynamics) is understanding kinematics, which is the description of motion. The learning of dynamics depends on a knowledge of kinematics (see Section 4.5). While most physicists would agree with this statement, few courses provide an effective environment for students to learn kinematics.

4.2.1 Student Understanding of Kinematics Concepts after Traditional Instruction

If all students were learning the fundamental concepts associated with motion and force, there might be less reason to consider changing the ways we teach kinematics and dynamics and less reason to explore effective means of evaluation. Unfortunately, students are not learning these concepts in traditional courses, at least in the United States. Figure 1 shows the result of asking more than two thousand students, mostly future engineers in calculus-based courses at major universities and colleges, 10 simple conceptual questions about velocity and acceleration. These questions, which test concepts and not just graphing, are shown in Figures 2 and 3. Note that error rates are shown. Physics professors unfamiliar with the results of this research predict error rates for their students of less than 5 to 10%, while the actual error rates

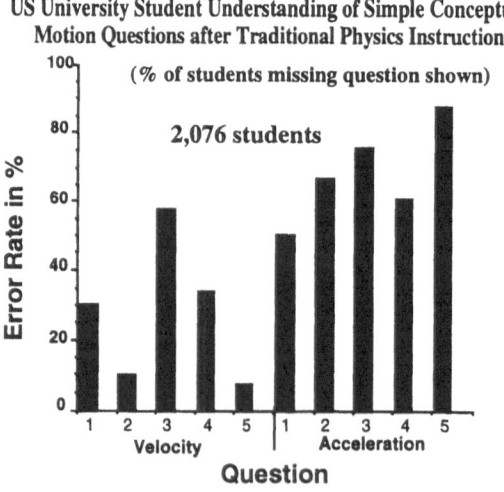

Figure 1   Percentage of US university students missing simple conceptual questions on kinematics after traditional instruction in first year calculus and algebra-based physics courses. The questions asked are shown in Figures 2 and 3. Note that velocity questions 2 and 5 require little understanding of kinematics.

Velocity-Time Graphs

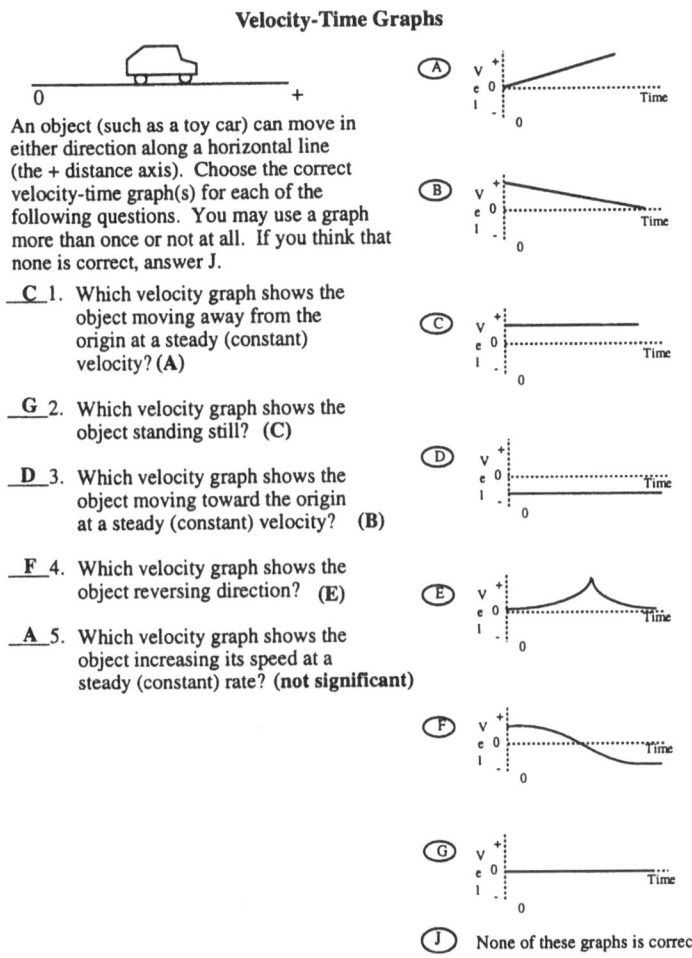

An object (such as a toy car) can move in either direction along a horizontal line (the + distance axis). Choose the correct velocity-time graph(s) for each of the following questions. You may use a graph more than once or not at all. If you think that none is correct, answer J.

__C__ 1. Which velocity graph shows the object moving away from the origin at a steady (constant) velocity? **(A)**

__G__ 2. Which velocity graph shows the object standing still? **(C)**

__D__ 3. Which velocity graph shows the object moving toward the origin at a steady (constant) velocity? **(B)**

__F__ 4. Which velocity graph shows the object reversing direction? **(E)**

__A__ 5. Which velocity graph shows the object increasing its speed at a steady (constant) rate? **(not significant)**

Figure 2   Some of the multiple choice velocity questions asked on the kinematics pre- and post-tests. Questions 1 through 5 are the velocity questions referred to in Figures 1 and 4. The most common "wrong" answer is shown in parentheses.

can be 40% or above for velocity concepts and 70–95% for acceleration concepts. The sample of students shown in Figure 1 show such an error rate. Our research also shows that students who cannot answer these simple questions will in general be at a great disadvantage in their subsequent study of kinematics and dynamics. Some evidence for this point of view will be presented in Section 4.5 of this paper.

The results shown are disquieting and do not speak well for traditional instruction. About 60% of the students seem to be learning the simplest velocity concepts and only 25% seem to be learning the simplest acceleration concepts in physics courses. Since everyone knows how hard physics is, some professors have felt that perhaps teaching 25% of the students to understand acceleration may be acceptable. This

Acceleration-Time Graphs

Questions 1-5 refer to a toy car which can move to the right or left along a horizontal line (the + distance axis).

Different motions of the toy car are described below. Choose the letter (**A** to **G**) of the acceleration-time graph which could correspond to the motion of the car described in each statement.

You may use a choice more than once or not at all. If you think that none is correct, answer choice **J**.

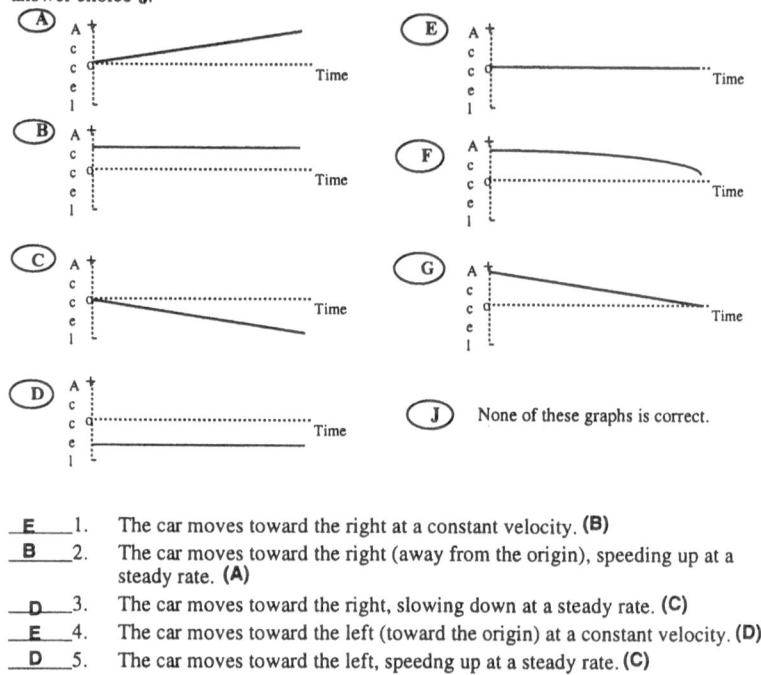

E	1.	The car moves toward the right at a constant velocity. **(B)**
B	2.	The car moves toward the right (away from the origin), speeding up at a steady rate. **(A)**
D	3.	The car moves toward the right, slowing down at a steady rate. **(C)**
E	4.	The car moves toward the left (toward the origin) at a constant velocity. **(D)**
D	5.	The car moves toward the left, speedng up at a steady rate. **(C)**

Figure 3   Some of the multiple choice acceleration questions asked on the kinematics pre- and post-tests. Questions 1 through 5 are the acceleration questions referred to in Figures 1 and 4. The most common wrong answers are shown in parentheses.

point of view has two problems. The first problem is that these are student responses to some of the simplest significant questions that indicate understanding. The students do less well on other questions. The second problem is that the large majority of students who demonstrate an understanding of the kinematics concepts are not learning them in the college physics classes. Most of the students who can answer the questions after traditional instruction knew them when they began the course.

To show that this is the case, we can look at students before and after traditional instruction at the University of Oregon in the environment described in the introduction. Figure 4 shows evidence that most students are not learning the concepts as a result of traditional instruction. This figure shows the result of asking the velocity

Oregon Non-Calculus Physics before and after Standard Instruction

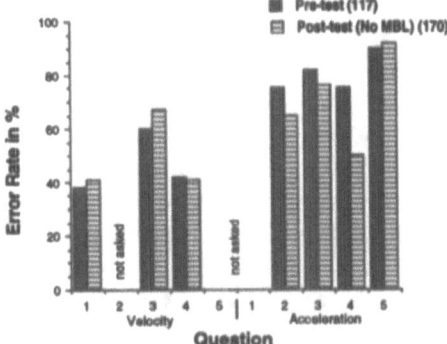

Figure 4 Results for introductory physics lecture students (non-calculus) at the University of Oregon, Fall, 1988—comparison of student error rates on seven velocity and acceleration questions given before and after standard instruction.

and acceleration questions shown in Figures 2 and 3 of students in the non-calculus physics class at the University of Oregon. The results are typical of other research we have done where we find less than 10% change in the error rate due to traditional instruction (for more information on student learning in colleges and high schools, see also references 11–16).

In this case, the post-test was given to all three sections, but it was only possible to give the pre-test to two of the lecture sections. Since the populations of these lecture sections were random (the only selection criterion was time of day) the pretests should have been similar for all three. There were no significant differences between the two sections that were pretested.

It was surprising to observe error rates as high as 40–60% on these simple velocity questions *after kinematics had been covered in lecture* and students had done standard problems. Such results are consistent with the data in Figure 1. As described above, most physics professors had predicted that fewer than 10% of their students would miss these questions. They also felt that students who were unable to answer such simple questions understood very little kinematics. All of the lecturers were aware of the testing, and all made a special effort to teach kinematics graphing and concepts in their lectures. The large error rates on questions 1 and 3 (41% and 67%, respectively, after instruction) are not simply the result of the wrong choice of sign. The most common error is the choice of the "distance analogs," graphs A and B. This is consistent with previous studies [8, 18] in which students used position models to interpret velocity graphs. The different error rates on these two questions show that students have significantly more difficulty interpreting negative velocities. (This conclusion is borne out by the results of additional testing.) Neither the results of this pretest nor the correct answers were shared with the students.

It should be noted that most students did not miss the questions because they were simply unable to read graphs. More than 95% could answer questions involving distance graphs correctly, and students interviewed were intentionally picking graphs consistent with their verbal or written explanations of velocity and acceleration. In 1991 at the University of Oregon we asked more than two hundred students in the non-calculus introductory physics course to explain why they chose particular graphs from the choices. The results confirm that student responses to the short answer questions are consistent with their written explanation of the phenomena. More than 97% of the time, their written explanations were consistent with their answers to the multiple-choice questions. As an example, consider question 2 in Figure 3 where students are to pick the appropriate acceleration graph for an object whose velocity is increasing at a steady rate as it moves to the right (positive direction, away from the origin). A typical student explanation given by those few students who think as physicists on the pretest is:

> A steadily increasing velocity means a constant acceleration and since it is moving to the right the acceleration is positive.

A typical student explanation for a student making the same "correct" choice later in the year is more detailed.

> Steadily increasing velocity toward the right means the velocity vs time graph looks like ^{v} ⟍. Since a=Δv/Δt equals the slope of the v-t graph, the acceleration is positive and constant.

A typical response by a student (in the majority) who picks graph A:

> Acceleration steadily increases as the velocity does.

Such students are not confusing velocity and acceleration but positing that acceleration behaves like velocity. They are using a velocity "model" for acceleration. The students who continue to view acceleration in this manner do not in general elaborate their responses after instruction in the manner of students choosing the physics explanation.

4.2.2 Why Use This Method of Evaluation?

In spite of the fact that physics professors thought initially that these questions were much too simple for their students and that very few would "miss" them, after seeing the results, some professors suggested that perhaps they are not significant or valid and reliable measures of knowledge. Our research does not support this point of view. The pre- and post-tests that we have used in these studies consist in part of multiple-choice questions. From earlier testing of students using free response questions requiring written answers and the drawing of graphs, we have constructed questions that seem to give a reasonable indication of students' basic knowledge of kinematics concepts and of graphical representation. Student answers to these questions correlate well with their written answers on these and earlier tests as the discussion above indicates. We find there are almost no random answers. Almost all students pick choices that we can associate with a small number of student models. This paper presents the results of only a few of more than 50 questions in different formats that are designed to distinguish among student models. Many of the multiple-choice questions require students to choose the correct graph from a group of up to nine graphs. Testing with smaller student samples shows that those who can pick the correct graph under these circumstances are almost equally successful at drawing the graph correctly without being presented with choices. The difficulties in convincing physics professors and high school teachers to give up course time for testing, our desire to make evaluation less subjective, and the effort involved in analyzing large samples moved us to use short answer questions for these studies. Although a more complete understanding of student learning can be gained by an open-ended questioning process, we decided to use short-answer questions in order to gather sufficient data at many different institutions to counter the common response that "my students do not have these difficulties you describe." From such questions we are also able to identify students with less common beliefs about motion and follow up with opportunities for open-ended responses to help us understand student thinking.

4.2.3 Student Understanding of Kinematics Concepts after MBL Instruction

A visit to an MBL classroom/laboratory illustrates the contrast with a traditional class. Students are actively involved in their learning. They are sketching predictions and discussing them in groups of two or three. They use MBL tools to collect physical data that are graphed in real time and then can be manipulated and analyzed. The discovery-based curricula take advantage of the fact that MBL tools present data in an immediately understandable graphical form. In the case of a motion laboratory, the students move in front of a motion detector that plots their motion. They appeal to features of the graphs they have just plotted to argue their points of view with their peers. They ask questions and, in many cases, either answer them themselves or find the answers with the help of fellow students. There is a level of student involvement, success, and understanding that is rare in physics and physical science courses. (For descriptions of the software tools, hardware probes, and the Tools for Scientific Thinking discovery-based curricula, see references 12–16.[3])

Student enthusiasm is wonderful and we feel that such a learning environment would have merit even if student learning of concepts was about the same as traditional instruction. We are pleased, however, to have made MBL tools and curricula that are very effective in helping students to learn motion concepts and have worked hard to evaluate such learning. We have been conducting studies of the effectiveness of traditional methods of teaching motion concepts, examples of which were given previously, and of those curricula we have developed that make use of MBL tools in the context of active and collaborative learning. As mentioned in the introduction, we have explored student understandings of physics concepts in middle schools and high schools. In colleges and universities we have studied the learning of students in calculus and algebra-based introductory physics courses, including those designed primarily for teacher preparation and enhancement programs. We have used pre- and

---

[3] These materials are available through Vernier Software, 2920 S.W. 89th Street, Portland, OR 97225.

post-testing and other forms of evaluation to examine the understandings of thousands of college and university physics students who have used the Tools for Scientific Thinking motion curriculum. We have also collected data for a large sample of secondary and middle school students. All of these contexts have provided strong evidence for significantly improved learning and retention by students who used the MBL materials, compared to those taught in lecture. Many examples of these learning results from university calculus and algebra-based physics courses, high schools, and teacher preparation programs have been published [12–16]. The results of this research show that in the case of kinematics, about 95% of university students understand velocity concepts and 80–95% understand acceleration concepts after using MBL curricula. Similar or better results have been achieved in high schools. To allow for more discussion of student learning of dynamics, we will not discuss these kinematics results in more detail and refer readers to the papers referenced above.

4.3 Evaluating Student Learning of Force and Motion (Dynamics) Concepts in Traditional and MBL Environments

4.3.1 Student Understanding of Dynamics Concepts after Traditional Instruction

Very few students entering universities understand force and motion from a Newtonian point a view. Unhappily, only a small additional percentage of students adopt a Newtonian framework after well-executed traditional instruction (see Figure 1). The results reported here reflect student understandings of Newton's 1st and 2nd Laws.

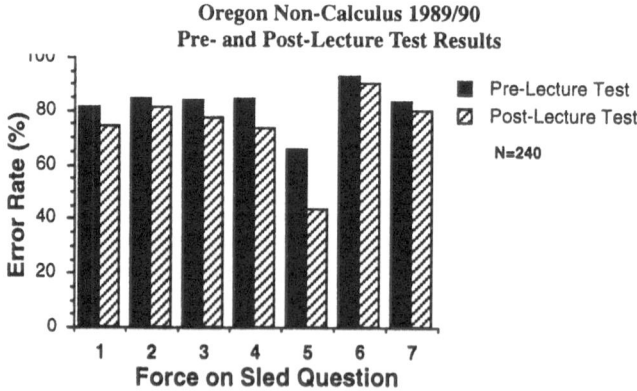

Figure 5   Error rates of 240 University of Oregon students in first year algebra-based physics on the simple dynamics questions shown in Figure 7. The dark bars show results before instruction and the striped bars after traditional instruction.

Figures 5 and 6 show that about 90% students in the introductory algebra-based physics course at the University of Oregon were unable to answer dynamics questions in ways that are consistent with a Newtonian view of the world either before or after traditional instruction.

These results are typical and not unique to the University of Oregon. The questions asked are shown in Figures 7 and 8. After standard instruction, this large error rate was reduced, on average, by an additional 7%. (Note that even if we assign none of the small gain to instruction but assign all 7% to asking the same questions twice, asking the same question twice does not have a large instructional effect. In fact our other research shows that asking the questions two or more times does not produce a measurable gain, certainly not 7%.)

These questions have been selected from the Force and Motion Conceptual Evaluation developed by the Center for Science and Mathematics Teaching at Tufts University. The questions have been asked of thousands of students. The fact that traditional instruction has little effect on student beliefs about force and motion, as shown by the results in Figures 5 and 6, is confirmed by considerable additional research. Note that although both sets of questions (force on a sled and force graph sequences) explore the relationship between force and motion, the format is very different. The force on sled questions make no overt reference to a coordinate system, they use "natural" language as much as possible, and they make no reference to graphs. Student responses to questions where there is an exact analog between the force on the sled question and the graphical questions are consistent in spite of these differences.

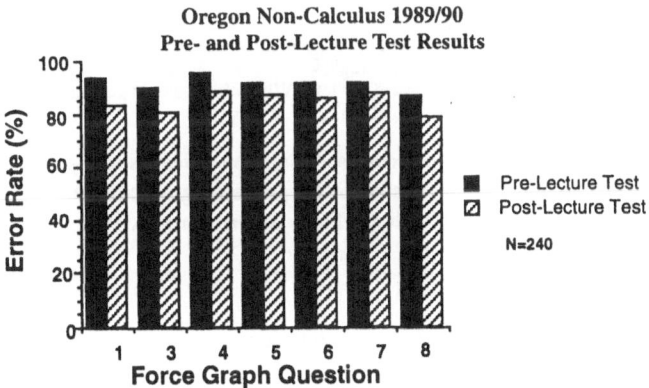

Figure 6 Error rates of 240 University of Oregon students in first year algebra-based physics on the simple dynamics questions shown in Figure 8. The dark bars show results before instruction and the striped bars after traditional instruction. Question 2 in this sequence from the Force and Motion Conceptual Evaluation was not asked.

Questions on Force and Motion

A sled on ice moves in the ways described in questions 1-7 below. Friction is so small that it can be ignored. A person wearing spiked shoes standing on the ice can apply a force to the sled and push it along the ice. Choose the one force (**A** through **G**) which would keep the sled moving as described in each statement below. You may use a choice more than once or not at all but choose only one answer for each blank. If you think that none is correct, answer choice **J**.

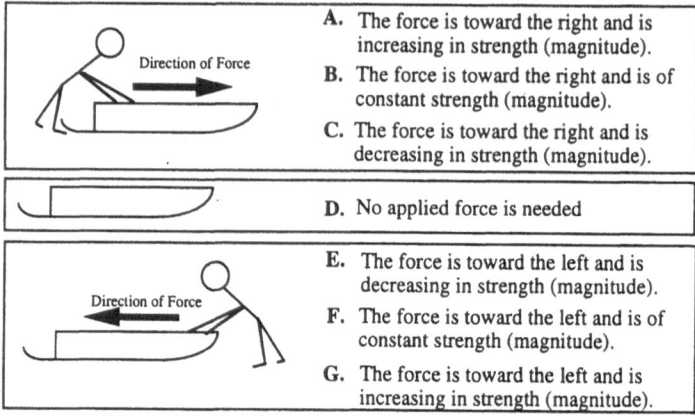

A. The force is toward the right and is increasing in strength (magnitude).

B. The force is toward the right and is of constant strength (magnitude).

C. The force is toward the right and is decreasing in strength (magnitude).

D. No applied force is needed

E. The force is toward the left and is decreasing in strength (magnitude).

F. The force is toward the left and is of constant strength (magnitude).

G. The force is toward the left and is increasing in strength (magnitude).

B 1. Which force would keep the sled moving toward the right and speeding up at a steady rate (constant acceleration)? **A**

D 2. Which force would keep the sled the sled moving toward the right at a steady (constant) velocity? **B**

F 3. The sled is moving toward the right. Which force would slow it down at a steady rate (constant acceleration)? **C**

F 4. Which force would keep the sled moving toward the left and speeding up at a steady rate (constant acceleration)? **G**

D 5. The sled was started from rest and pushed until it reached a steady (constant) velocity toward the right. Which force would keep the sled moving at this velocity? **B**

B 6. The sled is slowing down at a steady rate and has an acceleration in the positve direction. (The positive direction is to the right.) Which force would account for this motion? **C**

B 7. The sled is moving toward the left. Which force would slow it down at a steady rate (constant acceleration)? **E**

Figure 7   Force on a sled questions corresponding to the results in Figures 5 and 6. The most common "wrong" answer on the pre-test is given after the questions. These answers are consistent with a velocity implies force hypothesis.

4.3.2 Student Understanding of Dynamics Concepts after Using the MBL Curriculum

The great majority of students at the University of Oregon who completed the kinematics and dynamics laboratory curriculum mentioned above and described in reference 16 answered the questions in Figures 7 and 8 as most physicists would (see Figures 9 and 10). This point of view is often described as Newtonian. The Newtonian

Force Graph Questions

Questions 1-8 refer to a toy car which can move to the right or left along a horizontal line (the positive part of the distance axis).

Assume that friction is so small that it can be ignored.

A force is applied to the car. Choose the one force graph (A through H) for each statement below which could allow the described motion of the car to continue.

You may use a choice more than once or not at all. If you think that none is correct, answer choice J.

E 1. The car moves toward the right (away from the origin) with a steady (constant) velocity. **A**

E 2. The car is at rest. **not significant**

A 3. The car moves toward the right and is speeding up at a steady rate (constant acceleration). **C**

E 4. The car moves toward the left (toward the origin) with a steady (constant) velocity. **B**

B 5. The car moves toward the right and is slowing down at a steady rate (constant acceleration). **H**

B 6. The car moves toward the left and is speeding up at a steady rate (constant acceleration). **D**

G 7. The car moves toward the right, speeds up and then slows down. **F**

E 8. The car was pushed toward the right and then released. Which graph describes the force after the car is released.   **H,F,A**

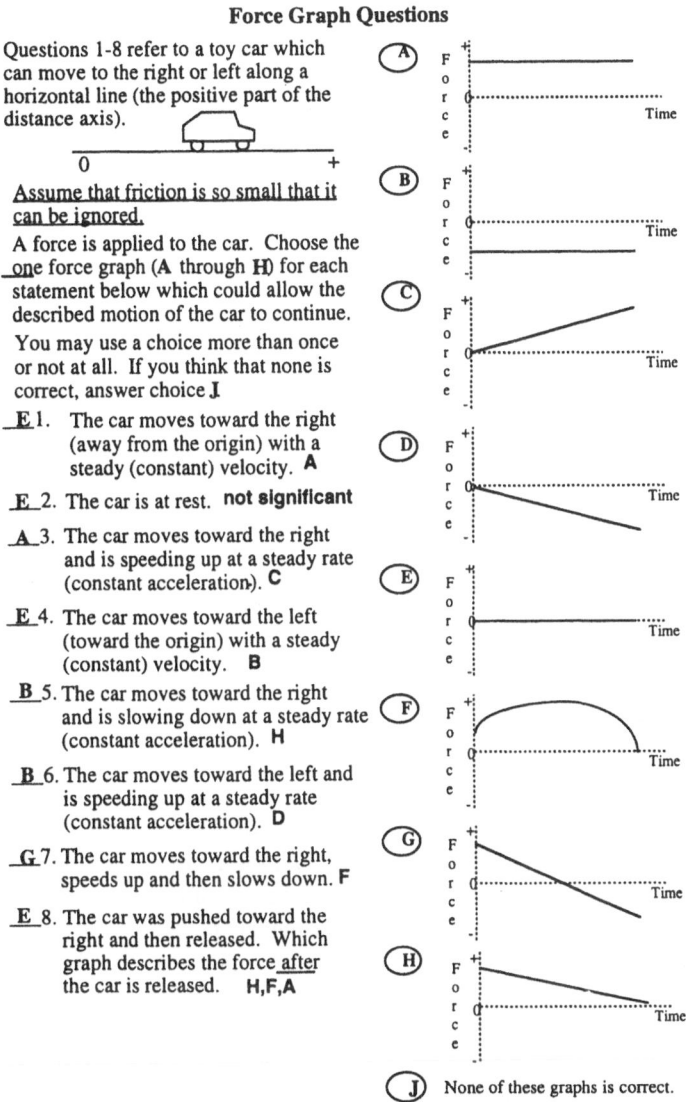

J   None of these graphs is correct.

Figure 8 Force graph questions corresponding to the results in Figures 5 and 6. The most common "wrong" answer on the pre-test is given after the questions. These answers are consistent with a velocity implies force hypothesis.

model accurately describes the behavior of objects of everyday sizes moving at ordinary velocities. Objects that are moving at a constant velocity or at rest have no net force acting upon them and objects that are speeding up at a constant rate (undergo-

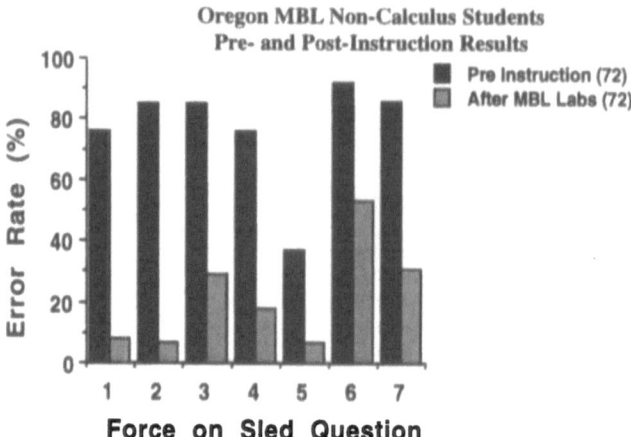

Figure 9   Student performance on the force on a sled questions shown in Figure 7 before
instruction (dark bars) and after lectures and MBL laboratories (lighter bars).
The same 72 students all took the pre-test, listened to lectures, did the MBL
laboratories, participated in an interactive MBL lecture demonstration, and took
a quiz after instruction.

ing constant acceleration) are acted upon by a constant force. Force is proportional to
acceleration.

This model held by physicists is in contrast to the almost universal student model
that while agreeing that an object at rest is acted upon by a net force of zero, proposes
that motion (or more specifically velocity) implies a force. Thus if an object is moving at constant velocity, it experiences a constant force while an object whose velocity is uniformly increasing must be acted upon by an uniformly increasing force.
Both views disagree with a Newtonian model. Such a model is sometimes called an
"impetus" model and, less accurately, an Aristotelian model.

It may be valuable to discuss a few individual questions. The largest error rate
after MBL instruction is question 6 of the force on a sled sequence. About 50% of
students missed this question. We know from additional research that 40% of
physics professors and high school teachers also miss this question but are then
unable to suggest a change in wording. Since some people who very consistently
answer questions from a Newtonian viewpoint still miss question 6, we must interpret the results cautiously. Such results confirm the value of asking a number
of questions to probe understanding of particular concepts and the value of asking
them of diverse audiences.

Some questions are asked to make sure that students understand the format and
(in some cases) can read English. Question 2 in the force graph sequence is the only
question where the most common student views and Newtonian view are the same
since the object is at rest (see above). Consequently, it is not "missed" often, even
before instruction.

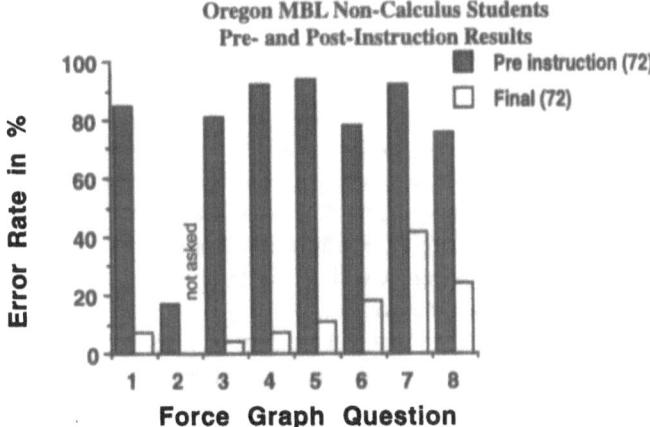

Figure 10 Student performance on the force graph questions shown in Figure 8 before instruction (dark bars) and after lectures and MBL laboratories (lighter bars). The same 72 students all took the pre-test, listened to lectures, did the MBL laboratories, participated in an interactive MBL lecture demonstration, and answered these questions on the final exam.

Like the earlier kinematics questions discussed in this paper, students do not answer these questions randomly. The choices were derived from open answer responses of students and from interviews. Most students can find an answer that matches the relationship between force and motion that they have in mind. More than 95% of the responses are either Newtonian or the most common student "model," which is the impetus model. The fact that most students are using "models" (even if they are only applied in very limited circumstances and give results that are in conflict with a more careful examination of physical phenomena) is a good beginning for instruction.

The contrast between 85% of students answering the above questions from a Newtonian perspective after the Tools for Scientific Thinking Curriculum while less than 20% do so after traditional instruction has led some to question the evaluation methods being used. After looking further at the evolution of student understanding of motion and force concepts and after exploring a new teaching technique useful in large lecture sections, we will examine in more detail the correlation between these questions and additional probes of students' conceptual knowledge of force and motion.

4.4 Hierarchical Understanding of Motion and Force Concepts

The use of the Tools for Scientific Thinking Force and Motion Conceptual Evaluation in actual university and high school classrooms has allowed us to study the learning of mechanics (and the mechanics of learning) by students in many different environments and over an entire course. We have strong statistical evidence that students who do not answer the simple velocity concept questions (Figure 2) as a physicist would will not answer the simple conceptual questions on acceleration or the questions on force and motion as a physicist would. Conversely, students able to answer the acceleration questions as a physicist would will answer the velocity questions "correctly." The same hierarchical relationship exists between acceleration and force. We can illustrate the hierarchy using data from the non-calculus course at Oregon in the fall of 1991. These are the learning data from which we drew many previous conclusions. In the following we will use the term "correct" to indicate the way most physicists would answer the questions.

From the results of the pre-test we find that 65 students miss two or more of the velocity questions shown in Figure 2. These same students miss on average 88% of the five acceleration questions (Figure 3) and over 90% of the force questions in Figures 7 and 8. To establish a hierarchy, we must also show that students who answer the acceleration questions correctly answer the velocity questions correctly. In this pre-instruction sample of 230 students only 13 are able to answer all 5 of the acceleration questions correctly. These 13 students also answer all of the velocity questions correctly. To improve the statistics by finding more students able to answer the acceleration questions correctly, we can look later in the semester. After approximately half of the students have taken the Tools for Scientific Thinking motion labs and all students have seen an interactive lecture demonstration on kinematics (see Section 4.5), there are 102 students who answer all 5 acceleration questions correctly. These same students answer the velocity questions correctly 98% of the time, as expected from the smaller sample.

Having established the hierarchical relationship between the learning of velocity and acceleration, we use additional data from the same source to establish the hierarchical relationship between the learning of acceleration and force concepts. On the pre-test, the 142 students who missed at least 4 out of 5 of the acceleration questions missed 94% of the force graph questions shown in Figure 7. By the final exam, 123 students missed no more than one force graph question. These same students also answered the acceleration questions correctly 96% of the time. Only 20 students missed a substantial number (at least 4 out of 5) of the acceleration questions on the final, and they also missed most of the force graph questions (75%).

There is a clear hierarchical learning relationship among simple concepts that indicate understanding of velocity, acceleration, and force. By examining the Oregon and additional learning data, we find that this hierarchy is true for traditional and MBL instruction. There remains a less likely possibility that some unexamined method of mechanics instruction would not result in such an hierarchy. It is also true that some students can answer some simple conceptual questions about acceleration, for example, as a physicist would while missing more sophisticated questions about velocity concepts.

4.5 MBL Interactive Lecture Demonstrations: Using Non-Traditional Methods to Improve Traditional Lecture Instruction

The data from the Oregon non-calculus course displayed in Figures 5 and 6 from 1989 and 1990 show that standard instruction resulted in an additional 7% of the students in the class answering the conceptual questions on force and motion in a Newtonian manner. The disappointing result becomes somewhat more understandable in light of the hierarchy presented above. At the end of all traditional kinematics and dynamics instruction, a given acceleration question was answered correctly (on average) by 28% of the students. (It is probable that this number was somewhat lower during the time the dynamics lectures were being given.) Since only students who understand acceleration concepts are likely to learn dynamics, the traditional instruction on dynamics could be of benefit to very few and partially accounts for the small number of students who develop a Newtonian view.

In an effort to improve learning of dynamics by students who did not have access to MBL laboratories, David Sokoloff and I developed further the concept of an interactive MBL lecture demonstration that had shown some promise in smaller scale experiments in previous years. In the fall of 1991, students in a non-calculus lecture course were given 40 minutes of interactive lecture demonstration on kinematics, based on the learning sequence used in the MBL motion labs. A protocol developed for effective implementation of the interactive lecture demonstrations is shown in Figure 11. The protocol encourages students to become actively engaged in the learning.

Interactive Lecture Demonstration Protocol

1. Describe the experiment/demonstration and do it for the class without MBL measurements.
2. Ask each student to record an individual prediction on the handout sheet.
3. Ask the class to engage in small group discussions in order to decide on a group prediction.
4. Ask each student to sketch a final prediction on the handout sheet (the group prediction if they came to an agreement). The prediction sheet will be collected at the end of the class.
5. Carry out the experiment/demonstration with MBL measurements displayed.
6. Ask a few students to describe the result and discuss results in the context of the demonstration. Students fill out results sheet which they keep.
7. Discuss analogous physical situations that produce a similar physical result but have different "surface" features.

Figure 11 This is the procedure to follow for each of the lecture demonstrations/experiments such as the ones shown in Figure 12. The students are given two sheets, a prediction sheet to hand in so they don't lose credit and a results sheet to fill out for their own use. (Filling out the results should also improve the learning.) The students are reminded that there are no wrong predictions.

As a result of this series of interactive demonstrations of kinematics, which were given after kinematics lectures, the acceleration questions were answered correctly by 63% of the students before the traditional lectures on dynamics. This is in contrast to previous years where approximately 25% of the students understood the acceleration concepts. After this preparation, the traditional lectures on dynamics resulted in a 28% ($\pm$ 6%) average improvement in the force graph questions compared to the 7% ($\pm$2 %) from the previous years. We attribute this improvement to the fact that more students understood kinematics, since nothing else was changed. The 40 minutes of time invested was well worthwhile since four times as many students benefited from

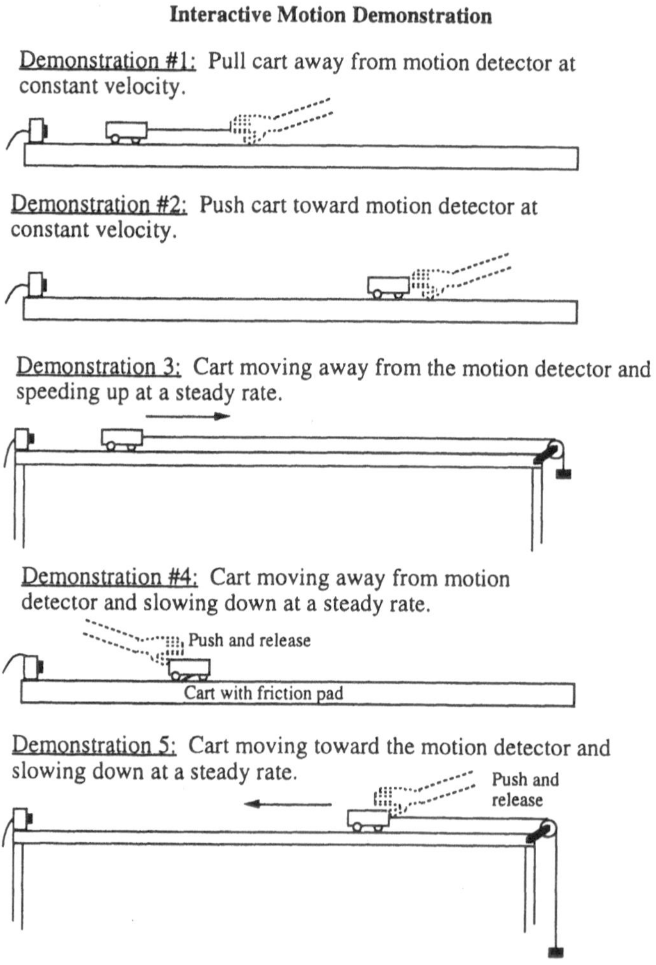

Interactive Motion Demonstration

Demonstration #1: Pull cart away from motion detector at constant velocity.

Demonstration #2: Push cart toward motion detector at constant velocity.

Demonstration 3: Cart moving away from the motion detector and speeding up at a steady rate.

Demonstration #4: Cart moving away from motion detector and slowing down at a steady rate.

Push and release

Cart with friction pad

Demonstration 5: Cart moving toward the motion detector and slowing down at a steady rate.

Push and release

Figure 12 First five of eight demonstrations/experiments for interactive motion demonstrations as examples. Student sheets, which are not shown, often ask students to predict both the velocity and acceleration graphs for each demonstration/experiment.

the traditional dynamics instruction. On the other hand, on average, 72% of the students still answer these questions from a non-Newtonian point of view. Even with the enhanced learning of kinematics, attributable in the interactive demonstration, the traditional lectures were not very effective at teaching a Newtonian point of view. The students who used the Tools for Scientific Thinking lab curricula did considerably better; only 15% of the students answer these questions in a non-Newtonian way (Figure 9).

We employed further interactive lecture demonstrations on force and motion after the traditional lectures. The results were gratifying, and approximately 65% of the non-laboratory students answered the force graph questions from a Newtonian point of view in contrast to previous years where only 15–20% did so (see Figure 6). We are examining the efficacy of interactive lecture demonstrations at other universities and high schools. (For a more complete description of materials and results, contact the author.)

4.6 Exploring the Significance of the Dynamics Questions on the Force and Motion Conceptual Evaluation

The success of students in answering the dynamics questions from the Force and Motion Conceptual Evaluation after using the Tools for Scientific Thinking (or Workshop Physics) curricula leads some researchers to ask if the questions are significant indicators of students' understanding of dynamics. To some extent, this paper has already addressed important questions that might be raised. Student answers to the multiple-choice graphical format questions of the type shown in Figure 7 correlate with answers to questions probing the same concepts when asked in the very different format of the questions shown in Figure 8, which neither use graphs nor refer explicitly to coordinate systems. The correlation holds both before and after traditional or MBL instruction. We have, however, found the graphical questions easier to formulate and to provide a more explicit indication of student understanding. Also, such questions can be asked a number of times without any significant learning taking place from the questions.

4.6.1 Student Responses to Additional Force and Motion Questions

To explore further the significance of the Newtonian student responses to the above graphical multiple-choice questions, we also asked a set of simple conceptual questions on the final exam that had never been asked of the students at the University of Oregon. These questions are of a different format and set in rather different contexts than the questions discussed above. Figure 13 shows ten of these questions. The research question we explored was, if students answered the graphical force questions discussed above (Figure 8) from a Newtonian point of view, what percentage would answer different and unfamiliar questions from the same point of view? The numbers in parentheses after the Newtonian answer indicate the percentage of students giving this answer. The results are very good, with six questions answered correctly by approximately 90% or more of the students. It is interesting that 20% of the students missed question 2 about an automobile moving at constant velocity, while only 1% of students missed question 3 where a skater moves at constant velocity. The large number of students who missed question 7 were equally split between choosing a constant force in the direction of and opposite to the motion. It may be the case that the concept of net force is not well understood. We are continuing to explore student responses to questions such as these.

Our expectations for students answering these questions using a Newtonian point of view were somewhat more modest than the actual results. Previous work had shown us that students often do not generalize in ways that seemed obvious to physicists without specific instructional effort to show such generalization is valid. Because of

New Force Questions on Final

In each of the following examples of motion of an object (1 - 10), choose the one description below (**A - J**) of the net (resultant) force on the object which could keep the object moving as described. You may use a choice more than once or not at all.

A. The net force is in the direction of the motion and is increasing in strength (magnitude).
B. The net force is in the direction of the motion and is of constant strength (magnitude)
C. The net force is in the direction of the motion and is decreasing in strength (magnitude).
D. The net force is zero.
E. The net force is in the direction opposite the motion and increasing in strength (magnitude).
F. The net force is in the direction opposite the motion and is of constant strength (magnitude)
G. The net force is in the direction opposite the motion and decreasing in strength(magnitude).
J. None of the net force descriptions is correct.

( ) **show % of students giving physics answer**

B (81)1. What net force will cause an automobile moving on a highway to speed up at a
      steady (constant) rate.
D (80)2. What net force will cause an automobile moving on a highway to maintain a constant
      speed of 55 miles per hour.
D (99)3. What is the net force on an ice skater gliding across a frozen lake at a constant speed.
E (95)4. A ball was thrown upward. What is the net force on the ball right after it is released
      and is moving upward, slowing down at a steady rate?
B (82)5. What is the net force on the same ball as it is falling downward after reaching its highest point?
F (93)6. An automobile moving at 55 miles per hour has the brakes applied suddenly to avoid a deer.
      What is the net force on the car as it slows down at a quick but steady (constant) rate?
D (54)7. What is the net force on a bicycle that is being pedaled up a hill at a steady (constant) speed?
B (91)8. What is the net force on a bicycle that is speeding up at a steady (constant) rate as it rolls
      down a hill?
F (89)9. A bicycle after coasting along level ground comes to a hill. What is the net force on the
      bicycle as it rolls up the hill slowing down at a steady (constant) rate.
F (89)10. What is the net force on an airplane as it moves down the runway slowing down at a
      steady (constant) rate after landing.

Figure 13 These questions, which test understanding of Newton's 1st and 2nd Laws in different contexts, were given on the final to students in the University of Oregon introductory non-calculus course. The students had not seen these questions previously. On the final, 123 students answered all or all but one of the force graph questions in Figure 8 in a Newtonian manner. The numbers in parenthesis show the % of these students also answering the questions above "correctly." The results show most students consistently apply a Newtonian point of view to unfamiliar questions.

the limited time devoted to the Tools for Scientific Thinking dynamics curriculum (two three-hour laboratories) and the lack of any discussion time to introduce different contexts, we expected higher error rates.

4.6.2 Coin Toss Problems and Analogs

Students most commonly use a motion-implies-force model when they are asked the traditional coin toss problem (e.g., A coin is tossed up into the air. What is the force on the coin on the way up (after release)? At the highest point? and on the way down?). After traditional instruction only 5% of the students in the University of Oregon sample answer the coin toss questions shown in Figure 14 as a physicist would. After the MBL curriculum over 90% of the students answer the coin toss as a physicist would.

The coin toss problem and its analogs provide more evidence that students who answer the graphical force questions in Figure 8 from a Newtonian point of view have made a fundamental belief change. If we look again at the sample of 123 students (see also Figure 14 and discussion) who answered all or all but one of the questions in Figure 8 "correctly," we find that 93% of these students answered the cart on the ramp questions shown in Figure 14 "correctly," i.e., from the Newtonian point of view. They had, however, seen this problem previously. Figure 15 shows a coin toss analog they had not seen, a block sliding into a spring. 92% of the students in this sample answered from the Newtonian point of view.

As with the other multiple-choice questions on the Force and Motion Conceptual Evaluation, students who answered correctly were also able to describe in words why they picked the answer they did. Students were asked to explain how they determined the force on a cart in Figure 14 just as it reached the highest point (question 2). Typical answers from students who answered these short answer questions about the forces from a Newtonian point of view were:

> After the car is released the only net force acting on it is the x-component of its weight which has a net force down the ramp in the positive direction.

> When the car is at the top of the ramp, its velocity is 0 for just an instant, but in the next instant it is moving down the ramp, v2-v1 = a pos number so it is accel. down. Also, gravity is always pulling down on the car no matter which way it is moving.

> The only two forces involved were gravity and friction. At the top of the ramp the net force was downward because gravity is higher in magnitude than friction (unless the tires & the ramp were sticky).

Typical student answers for those who answered as if motion implies force were:

> At the highest point, the toy car's force is switching from one direction to another and there are no net forces acting upon it, so it is zero.

> Because at the one instant the car is at its highest point it is no longer moving so the force is zero for that one instant it is at rest = net force = 0

Cart up and down Ramp and Coin Toss Questions

Questions 1-3 refer to a toy car which is given a push up an inclined ramp. After it is released, it rolls up, reaches its highest point and rolls back down again.

Use one of the following choices (A through C) to indicate the net force acting on the car for each of the cases described below. Answer choice J if you think that none is correct.

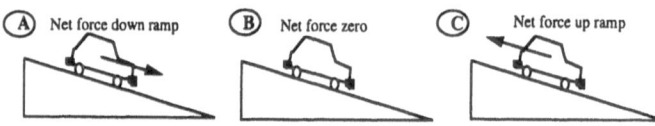

_A_1. The car is moving up the ramp after it is released. **(C)**
_A_2. The car is at its highest point. **(B)**
_A_3. The car is moving down the ramp. **(A)**

Questions 4-6 refer to a coin which is tossed straight up into the air. After it is released it moves upward, reaches its highest point and falls back down again. Use one of the following choices (A through C) to indicate the force acting on the coin for each of the cases described below. Answer choice J if you think that none is correct.

A. The force is downward.
B. The force is zero.
C. The force is upward.

_A_4. The coin is moving upward after it is released. **(C)**
_A_5. The coin is at its highest point. **(B)**
_A_6. The coin is moving downward. **(A)**

Figure 14 Coin toss question and an analog from the Force and Motion Conceptual Evaluation. Questions 4 through 6 are one version of the classic coin toss question. The most common pre-test "wrong" answer (shown after question) is consistent with the motion implies force model. Questions 1-3 involve the same knowledge of physics but seem slightly more difficult for students than the coin toss.

The agreement between the multiple-choice and open-answer responses is almost 100%.

In summary, most students who answer the force graph or force on a sled questions from a Newtonian point of view are able to answer other questions that they have never seen from the same Newtonian point of view and are also able to answer coin toss and coin toss analog questions from the same point of view. In addition,

Spring and Block—Coin Toss Analog

Questions 23-25 refer to a block on a table with negligible friction. The block is initially moving toward the left, when it crashes into a spring.

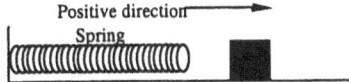

Positive direction

Spring

For each of the cases described below, use one of the following choices **A - C**) to indicate the force acting on the block. Answer choice **J** if you think that none is correct.

    A. The force is positive.

    B. The force is zero.

    C. The force is negative.

A 23. The block is in contact with the spring, and is moving toward the left and slowing down

A 24. The block is in contact with the spring, and has momentarily come to rest

A 25. The block is in contact with the spring, and is moving toward the right and speeding up

Figure 15 Coin toss analog from the Force and Motion Conceptual Evaluation.

students' written explanations agree with their choices on these carefully constructed multiple-choice questions. These results support the usefulness of the questions on the Force and Motion Conceptual Evaluation for evaluating student understanding.

4.7 Conclusions

We have used large-scale classroom research to study student conceptual learning in mechanics (kinematics and dynamics) and its use in developing new approaches to learning. This paper has combined presentation of actual conceptual learning results as a result of traditional and non-traditional instruction, discussion of the evaluation methods, and evidence for the significance of the evaluation.

Our classroom research on kinematics concepts, based on data from more than 2000 students, has shown that traditional instruction in kinematics is not very effective in US universities. Additional data on kinematics learning shows that most students who understand kinematics concepts are in fact not learning the concepts during their introductory courses but know them when they enter the course. Such data lead to the conclusion that traditional instruction in kinematics has little effect on students' conceptual understanding of this topic. In contrast, the Tools for Scientific Thinking Motion Curricula with MBL laboratories are shown to be effective in teaching kinematics concepts. The questions used to measure conceptual knowledge in kinematics include multiple-choice graphical questions. Results from more than two-hundred students in an introductory physics class at the University of Oregon show that students' written explanations of kinematics concepts correlate with their choices on these short-answer kinematics questions which are part of Tools for Scientific Thinking Force and Motion Conceptual Evaluation.

Large-scale classroom research to study student learning in dynamics shows that traditional instruction has even less effect on student beliefs. Questions on dynamics from the Force and Motion Conceptual Evaluation that probe the simplest concepts related to Newton's 1st and 2nd laws show that in many universities, only 10% of students understand such concepts when they enter the class. After high quality traditional instruction at the University of Oregon, only an additional 7% in a sample of more than two hundred students understood force and motion from the Newtonian point of view. However, more than 85% of students who have used the Tools for Scientific Thinking Motion and Force laboratory curriculum answer the same questions from a Newtonian point of view. Only 5% of students answer the coin toss question and its analogs in a Newtonian manner after traditional instruction, yet more than 85% of the students using the MBL curriculum do so.

Dramatic results might lead one to question the validity of the measurement process. This paper presented evidence that student responses were consistent on different format questions, graphical multiple choice and natural language questions with no overt references to coordinate systems. Students give almost no random answers. 95% of all responses are consistent with the most common student impetus model or with a Newtonian model. Approximately 90% of students who answered the graphical multiple-choice questions are able to answer most of the additional questions on the final exam with different surface features and format (questions they have not previously been asked). Written answers of more than 200 students were more than 98% consistent with their multiple-choice answers. More than 90% of students who answered the graphical multiple choice in a Newtonian manner also answered the coin toss questions and analog problems from the same point of view. Written answers for the coin toss problems are consistent with their short-answer choices.

Further evidence that the questions are probing significant understanding is the strong evidence for a learning hierarchy among velocity, acceleration, and force concepts. Students unable to answer the simple conceptual velocity questions as a physicist would are unable to answer the acceleration questions in the same manner. Statistically, students must know the acceleration concepts to answer the force and motion questions from a Newtonian point of view. If students do answer the force and motion questions in a Newtonian manner, then they also answer the acceleration and velocity questions as a physicist would. The evaluation of student conceptual learning in many different contexts has shown that curricula employing microcomputer-based laboratory tools allow students to develop a solid conceptual basis for understanding the world around them. Through the use of these tools, techniques and curricula, students connect their interactions with the physical world to the theories that constitute scientific knowledge.

We have used MBL Interactive Lecture Demonstrations to improve traditional instruction in large classrooms with over 200 students. Because of the hierarchical nature of learning in mechanics discussed above, most students are unable to learn force and motion concepts if they do not know kinematics concepts. As a result of a 40-minute series of interactive demonstrations of kinematics (using a protocol to encourage interactive learning), in which students participated after kinematics lectures, approximately 63% of the students understood acceleration concepts before the traditional lectures on dynamics. In previous years when approximately 25% of the students understood the acceleration concepts, traditional instruction in dynamics resulted in only 7% ($\pm$2%) of students answering force and motion questions from a Newtonian point of view. Because more students understood kinematics concepts as a result of the interactive lecture demonstrations, 28% ($\pm$ 6%) of students adopted a Newtonian point of view after traditional dynamics instruction. When additional MBL Interactive Lecture Demonstrations in dynamics (another 50 minutes) were given after the lectures, approximately 65% of the non-laboratory students answered the force graph questions from a Newtonian point of view in strong contrast to previous years where only 15–20% did so after all instruction.

The evaluation of student conceptual learning in many different contexts has shown that, while traditional instruction has little effect, curricula based on classroom research and employing MBL tools allow students to develop a solid conceptual basis for understanding the world around them. There is evidence that the Force and Motion Conceptual Evaluation is a useful measure of conceptual understanding on the part of the students.

References

1. Arons, A.: Achieving wider scientific literacy, Daedalus **112**, 91–117 (1983)
2. Brassell, H.: The effect of real-time laboratory graphing on learning graphic representations of distance and velocity, J. Res. Sci. Teaching **24**, 385–395 (1987)
3. diSessa, A.A.: The third revolution in computers and education, J. Res. Sci. Teaching **24**, 343–367 (1987)
4. Halloun, I.A. and Hestenes, D.: The initial knowledge state of college physics students, Am. J. Phys. **53**, 1043–1055 (1985) and Common sense concepts about motion, Am. J. Phys. **53**, 1056–1065 (1985)
5. Laws, P.: Workshop physics: replacing lectures with real experience. In: Redish, E. and Risley, J. (eds.): *Proc. Conf. Computers in Phys. Instruction*, Reading, MA: Addison Wesley 1989, pp. 22–32
6. Laws, P.: Calculus-based physics without lectures, Phys. Today **44**, 24–31 (Dec., 1991)
7. McDermott, L.C.: Research on conceptual understanding in mechanics, Phys. Today **37**, 24–32 (July, 1984)
8. McDermott, L.C., Rosenquist, M.L. and van Zee, E.H.: Student difficulties in connecting graphs and physics: Examples from kinematics, Am. J. Phys. **55**, 503–513 (1987)
9. Peters, P.C.: Even honors students have conceptual difficulties with physics, Am. J. Phys. **50**, 501–508 (1982)
10. Rosenquist, M.L. and McDermott, L.C.: A conceptual approach to teaching kinematics, Am. J. Phys. **55**, 407–415 (1987)
11. Thornton, R.K.: Access to college science: Microcomputer-based laboratories for the naive science learner, Collegiate Microcomputer **V** (1), 100–106 (1987)
12. Thornton, R.K.: Tools for scientific thinking: Learning physical concepts with real-time laboratory measurement tools. In: Redish, E. and Risley, J. (eds.): *Proc. Conf. Computers in Sci. Teaching*, Reading, MA: Addison Wesley 1989, pp. 177–189
13. Thornton, R.K.: Tools for scientific thinking—microcomputer-based laboratories for teaching physics, Phys. Ed. **22**, 230–238 (1987)
14. Thornton, R.K.: Enhancing and evaluating students' learning of motion concepts. Chapter in: A. Tiberghien and H. Mandl (eds.): *Intelligent Learning Environments and Knowledge Acquisition in Physics*. NATO ASI Series F, Vol. 86. Berlin: Springer-Verlag, 1992

15. Thornton, R.K. and Sokoloff, D.: Learning motion concepts using real-time microcomputer-based laboratory tools, Am. J. Phys. **58** (9), 858–66 (Sept. 1990)
16. Thornton, R.K.: Changing the physics teaching laboratory: Using technology and new approaches to learning to create an experiential environment for learning physics concepts, *Proc. of the Europhysics Conference on the Role of Experiment in Physics Education*, University of Ljubljana Slovenia (in press)
17. Tinker, R.K. and Thornton, R.K.: Constructing student knowledge in science. In: Scanlon, E. and O'Shea, T. (eds.): *New Directions in Educational Technology.* NATO ASI Series F, Vol. 96. Berlin: Springer-Verlag, 1992
18. Trowbridge, D.E. and McDermott, L.C.: Investigation of student understanding of the concept of velocity in one dimension, Am. J. Phys. **48**, 1020–1028 (1980) and Investigation of student understanding of the concept of acceleration in one dimension, Am. J. Phys. **49**, 242–253 (1981)

5. A New Mechanics Case Study: Using Collisions to Learn about Newton's Third Law

Priscilla W. Laws

Dickinson College

Abstract: Several researchers have reported on conceptual difficulties students encounter in the study of Newton's Laws, especially Newton's Third Law. This paper describes a project to restructure the introductory physics mechanics curriculum to present Newton's Laws in a more logical sequence. This curriculum is based on the use of direct experience coupled with Microcomputer-Based Laboratory (MBL) tools. This paper gives particular attention to the sequence of learning experiences developed to improve student understanding of Third Law concepts applied to collision processes. The results of pre- and post-testing show significant gains in student ability to apply the Third Law to different types of interactions.

5.1 Introduction

In the study of introductory mechanics, acquiring a conceptual understanding of Newton's Laws has proven to be one of the most difficult challenges faced by students. Recent surveys by Hestenes, et.al.,[1] of student conceptual gains before and after traditional instruction have been disappointing. Arons and Rothman[2] have observed that many popular textbooks have treatments of Newtonian dynamics that are logically inconsistent. In addition a number of science education researchers have discovered that many students begin the study of mechanics with misleading conceptions about the nature of motion which are extremely hard to overcome[3].

---

[1] David Hestenes, Malcolm Wells, and Gregg Swackhamer, "Force Concept Inventory," *The Physics Teacher,* Vol. 30, 141-158. (March, 1992).

[2] Arons, Arnold. *A Guide to Introductory Physics Teaching,* (John Wiley, New York, 1990) Chapter 3; and Rothman, Milton A. *Discovering the natural laws; the experimental basis of physics,* (Doubleday, New York, 1972) Chapter 2.

[3] Lillian McDermott, "Research on conceptual understanding in mechanics," *Physics Today,* 2-10. (July, 1984); and David Hestenes, Malcolm Wells, and Gregg Swackhamer, "Force Concept Inventory," *The Physics Teacher,* Vol. 30, 141-158. (March, 1992).

Research has shown that microcomputer-based laboratory tools are effective in enhancing student learning in kinematics and dynamics.[4] Recent improvements in microcomputer-based laboratory systems for the study of force and motion[5] and the availability of new low friction dynamics carts[6] have made it possible to design new activities in which students can observe relationships between force and motion quickly and easily.

David Sokoloff, Ronald Thornton and the author outlined a sequence of laboratory activities to be used for teaching mechanics concepts in the Workshop Physics program[7] and the RealTime Physics project[8] involving the adaptation of curricular materials from the Workshop Physics and Tools for Scientific Thinking Programs[9] to introductory laboratory sequences in Mechanics, Heat and Temperature, and Circuits. In the summer of 1992 a small conference attended by individuals active in physics education research and curriculum development[10] was held at Tufts University. Participants were asked to critique the ideas for the new mechanics sequence. Thus, the outcomes of research on student learning, insights offered by Arons and Rothman on logical development, new MBL tools, activities designed for Workshop Physics and Tools for Scientific Thinking programs, and ideas generated by participants in the new mechanics conference were used as a basis for the development of a new activity-based mechanics curriculum. During the 1992-93 academic year this curriculum was tested in the Workshop Physics programs at Dickinson College and Gettysburg High School and in activity-centered RealTime Physics laboratories at the University of Oregon and Arizona State University.

---

[4] c.f. The article in this volume by Ronald K. Thornton; and Ronald K. Thornton and David Sokoloff, "Learning Motion Concepts Using Real-Time Microcomputer-Based Laboratory Tools," *Am. J. Phys. 58* (9) 858-867 (Sept., 1990).

[5] A motion detector, force probe, interface and motion software for Macintosh and MS DOS computers can be purchased from Vernier Software Company, 2920 S.W. 89th Street, Portland, OR 97225.

[6] Low friction dynamics carts can be purchased from PASCO Scientific Company, 10101 Foothills Blvd., PO Box 619011, Roseville, CA 95678-9011.

[7] Priscilla Laws, "Calculus-Based Physics Without Lectures," *Physics Today,* Vol. 44, No. 12, (Dec. 1991); and Priscilla Laws, "Workshop Physics—Learning Introductory Physics by Doing It," *Change,* 20-27 (July/Aug. 1991); and Priscilla Laws, "Workshop Physics-Replacing Lectures with Real Experience," *Proceedings of the Conference on Computers in Physics Instruction,* Addison-Wesley, Reading, MA, 1989.

[8] The RealTime Physics project, directed by David Sokoloff, is funded by the National Science Foundation ILI Laboratory Leadership program, #USE-9054224, at the University of Oregon.

[9] FIPSE Comprehensive Program Grant #'s: 1) G008642149, *Tools for Scientific Thinking — Microcomputer-Based Laboratories;* 2) G008642146, *Workshop Physics* from 10/86 - 9/89; and 3) P116B90692, *Interactive Physics: Using Workshop Physics and MBL in the University Classroom and Laboratory* from 10/89-9/92; and NSF Undergraduate Curriculum Development Program, USE-9150589, *Student-Oriented Science (SOS): Curricula, Techniques & Computer Tools for Interactive Learning* from 9/91 to 2/93.

[10] The New Mechanics conference which was held on August 6-7, 1992 in Medford, MA was attended by Pat Cooney, Dewey Dykstra, David Hammer, David Hestenes, Priscilla Laws, Suzanne Lea, Lillian McDermott, Robert Morse, Hans Pfister, Edward F. Redish, David Sokoloff, and Ronald Thornton.

5.2 The New Mechanics Sequence

The New Mechanics sequence differs from the traditional sequence in several ways:

(a) The order in which Newton's three laws are presented is based on the difficulty students appear to encounter in understanding them. Students begin with Second Law activities before they consider First Law phenomena.[11] Finally they work with Third Law concepts which appear to be the hardest to master.[12]

(b) Activities using MBL force and motion sensors and low friction dynamics carts are designed to enable students to make direct observations of basic elements of Newtonian dynamics without recourse to textbooks.

(c) Extra efforts are made to help students look at the elements of Newton's laws and be able to distinguish definitions such as acceleration, force, and inertial mass from observed phenomena; for example, more "pull" causes more acceleration and more "stuff" causes in less acceleration.

(d) Concepts in kinematics and dynamics are initially developed for one dimensional horizontal motion with visible applied forces (pushes or pulls) with little friction present.

(e) Students are then asked to make additional observations which lead them to invent invisible forces (i.e., friction forces, gravitational interaction forces, normal forces, and tension forces) in order to maintain the viability of the Newtonian schema for predicting motions. This process of modifying mental schema to apply to new situations is labeled by cognitive scientists as accommodation.[13]

(f) The study of kinematics and dynamics is finally extended to two dimensional phenomena such as projectile motion, circular motion and motion on inclines.

(h) Students work with the impulse-momentum theorem, forces in collisions, the Law of Conservation of Momentum, and center-of-mass concepts before dealing with the conservation of energy. The inversion of momentum and energy topics was suggested by Arons[14] on the basis that (1) the momentum concept is simpler than the energy concept, in both historical and modern contexts and (2) the study of momentum conservation entails development of the concept of center-of-mass which is needed for a proper development of energy concepts.

---

[11] James Minstrell, "Teaching for the Development of Understanding of Ideas: Forces on Moving Objects" from the *1984 Yearbook of the Association for Supervision and Curriculum Development*, Editor Charles W. Andersen, December 1984.

[12] David Hestenes, Malcolm Wells, and Gregg Swackhamer, "Force Concept Inventory," *The Physics Teacher*, Vol. 30, 141-158. (March, 1992).

[13] David Ausubel, "Learning as Constructing Meaning," from *New Directions in Educational Psychology: 1–Learning and Teaching*, D. Entwhistle, Ed. (Falmer Press, London, 1985).

[14] Private Communication with Arnold Arons, "Preliminary Notes and Suggestions," August 19, 1990; and Arnold Arons, *Development of Concepts of Physics* (Addison-Wesley, Reading MA, 1965).

New Mechanics	Traditional Mechanics
I. D Kinematics w/ MBL	**I. D Kinematics**
• body motion	• lectures
• constant velocity	• textbook problems
• constant acceleration	• lab w/ const. accel.
II. D Dynamics (low friction w/ visible applied forces)	**II. D Kinematics**
• define constant F_{app}	• projectiles
• observe a∝ F_{app}	• centripetal acceleration
• define F scale	
• observe more stuff ' < a	
• define static mass scale	
• observe a∝(1/m)	
• observe $F_{net} \equiv \Sigma F_{app} = ma$	
• observe First Law, i.e., if $\Sigma F_{app} = 0$, then v=const.	
III. D Dynamics (invisible Fs)	**III. D and 2D Dynamics**
• observe friction w/ visible drag as a "passive force"	• state First Law
• postulate friction as force	• state Second Law F=ma
• observe vertical fall	• state superposition i.e., $F_{net} \equiv \Sigma F$
• postulate constant gravitational force	• describe action of forces due to gravity, friction, surfaces and strings
• observe effects of strings and surfaces	• state third law as "action/reaction"
• postulate Newton's third law to explain a= 0 cases	• state an elaborate rules for problem solutions w/ free body diagrams
• postulate T and N as passive forces	
IV. 2D Dynamics	**IV. Mechanical Energy Conservation**
• observe vector combination of forces	
• discover rules for freebody diagrams and 2D problem solving	
• apply Newton's laws to projectile, circular and inclined plane motion	
V. Momentum Conservation and Collisions	**V. Momentum Conservation and Collisions**
• Observe impulses and p-change	• Impulse
• Derive FΔt=Δp theorem from Newton's 2nd Law	• p-conservation
• Use MBL to verify FΔt=Δp theorem experimentally	• 1D collisions
• predict interaction forces during collisions if $m_1 > m_1$ or $v_1 > v_2$	• 2D collisions
. Use MBL to observe interaction forces ($F_{12} = -F_{21}$ always)	• Center-of-mass
• Combine dynamic Third Law w/FΔt=Δp to get 1D p-conservation	• Particle systems
• Center-of-mass, 1D collisions, 2D collisions, and particle systems	
VI. Mechanical Energy Conservation	

Table 1:   Outline of the New Mechanics Sequence

Although this paper focuses on elements of the sequence designed to help students acquire an understanding of Newton's Third Law and collision processes, key elements of the New Mechanics Sequence are summarized in Table 1. A more detailed description of the activities developed to help students understand these elements will be published in the near future in an article currently being prepared by the author in collaboration with David Sokoloff and Ronald Thornton.

5.3 Helping Students Understand the Third Law

5.3.1 Overview

One of the most challenging and interesting parts of the New Mechanics sequence involves the application of Newton's Third Law to one-dimensional collision processes. Many students can apply Newton's Third Law to the construction of free body diagrams when two interacting objects are in equilibrium. Virtually all introductory physics students can recite Newton's Third Law in the form "for every action there is an equal and opposite reaction" or "forces are always equal and opposite". However, the majority of students who complete introductory mechanics either do not understand the meaning of these phrases or don't really believe them when considering contact forces in collisions. For example, traditional instruction in two high school classes in Arizona, one regular and one honors, reduced average error rates from 90% to only 72% on conceptual questions requiring an understanding of Third Law concepts.

The existence of common misconceptions about interaction forces in collisions is not surprising for two reasons. First, when students observe elastic collisions between a rapidly moving object (i.e. an active agent) and a stationary object having the same mass, a dramatic momentum transfer seems to take place. Pretest scores indicate that about 80% to 90% of students begin the study of introductory physics with the belief that in a collision there are circumstances under which one object exerts more force on another. Second, when students observe a head-on collision between a heavy object and a light one moving at the same speed, the light object undergoes a more dramatic acceleration than the heavy one. This leads to the belief that the object with the greatest mass exerts the most force in a collision. Only students with good physics training and intuition recognize that Newton's Second Law reveals that momentum changes are essential to determining relative magnitudes of interaction forces.

Arons asserts that an understanding of Newton's Third Law requires students to recognize that "all interacting objects exert equal and opposite forces on each other instant by instant, and this applies to widely separated gravitating bodies as well as to those exerting contact forces on each other, and that zero time elapses between a change occurring at one body and the effect of the change being felt by the other."[15]

---

[15] Arons, Arnold. *A Guide to Introductory Physics Teaching*, (John Wiley, New York, 1990), p. 67.

From the perspective of modern physics, we now understand that the requirement that zero time elapse between a change in one body and a change in another cannot be met. Thus, although Newton's Third Law does hold for mechanical contact forces between objects, modern physics ultimately gives primacy to Conservation of Momentum in the hierarchy of physical law. In fact, one of Newton's many brilliant insights was that the experimental fact that momentum (quantity of motion) is conserved in collisions implies that the interaction forces between two objects must have the same magnitude.

5.3.2 The Sequence of Activities on Collisions and the Third Law

A major goal of this sequence of activities is to help students understand how Newton's Laws lead naturally to the Law of Conservation of Momentum in the description of collision processes. As we explained in section 5.2 we decided to introduce momentum and its conservation before exposing students to energy concepts.

The sequence of activities was designed to help students to: (1) understand the relationship between forces experienced by a single object and its change in momentum, (2) consider mutual interaction forces between two bodies undergoing a collision, and (3) realize that the Law of Conservation of Momentum is a consequence of Newton's Second and Third Laws. Students perform the following activities:

Recasting Newton's Second Law in Momentum Form

a. Using a Thought Experiment to Define Momentum: Students perform a thought experiment and try to predict at what speed, V, a small car of mass m must move in order to stop a truck of mass M moving at a slower speed, v. The outcome of this discussion is used as a basis for defining momentum as $\vec{p} = \mu\vec{v}$.

b. Deriving Newton's Second Law in terms of momentum: Students show mathematically that $\Sigma\vec{F} = m\vec{a} = \dfrac{d\vec{p}}{dt}$

The Impulse-Momentum Theorem

a. Reviewing the mathematical definition of p-change: Students need help realizing that a super ball undergoes more momentum change than a clay blob. They are asked to practice calculating momentum changes.

b. Gaining intuition about impulse, average force and momentum change: Students discuss why they tend to catch raw eggs more slowly than a ball. This discussion makes the definition of impulse as the time integral of force a bit more plausible.

c. Measuring Impulses: Quantitative data on forces during a collision is performed using the MBL system set up with motion software and a force probe. A force probe is attached to a cart and allowed to collide gently with a wall or another object. The data analysis feature of the motion software is used to determine the impulse resulting from the collision. A sketch of the apparatus is shown in Figure 1.

c. Deriving the Impulse-Momentum Theorem from Newton's Second Law: Students are asked to perform a mathematical derivation to show that

$$\int_{ti}^{tf} \Sigma\vec{F}\,dt = \int_{ti}^{tf} \frac{d\vec{p}}{dt}\,dt = (\vec{p_f} - \vec{p_i}) = \Delta\vec{p}$$

d. Verifying experimentally that the impulse-momentum theorem holds: A quantitative experiment is performed using the MBL system set up with motion software, a force probe, and a motion detector. A force probe is attached to a cart so that it can undergo a relatively slow collision with something soft such as a piece of foam rubber. Force readings are taken during the collision while the motion detector is used to determine the velocity of the cart just before and just after the collision. Students find that the quantitative verification of the impulse-momentum theorem is good to within 5 or 10% if they take careful measurements.

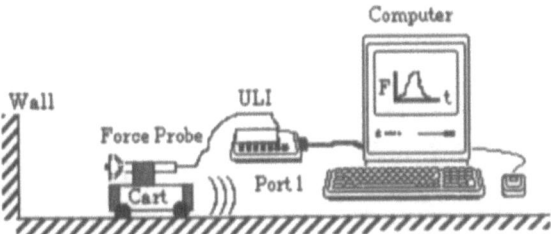

Figure 1: MBL apparatus set up for measuring collision forces on a force probe mounted on a cart

Mutual Interaction Forces

a. *Predicting relative force magnitudes in a collision:* Students are presented with several collision scenarios such as two cars of equal mass undergoing a collision, a moving car hitting a stationary truck, and a school bus smashing a fleeing mosquito. They are asked to predict the relative forces and discuss the circumstances under which one object might exert a greater magnitude of force on another object.

b. *Observing interaction forces in "real time":* Students are given two low friction carts with force probes mounted on them and some extra masses. The force probes are hooked into an MBL system, and they are provided with a special version of the motion software which can record data from two probes simultaneously. Students are then asked to use this equipment to investigate the circumstances under which one object exerts more force on another. A typical setup is shown in Figure 2. When these observations are carefully done, students discover that contact forces of interaction are equal in magnitude and opposite in direction on an instant-by-instant basis for all circumstances including that of a heavily loaded cart bearing down on a light cart which is at rest. Many students are surprised to see the force vs. time graphs for the two force probes looking equal and opposite! Sample graphs from these types of experiments are shown in Figures 3 and 4 for slow and fast encounters respectively.

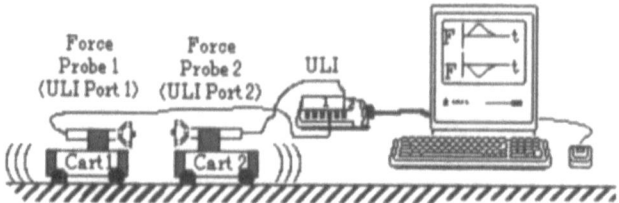

Figure 2: MBL set up for reading two forces at once during a gentle collision

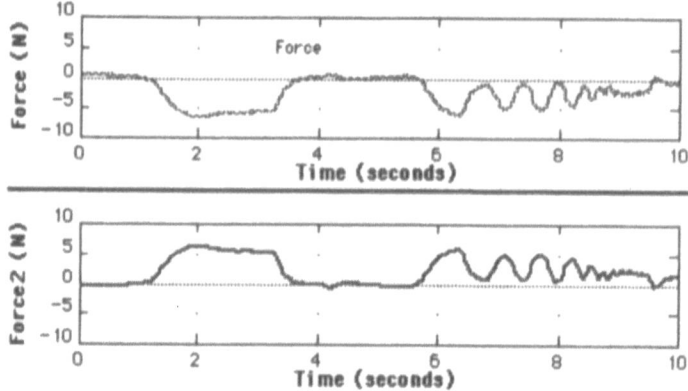

Figure 3: Two carts undergo slow collisions. Sometimes the first cart does the pushing and other times the second cart does the pushing. These graphs are made with MBL software, a ULI, and two force probes.

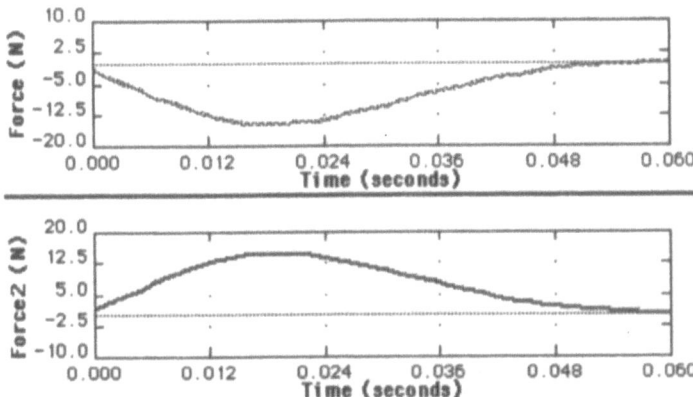

Figure 4: A 1.0 kg cart which is moving collides gently with a 0.5 kg cart which is at rest. These graphs are made with MBL software, a ULI, and two force probes. The data rate was set for 1000 readings per second.

Momentum Conservation

a. *Deriving Momentum Conservation as a Consequence of the Second and Third Laws:* Students combine the impulse-momentum theorem which is a form of the Second Law and the Third Law to predict mathematically that momentum ought to be conserved for collision processes.

b. *Observing Momentum Conservation Qualitatively for Simple Situations:* Students watch carts of the same mass interact in elastic collisions involving both contact forces and magnetic action-at-a-distance forces. They also observe inelastic collisions and "explosions."

c. *Deriving the Equivalence between Momentum Conservation and Constant Center-of-Mass Motion:* The idea of center-of-mass is introduced and students show mathematically that the center of mass of an isolated system always moves at a constant velocity.

d. *Observing Center-of-Mass motion in 1D and 2D Collisions:* Collisions between low friction carts having different masses and of pucks on an air table enable students to verify momentum conservation in 1D and 2D.

5.4 Assessing Learning Gains for Third Law Concepts

5.4.1 Initial Use of the Sequence at Three Institutions

Activity-based student worksheets using New Mechanics sequences were prepared in two slightly different formats, one for the Workshop Physics program and one for the RealTime Physics laboratory program. Preliminary versions of the New Mechanics curriculum were then tested at Dickinson College, the University of Oregon, and Arizona State University in the fall of 1992. The author introduced activities to three sections of the Workshop Physics calculus-based physics course at Dickinson College with a total enrollment of 71 students. These students had no formal lectures and met for three sessions of two hours in length each week for the semester. At the same time a RealTime Physics Laboratory program using the New Mechanics sequence was used under the direction of David Sokoloff in an algebra-based introductory laboratory course taken by 257 students. These students met in the laboratory once each week for three hours. In addition, the University of Oregon students were enrolled in a parallel lecture course which included recitation sessions in addition to lectures. Finally, Cheryl Claussen, a graduate teaching assistant at Arizona State University, introduced the new RealTime Physics Laboratory materials to students in one laboratory section of algebra-based physics at Arizona State University which met for two hours each week for a semester.

From an instructor's perspective the trials went as well as could be expected for a first time through. However, many changes are being made as a result of the classroom testing. For example, the preparation for the activities involving the Third Law was quite labor intensive because the software was so new that there was insufficient time to test it and some of the older force probes didn't work properly with the new software. Thus, while many of the students made observations on colliding carts that convinced them that forces between carts were always "equal and opposite," some students encountered technical difficulties. We expect that with fully tested software and procedures the reliability of the MBL observations made throughout the sequence will be improved significantly.

5.4.2 The Results of Pre and Post Testing on Third Law Concepts

The Force Concepts Inventory examination was administered to students in the calculus-based sections of Workshop Physics in the Fall of 1992 both before and after students worked with the New Mechanics activities. At the University of Oregon students were given a related Force and Motion Concepts Test[16] after completing the New Mechanics sequence as part of the RealTime Physics laboratory program. Since the testing of the curriculum at Arizona State University was done only on a pilot basis, no formal analysis of the ASU results was performed.

Three of the questions covering Third Law concepts on the Force and Motion Concepts Test were based on questions developed for the Force Concepts Inventory. Each of these three questions tests several important elements in student misconceptions about forces in collisions: (1) the notion that an object with a greater mass exerts a greater force even if the objects are moving at the same speed when they collide head on, (2) the notion that a larger active agent with more mass will exert

---

[16] For more information on the test, see Ronald K. Thornton, "Using Large-Scale Classroom Research to Study Student Conceptual Learning in Mechanics and to Develop New Approaches to Learning," this volume.

more force on a smaller passive agent, and (3) the question of whether more force is exerted during contact by a small active agent or by a large passive agent. In this article the error rates are reported for these three questions at both Dickinson College and the University of Oregon after students completed the New Mechanics activities. In addition, Hestenes, et. al.1 have reported results for those questions for some other groups including two high school classes, one honors and one regular, that had received traditional instruction and two high school honors sections that had received special instruction. The results are summarized in Table 2.

Misconception	FCI	FMT	Traditional Instruction HS honors & regular % Error	Special Instruction HS honors % Error	RTP New Mechanics N=257 Univ of Oregon % Error	WP New Mechanics N=71 Dickinson College % Error
1. Greater mass results in greater force when truck and car collide head on	Q2	Q36	65 (88)	08 (85)	11 (–)	14 (100)
2. Active agent w/ more mass exerts more force as a student pushes another	Q11	Q45	61 (89)	03 (86)	09 (–)	11 (73)
3. Active or most massive agent exerts more force when car pushes truck	Q13	Q42	89 (93)	22 (89)	39 (–)	30 (78)
Average Error Rate %			**72 (90)**	**11 (87)**	**20 (–)**	**18 (84)**

Table 2: Percentage Error on Post (Pre) test questions involving the application of Newton's Third Law to contact forces

5.4.3 Comments on the Results

Some conclusions can be drawn from the data in Table 2. Based on pretest error rates at Dickinson College and in the high school groups tested by Hestenes, et. al., between 80 and 90% of any class have significant misconceptions about interaction forces in collisions. After traditional instruction error rates are only reduced at best for the "easiest" of the questions to about 60%. The post instruction error rates on Third Law concepts for students using the New Mechanics sequence were very similar for students at Dickinson College and the University of Oregon. The lowest error rates were achieved for question 2 and were about 10% in each case with the average error rate on all three questions being about 20%.

Question 3 in which a small car was pushing a large truck with its engine turned off remained the hardest for all the classes tested. If students thought either that the more active agent exerts more force or that the more massive object exerts more force they could answer the question incorrectly. However, essentially all of the students at Dickinson and University of Oregon who still answered the question incorrectly after instruction did so because they believed that the small car as an active agent would exert more force on the large truck which was passive and pushed in the direction of the car.

This overall result for error reduction in Third Law concepts, although quite impressive, is not quite as impressive as the 11% post test error rate achieved by a high school honors class taught by Malcolm Wells. No details are reported on how Wells achieved the learning gains in this particular class. There are several possible factors that might explain the difference between his results and those obtains by students completing the New Mechanics activities. It could be that difficulties (which we expect to overcome) in keeping the equipment calibrated and working smoothly in the MBL based force probe collisions prevented a few students groups from discovering the Third Law for themselves. It may be that there are a larger proportion of students in classes at the University of Oregon and Dickinson College who were very slow learners than in the high school honors class. It may be that differences of 10% in error rates are simply not statistically significant when the unreported sample size in the high school honors class is probably quite small.

5.5 Conclusions

As a result of the pilot testing in the fall of 1992, instructors generally agreed that the New Mechanics sequence shows promise in helping students develop a deeper conceptual understanding of Newton's Laws. The curriculum needs further refinement and more classroom testing. We must do much more careful analysis of learning gains for elements of all three of Newton's Laws before reaching firm conclusions about the impact of the new curriculum on learning. By examining the teaching of Third Law concepts in more detail, perhaps some light has been shed on the educational potential of the New Mechanics activities.

5.6 Acknowledgments

This work would not have been undertaken without the help of David Sokoloff and Ronald Thornton, who played a major role in the development and implementation of the New Mechanics sequence. As collaborators we owe a debt of gratitude Arnold Arons for consulting with us over a period of a year and a half to suggest better ways to present mechanics. Dewey Dykstra spent time early in the conception of the new sequence helping us refine our ideas and reminding us that teaching mechanics to students requires much work prior to the presentation of logical sequences and quantification. We would especially like to thank participants in the New Mechanics Conference who served as sounding boards for our ideas. Special mention should go to Cheryl Clausen who is cooperating with us in the testing of the activities at Arizona State University and to Bob Morse of St. Albans School who was typically several steps ahead of us in the conceptualization and classroom testing of a number of activities we have incorporated into the New Mechanics curricula.

Appendix
Conceptual Questions on Newton's Third Law

Questions 36, 42, and 45 from Force and Motion Concepts Test developed by David Sokoloff and Ronald Thornton are reproduced below. These questions are adapted from Force Concepts Inventory1 questions 2, 13, and 11 respectively.

Questions 36–40 refer to collisions between a car and a truck. For each description of a collision (36–40) below, choose the one answer from the possibilities **A** though **J** that best describes the forces between the car and the truck. You may use a choice more than once or not at all.

A. The truck exerts a greater amount of force on the car than the car exerts on the truck.

B. The car exerts a greater amount of force on the truck than the truck exerts on the car.

C. Neither exerts a force on the other; the car gets smashed simply because it is in the way of the truck.

D. The truck exerts a force on the car but the car doesn't exert a force on the truck.

E. The truck exerts the same amount of force on the car as the car exerts on the truck.

F. Not enough information is given to pick one of the answers above.

J. None of the answers above describes the situation correctly.

In questions 36 through 38 the truck is much heavier than the car.

36. They are both moving at the same speed when they collide. Which choice describes the forces?

37. The car is moving much faster than the heavier truck when they collide. Which choice describes the forces?

38. The heavier truck is standing still when the car hits it. Which choice describes the forces?

Questions 41–43 refer to a large truck which
breaks down out on the road and receives a
push back to town by a small compact car.

Pick one of the choices **A** through **J** below which correctly describes the forces between the car and the truck for each of the descriptions (34–36). You may use a choice more than once or not at all.

A. The force of the car pushing against the truck is equal to that of the truck pushing back against the car.

B. The force of the car pushing against the truck is less then that of the truck pushing back against the car.

C. The force of the car pushing against the truck is greater then that of the truck pushing back against the car.

D. The car's engine is running so it applies a force as it pushes against the truck, but the truck's engine isn't running so it can't push back with a force against the car.

E. Neither the car nor the truck exert any force on each other. The truck is pushed forward simply because it is in the way of the car.

J. None of these descriptions is correct.

41. The car is pushing on the truck, but not hard enough to make the truck move.

42. The car, still pushing the truck, is speeding up to get to cruising speed.

43. The car, still pushing the truck, is at cruising speed and continues to travel at the same speed.

45. Two students sit in identical office chairs facing each other. Bob has a mass of 95 kg, while Jim has a mass of 77 kg. Bob places his bare feet on Jim's knees, as shown to the right. Bob then suddenly pushes outward with his feet, causing both chairs to move. In this situation, while Bob's feet are in contact with Jim's knees,

Bob   Jim

A. Neither student exerts a force on the other.

B. Bob exerts a force on Jim, but Jim doesn't exert any force on Bob.

C. Each student exerts a force on the other, but Jim exerts the larger force.

D. Each student exerts a force on the other, but Bob exerts the larger force.

E. Each student exerts the same amount of force on the other.

J. None of these answers is correct.

6. Teaching Electric Circuit Concepts Using Microcomputer-Based Current/Voltage Probes

David R. Sokoloff

University of Oregon

Abstract. A number of researchers have reported on the difficulties students have with simple electric circuit concepts. This paper reports on the use of microcomputer-based Current/Voltage probes in conjunction with a highly interactive *Electric Circuit* curriculum to teach these concepts in the introductory college physics laboratory. An *Electric Circuit Conceptual Evaluation* has been developed and has been used to assess student understanding of circuit concepts. The results of pre- and post-testing show dramatic gains in student understanding of current and voltage in simple series and parallel direct current circuits[1].

6.1 Introduction

Over the last dozen or so years, a number of studies involving introductory physics students have revealed serious misunderstandings of concepts which are essential to an understanding of the operation of simple direct current circuits. In a recent paper, Lillian McDermott and Peter Shaffer [1] report on student difficulties which have been observed by them and previous researchers. Most students come into the introductory course lacking concrete experiences with real circuits. They fail to differentiate among the important circuit quantities: current, potential difference, energy and power. Many think of current as being used up as it flows around a circuit, and therefore think that the amount of current reaching a circuit element depends on the order of the elements. An ideal battery is thought of by many students as a constant current source rather than a source of a constant potential difference. Such students believe that the current is unaffected by changes in the circuit to which the battery is connected. Potential and potential difference are confused, and therefore nearness to the positive terminal of the battery is often associated with a larger potential difference

[1] This work was supported in part by the Fund for Improvement of Post-secondary Education (FIPSE) of the U.S. Department of Education under the *Tools for Scientific Thinking* and *Interactive Physics* projects, and by the National Science Foundation under the *Student Oriented Science* and *The Workshop Physics Laboratory Featuring Tools for Scientific Thinking* projects at Tufts University, Dickinson College and University of Oregon.

across a circuit element. Also common is confusion in identifying and distinguishing series and parallel connections of circuit elements.

McDermott and Shaffer have reported some success in improving students' understandings of circuit concepts through the development of highly interactive materials designed for use within the constraints of a traditional introductory physics course [2,3]. These materials make use of increasingly more sophisticated observations using batteries and bulbs. They have been used in both a laboratory setting and, more recently, in interactive tutorials which replace traditional recitation sections.

6.2 The Electric Circuit Conceptual Evaluation

Over the last several years, as part of the *Tools for Scientific Thinking* project [4] the author and Ronald Thornton, Director of the Center for Science and Math Teaching at Tufts University, have developed diagnostic tests to assess student understandings of concepts in a number of areas of physics. For example, the *Force and Motion Conceptual Evaluation* has been used extensively as a pre- and post-test to examine student mastery of kinematics and dynamics concepts [5]. The multiple-choice format of these tests has been advantageous for several reasons:

1. It makes possible evaluation of large, varied populations for prevalence of conceptual errors.
2. The short testing time required makes faculty more willing to administer them.
3. Tracking of student progress and retention is possible by testing several times during a course.
4. The results are not biased by an interviewer or test evaluator.

Such a test can evaluate the conceptual models used by students if it is carefully designed using the results of more open-ended research into student understandings, and if questions are asked in a number of different contexts. The tests also include several open-ended questions, which can be correlated with the multiple-choice answers.

The *Electric Circuit Conceptual Evaluation* was developed to assess student understandings of simple circuit concepts. Several questions from the test are shown in Figure 1, and the complete test can be found in the Appendix. The questions assess student understandings of current and potential difference in simple series and parallel circuits. The test was based on the physics education research literature and on University of Washington student responses on long-answer questions administered by the *Physics Education Group* there. Every effort was made to include choices reflecting all known student and expert circuit models.

During Spring term, 1992, the diagnostic test was administered as both a pre- and post-test to the students in one lecture section of the algebra/trigonometry-based (non-calculus) general physics course (PHYS 203) at the University of Oregon. The structure of this four credit hour course at Oregon differs from the traditional setting in that:

All batteries are ideal (they have no internal resistance), and connecting wires have no resistance.

1. A bulb and a battery are connected as shown below.

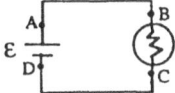

Which is true about the current at various points in this circuit?
- A. The current is largest at A
- B. The current is largest at B
- C. The current is largest at C
- D. The current is largest at D
- E. The current is the same everywhere
- F. The current is the same in AB and smaller than in CD
- G. The current is the same in AB and larger than in CD
- H. The current is the same everywhere except in the bulb
- I. The current is the same everywhere except in the battery
- J. None of these is true.

For questions 2–4, a second identical bulb is added to the circuit in (1) as shown below.

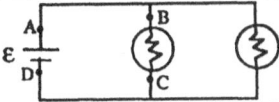

2. Compare the current at A now to the current at A before with only one bulb.
- A. The current at A is now twice as large as before
- B. The current at A is now larger than before but not twice as large
- C. The current at A is the same as before
- D. The current at A is now half as large as before
- E. The current at A is now smaller than before not half as large
- J. None of these is correct.

Briefly explain in the space below how you arrived at your answer to question 2.

3. Compare the brightness of the bulb connected between B and C to its brightness before when there was only one bulb.
- A. The bulb is brighter than it was before
- B. The bulb is just as bright as before
- C. The bulb is dimmer than it was before

4. Compare the potential difference across the bulb, V_{BC}, now to what it was before when there was only one bulb.
- A. The potential difference is now twice as large as before
- B. The potential difference is now larger than before but not twice as large
- C. The potential difference is the same as before
- D. The potential difference is now half as large as before
- E. The potential difference is now smaller than before but not half as large
- J. None of these is correct.

Figure 1   Excerpt from the *Electric Circuit Conceptual Evaluation*

1. There are four, one-hour lectures each week and no recitation session, and

2. The introductory physics laboratory is a separate two credit hour course (PHYS 206) in which only one half to two thirds of the lecture students are normally enrolled.

This structure makes correlation between lecture and laboratory impossible.

For assessment purposes, the lecture is conveniently divided into a LAB group—simultaneously enrolled in lecture and laboratory—and a NOLAB group—enrolled in lecture only. For the section assessed during Spring, 1992, there were 43 students in the matched LAB group of students who took both the pre- and post-tests, and 31

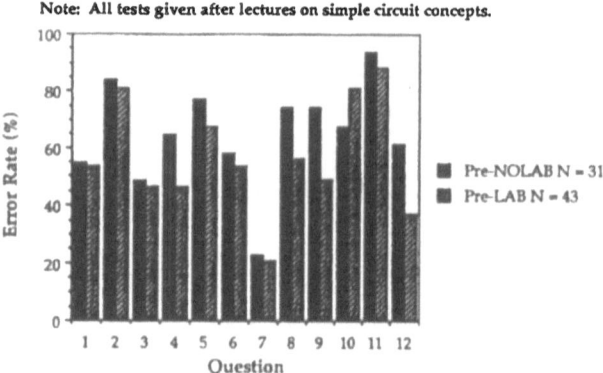

Note: All tests given after lectures on simple circuit concepts.

Figure 2   Comparison of error rates of LAB and NOLAB groups on the *Electric Circuit Conceptual Evaluation* given as a pre-test after lectures on simple circuit concepts

students in the matched NOLAB group. The pre-test was given roughly three weeks into the term after all the material on electrostatics, capacitance and simple direct current circuits had been covered in lecture [6], but before any laboratory experiments on circuits. The instructor had seen the test in advance and made a special effort in lecture to discuss simple circuit concepts. Figure 2 shows the error rates for the two student groups on the twelve test questions. It can be seen that the LAB and NOLAB groups are fairly comparable, and that both groups had error rates of 50–80% on most of the questions, *after all lecture and text instruction had been completed.* These results support the conclusions of previous studies—students do not master simple circuit concepts through a traditional instructional approach [1].

6.3 The Microcomputer-Based Current/Voltage Probes and Electric Circuit Curriculum

Since 1986, faculty at Tufts University's Center for Science and Mathematics Teaching and members of the Department of Physics at Dickinson College have been developing microcomputer-based (MBL) tools of the style first developed at TERC [7]. More recently, tools for Macintosh and MS-DOS computers have been developed [8]. They make use of inexpensive probes, connected to a Macintosh or MS-DOS computer through a Universal Laboratory Interface (ULI), to measure such physical quantities as temperature, position, velocity, acceleration, force, sound pressure, light intensity, or magnetic field.

Students are not required to know anything about computers to use these MBL tools. Menu-driven, self-explanatory software is friendly, even for first-time users, and encourages underprepared and anxious students. Students are in control of their learning since they select the measurements to be made and the way the data are displayed. Data are displayed in digital and graphical form on the computer monitor as the measurements are taken. Students can transform and analyze the data, print graphs or tables, or save data to disks for later analysis. The tools do not simulate physical phenomena, but instead are a means of changing inexpensive computers into instruments for student-directed exploration of the physical world.

The following characteristics of these tools are important to student learning:

1. The tools allow student-directed exploration but free students from most of the time-consuming drudgery associated with data collection and display.

2. The data are plotted in graphical form in real time so that students get immediate feedback and see the data in an understandable form.

3. Because data are quickly taken and displayed, students can easily examine the consequences of a large number of changes in experimental conditions during a single laboratory period. The students spend a large portion of their laboratory time observing physical phenomena and interpreting, discussing, and analyzing data.

4. The hardware and software tools are general—independent of the experiments. The variety of probes use the same interface and the same software format. Students are able to focus on the investigation of many different physical phenomena without spending a large amount of time learning to use complicated tools.

5. The tools dictate neither the phenomena to be investigated, the steps of the investigation, nor the level or sophistication of the curriculum. Thus a wide range of students from elementary school to university level are able to use this same set of tools to investigate the physical world.

The Current/Voltage probes for the ULI and *Electricity* software for the Macintosh were first available in prototype form during Spring, 1992. Connection as a current probe in a simple battery and bulb circuit is shown schematically in Figure 3. The probe is connected to the ULI through a programmable amplifier which allows the range to be selected through software settings. The ULI senses an identification resistor which tells it if the toggle switch on the probe is in the current or voltage position. Because it is desirable to be able to measure current and/or voltage simultaneously at several different locations in a circuit, it is necessary to have probes which don't need to be grounded. The current and voltage probes are effectively isolated. This is accomplished through signal addition and subtraction programmed into the EPROM of the ULI. The software has all of the standard features listed above, so students can observe real time digital and/or graphical displays of current and/or voltage.

During 1992, the author and Priscilla Laws of Dickinson College conducted parallel curriculum development projects. For several years, some of Laws' *Workshop Physics* [9] materials have been modified for use in the more traditional setting at the University of Oregon. The *Electric Circuit* laboratory curriculum makes heavy use of observations with batteries and bulbs and is based on *Physics by Inquiry* [3]. As with all of the MBL laboratory curricula developed as part of the *Tools for Scientific Thinking* and *Interactive Physics* projects, the *Electric Circuit* curriculum allows stu-

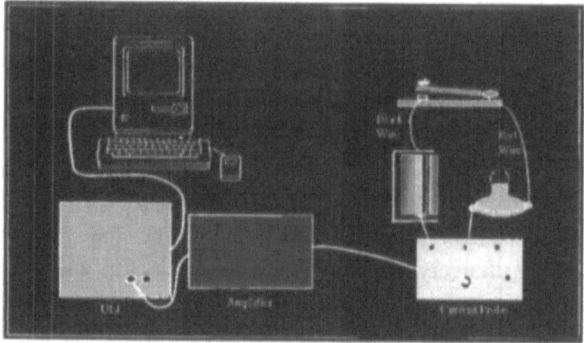

Figure 3 Schematic diagram of connection of Current/Voltage probe, Programmable Amplifier and ULI to measure current in a simple series circuit

dents to take an active role in their learning and encourages them to construct physical knowledge from actual observations through the following features:

1. A guided discovery approach is used with groups of two to four students.
2. Peer learning is supported by presenting data immediately in an understandable form.
3. Predictions are used to engage the student and provide a vehicle for discussion.
4. Attention is paid to student alternative understandings that have been documented in the research literature.
5. Students are encouraged to construct knowledge for themselves.

With the Current/Voltage probes available at Oregon, the author decided to test their effectiveness by incorporating them into the curriculum. It was hoped that observing the behavior of current and/or voltage graphically at different locations in simple circuits would be an effective learning tool which would complement observations of the brightnesses of bulbs. Three laboratory units requiring three, three-hour sessions were developed for use at Oregon. The units were *Batteries, Bulbs and Current*; *Simple Direct Current Circuits*; and *Ohm's Law*. Each lab unit includes a Pre-Lab Preparation Sheet to be filled in before coming to class and a Homework to be turned in several days after completing the lab. Figure 4 shows the cover page of *Simple Direct Current Circuits*.

Only a brief summary of the three laboratory units will be given here. Complete copies are available from the author. *Batteries, Bulbs and Current* begins with observations that static and current electricity involve the same "stuff," using an electroscope, parallel metal plates, a foil-covered pith ball and various methods of charging the ball. Then students are given a battery, bulb, and wire and challenged to find different connections which will cause the bulb to light. Adding a switch and putting a variety of materials into the circuit leads to a study of conductors and insulators. Students are then asked to choose among several possible models for current flow in their battery and bulb circuit. They then explore these models by connecting current probes at two differ-

Name _____ Date _____ Partners _____

UNIT 5: SIMPLE DIRECT CURRENT CIRCUITS

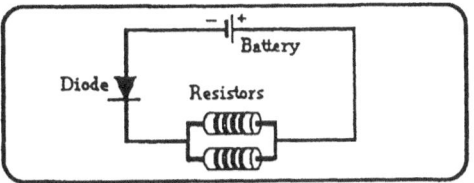

If it's green and it wiggles, it's biology.
If it stinks, it's chemistry.
If it doesn't work, it's physics.
If it's incomprehensible, it's mathematics.
If it doesn't make sense, it's either economics or psychology.
                                          From A. Bloch's
                                          Murphy's Law Bk 3

"Each time a person stands up for an ideal, or acts to improve the lot of others. . .he sends forth a tiny ripple of hope, and crossing each other from a million different centers of energy and daring. Those ripples build a current that can sweep down the mightiest walls of oppression and resistance."
                                          Robert F. Kennedy

OBJECTIVES

1. To understand how current flows through a light bulb as the result of the action of a battery.

2. To understand the meaning of series and parallel connections in an electric circuit.

3. To understand the relationship between the current through different parts of a series circuit.

4. To understand the relationship between the current through different branches of a parallel circuit.

5. To understand the concept of resistance.

6. To learn to apply the concept of potential difference to explain the action of a battery in circuits.

7. To understand the relationship between the potential difference across different parts of a series circuit.

8. To understand the relationship between the potential difference across different branches of a parallel circuit.

© 1992 Dickinson College, Tufts Univrsity , University of Oregon
Supported by National Science Foundation and the U.S. Dept. of Education (FIPSE)

Figure 4   Cover page of *Simple Direct Current Circuits* laboratory unit

ent locations in the circuit and simultaneously observing the currents at these two locations. Figure 5 shows the circuit and typical graphs when the switch is opened and closed several times during the graphing interval. Designing several useful circuits like Christmas tree lights, exploring symbolic circuit diagrams and the connection of cur-

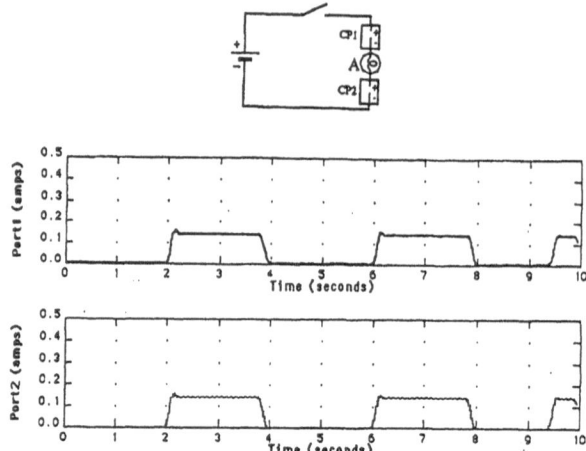

Figure 5 Circuit diagram showing current probes connected to measure current at two different locations in a simple series circuit. Graphs show the currents measured by the probes as the switch is opened and closed.

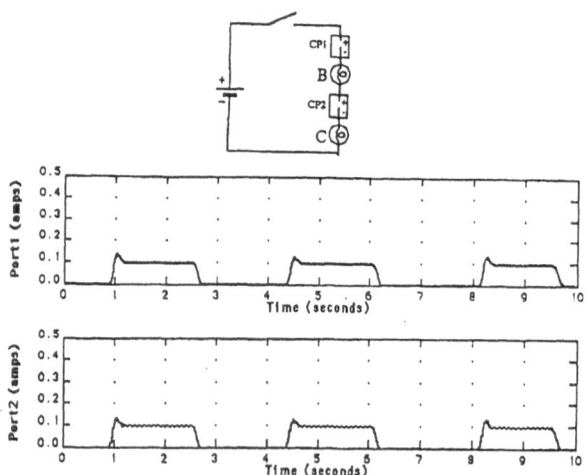

Figure 6 Circuit diagram showing current probes connected to measure current at two different locations in a circuit with two identical bulbs connected in series. Graphs show the currents measured by the probes as the switch is opened and closed.

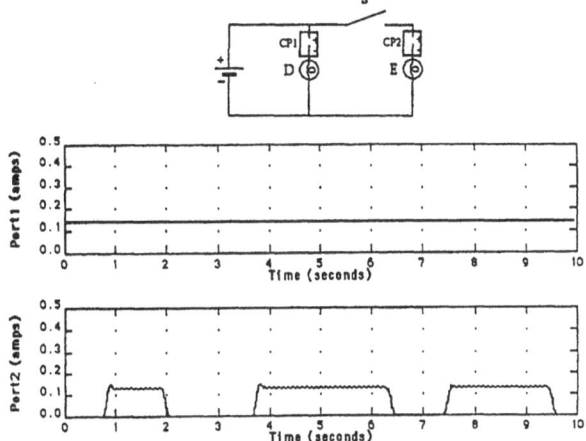

Figure 7 Circuit diagram showing current probes connected to measure the currents through identical bulbs connected in parallel. Graphs show the currents measured by the probes as the switch is opened and closed.

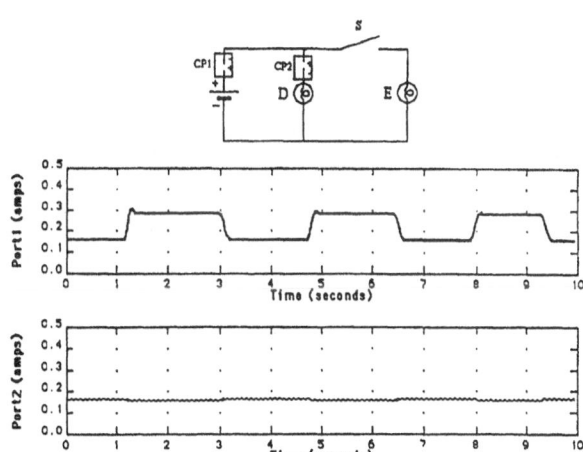

Figure 8 Circuit diagram showing current probes connected to measure the current through the battery and the current through one of the two identical bulbs connected in parallel. Graphs show the currents measured by the probes as the switch is opened and closed.

rent and voltage probes are then followed by observation of a current flow analog fabricated from an inclined board, a bunch of nails and several marbles.

In *Simple Direct Current Circuits* students begin to explore series and parallel circuits. Conclusions are based on observations of relative brightness of bulbs and graphs made with the MBL system. First, students connect two identical bulbs in series, and compare their observations to those with a single bulb. Figure 6 shows the circuit diagram and typical graphs. The next step is to compare this to a parallel connection of two identical bulbs. Two sets of observations are made with different connections of the two current probes. These are shown in Figures 7 and 8. Next, currents are observed in somewhat more complex series and parallel networks. The second half of the lab shifts the point of view to potential differences. Qualitative and quantitative observations are made on the circuit with two bulbs in series, first with one battery and then with two batteries in series. Students observe that potential differences add in a series circuit. Then the voltage probes are used to explore potential differences in a parallel connection. The circuits and graphs are shown in Figures 9 and 10. Finally, a more complex network is examined from a potential difference point of view.

Ohm's Law begins with semi-quantitative observations of the relationship between the current through a bulb and the potential difference across it. The circuit is the same as that shown in Figure 11, but with a bulb substituted for the 10 ohm resistor. This is followed by a more quantitative look using current and voltage probes and the 10 ohm resistor. Figure 11 also shows graphs plotted by the *Electricity* software. The top graph is potential difference across the resistor as a function of time.

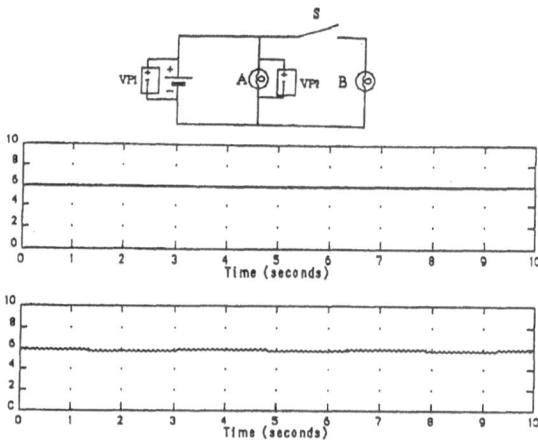

Figure 9 Circuit diagram showing voltage probes connected to measure the potential differences across the battery and across one of the two identical bulbs connected in parallel. Graphs show the potential differences measured by the probes as the switch is opened and closed.

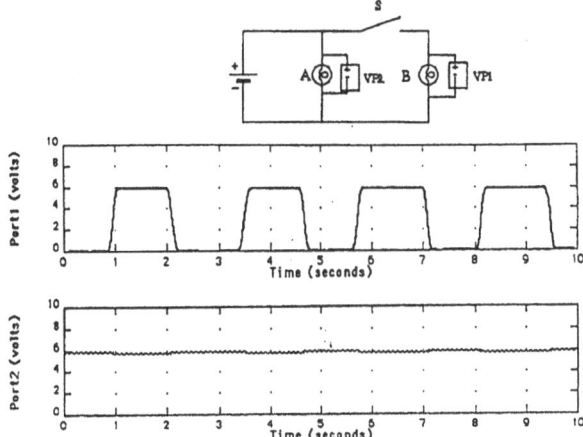

Figure 10 Circuit diagram showing voltage probes connected to measure the potential differences across each of the two identical bulbs connected in parallel. Graphs show the potential differences measured by the probes as the switch is opened and closed.

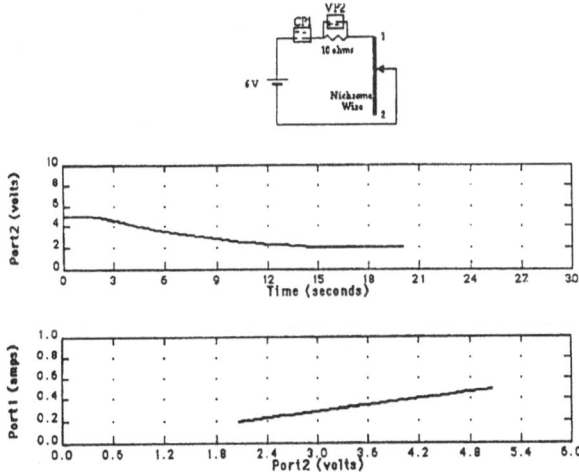

Figure 11 Circuit diagram showing current and voltage probes connected to measure the current through the resistor and the potential difference across the resistor. The upper graph shows the potential difference plotted as a function of time as the alligator clip is moved along the nichrome wire from 1 to 2. The lower graph shows the current plotted as a function of the potential difference.

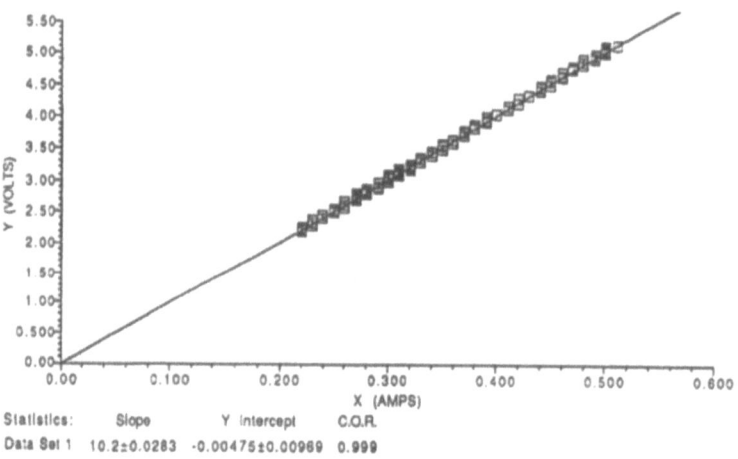

Figure 12 Voltage and current data from Figure 11 plotted using Vernier's *Graphical Analysis*

Its shape depends on the manner in which the alligator clip was moved along the nichrome wire. The bottom graph is current vs. voltage which is a straight line for this ohmic resistor regardless of how the clip was moved. The data may be copied from the table and pasted into a graphing program. Figure 12 shows the data plotted using Vernier's *Graphical Analysis* [8]. The proportionality of current and potential difference is clearly seen, and the value of the resistance from the slope is in excellent agreement with the marked value. The remainder of this lab unit introduces students to the use of multimeters to measure current and voltage, and then guides them as they explore the laws for finding the equivalent resistance of resistors connected in series and in parallel.

6.4 Evidence for Learning Gains

The three *Electric Circuit* laboratory units were completed by the LAB group during the three weeks following the pre-test. During this time, the material on direct current circuits (Tipler [6], Chapter 20) was completed in lecture, and all students were given a one-hour midterm examination on this material. Then the post-test was given. Figure 13 compares the error rates of the NOLAB group on the pre-test and post-test. There was essentially no change, despite completion of all instruction on these concepts, and review for the midterm examination! Figure 14 compares the error rates for the LAB group on the pre-test and post-test. The improvement is significant on all questions. To test retention, the *Electric Circuit Conceptual Evaluation* was included on the laboratory Final Examination given four weeks after the post-test. Figure 15

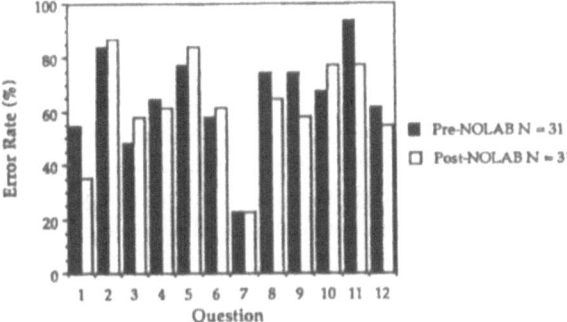

Figure 13 Comparison of error rates of NOLAB group on pre-test and post-test.

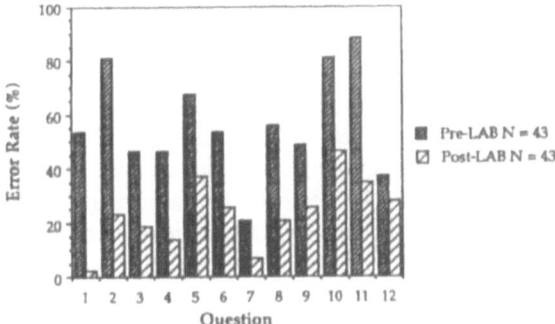

Figure 14 Comparison of error rates of LAB group on pre-test and post-test.

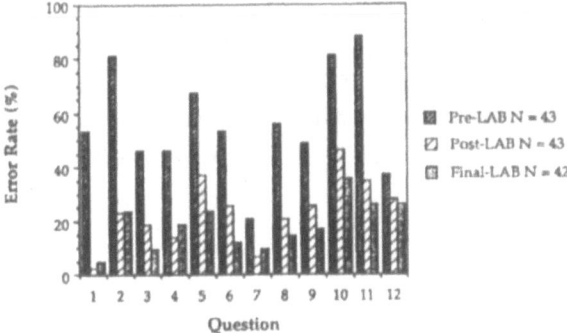

Figure 15 Comparison of error rates of LAB group on pre-test, post-test and final examination.

compares the error rates for the LAB group on the three tests. Retention can be seen to be excellent.

Was the use of the Current/Voltage probes significant in achieving these learning gains? Some insight can be gained by comparing the results on the diagnostic test at Oregon to those at Dickinson College. Most of the circuit laboratory exercises done by *Workshop Physics* students at Dickinson during Spring, 1992, were the same or very similar to those at Oregon. However, the Current/Voltage probes were not used at Dickinson. The *Workshop Physics* course is calculus-based, and includes physics and other science and engineering majors. Compared to the larger class size and traditional approach at Oregon, Dickinson students get a lot more personal attention—from highly motivated professors rather than teaching assistants—in a much more focused learning environment. Dickinson students were pre- and post-tested with a test including questions 1–4 and 8–12. Their average pre-test error rate on these nine questions prior to any instruction on circuits was 66% compared to 60% for the Oregon LAB group after lecture instruction. The average error rates after workshop instruction at Dickinson was 33% compared to 24% after laboratory and lecture instruction at Oregon. The graphical observation in real time available with the MBL Current/Voltage probes may have contributed to the better performance of the Oregon students.

After the post-test, Dickenson students had further discussions on these concepts. They were then tested on a smaller subset of these questions, 2–4 and 10–12. The average error rate on these six questions at Dickenson was 76% before instruction, 39% after lab instruction and 17% percent after these additional discussions, compared to 64%, 28% and 23% at Oregon. Clearly more research is needed to assess the effectiveness of the graphical representations in teaching circuit concepts.

6.5 Conclusions

Extensive research into student understandings of electric circuit concepts was used to design an inquiry-based *Electric Circuit* laboratory curriculum and an *Electric Circuit Conceptual Evaluation.* The curriculum made use of real time graphing of current and voltage at various locations in a circuit available with MBL Current/Voltage probes and *Electricity* software. Significant conceptual learning gains as measured by the conceptual evaluation were achieved by students in an otherwise traditional general physics course at the University of Oregon. While an unacceptable number of students still made errors on simple circuit concepts, the results of this study can now be used to improve the curriculum to tackle these problems.

Acknowledgements

The author is indebted to Ronald Thornton and Priscilla Laws for their efforts to create MBL tools, and especially for making the prototype Current/Voltage probes and *Electricity* software available for use at the University of Oregon. Priscilla Laws' collaboration on curriculum development was invaluable. The author also thanks Lillian McDermott, Peter Shaffer and the Physics Education Group at the University of Washington for their assistance in developing the *Electric Circuit Conceptual Evaluation*, and for their development of *Physics by Inquiry*. Elena Sassi of the University of Naples also made a number of useful suggestions for the test. Finally, this work could not have been done without the tireless efforts of Michael Chmelik, graduate teaching fellow at the University of Oregon and head teaching assistant for the Introductory Physics Laboratory during Spring, 1992.

References

1. McDermott, Lillian C. and Shaffer, Peter S.: Research as a guide for curriculum development: An example from introductory electricity. Part I: Investigation of student understanding, *Am. J. Phys.* 60, 994–1002 (1992).
2. Shaffer, Peter S. and McDermott, Lillian C.: Research as a guide for curriculum development: An example from introductory electricity. Part II: Design of instructional strategies, *Am. J. Phys.* 60, 1003–1013 (1992).
3. McDermott, L.C. et al., *Physics by inquiry* (Physics Education Group, University of Washington, Seattle, Washington, 1982–1992).
4. Thornton, Ronald K. and Sokoloff, David R.: Learning motion concepts using real-time microcomputer-based laboratory tools, *Am. J. Phys.* 58, 858–867 (1990).
5. Some kinematics results have been published. See reference 4. Dynamics results have not yet been published. For more information on the test and results, contact the author or Ronald K. Thornton, Center for Science and Math Teaching, Tufts University, 4 Colby Street, Medford, MA 02155.
6. This corresponds to chapters 18, 19 and 20 (through section 20-4) of Paul Tipler's *College Physics* (Worth, New York, 1987).
7. Technical Education Research Center, 2067 Massachusetts Avenue, Cambridge, MA 02140. These original Apple II-based tools, originally developed for use with middle school students, are available as HRM Motion, Heat and Temperature and Sound Microcomputer-Based Laboratories from Queue, Inc., 338 Commerce Drive, Fairfield, CT 06430.
8. For more information write to Ronald Thornton (see reference 5) and Priscilla Laws, Department of Physics and Astronomy, Dickinson College, Carlisle, PA 17013. These materials are available through Vernier Software, 2920 S.W. 89th Street, Portland, OR 97225.
9. Laws, Priscilla W.: Calculus-based physics without lectures, *Phys. Today* 44, 24-31 (December, 1991).

144   D.R. Sokoloff

Appendix: Electric Circuit Conceptual Evaluation

Name_____

1. A bulb and a battery are connected as shown below.

Which is true about the current at various points in this circuit?
   A. The current is largest at A
   B. The current is largest at B
   C. The current is largest at C
   D. The current is largest at D
   E. The current is the same everywhere
   F. The current is the same in AB and smaller than in CD
   G. The current is the same in AB and larger than in CD
   H. The current is the same everywhere except in the bulb
   I. The current is the same everywhere except in the battery
   J. None of these is true.

For questions 2-4, a second identical bulb is added to the circuit in (1) as shown below.

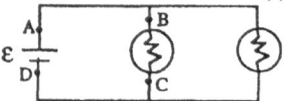

2. Compare the current at A now to the current at A before with only one bulb.
   A. The current at A is now twice as large as before
   B. The current at A is now larger than before but not twice as large
   C. The current at A is the same as before
   D. The current at A is now half as large as before
   E. The current at A is now smaller than before not half as large
   J. None of these is correct.

Briefly explain in the space below how you arrived at your answer to question 2.

3. Compare the brightness of the bulb connected between B and C to its brightness before when there was only one bulb.
   A. The bulb is brighter than it was before
   B. The bulb is just as bright as before
   C. The bulb is dimmer than it was before

Figure 16 Page 1 of the *Electric Circuit Conceptual Evaluation*

4. Compare the potential difference across the bulb, V_{BC}, now to what it was before when there was only one bulb.

     A. The potential difference is now twice as large as before
     B. The potential difference is now larger than before but not twice as large
     C. The potential difference is the same as before
     D. The potential difference is now half as large as before
     E. The potential difference is now smaller than before but not half as large
     J. None of these is correct.

> Briefly explain in the space below how you arrived at your answer to question 4.

For questions 5-7, a second bulb is added to the circuit in (1) as shown below.

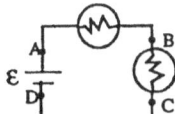

5. Compare the current at A now to the current at A in circuit (1) with only one bulb.

     A. The current at A is now twice as large as before
     B. The current at A is now larger than before but not twice as large
     C. The current at A is the same as before
     D. The current at A is now half as large as before
     E. The current at A is now smaller than before but not half as large
     J. None of these is correct.

> Briefly explain in the space below how you arrived at your answer to question 5.

6. Compare the potential difference across the bulb, V_{BC}, now to what it was before when there was only one bulb, as in circuit (1).

     A. The potential difference is now twice as large as before
     B. The potential difference is now larger than before but not twice as large
     C. The potential difference is the same as before
     D. The potential difference is now half as large as before
     E. The potential difference is now smaller than before but not half as large
     J. None of these is correct.

> Briefly explain in the space below how you arrived at your answer to question 6.

7. Compare the brightness of the bulb connected between B and C to its brightness before when there was only one bulb, as in circuit (1).

     A. The bulb is brighter than it was before
     B. The bulb is just as bright as before
     C. The bulb is dimmer than it was before

Figure 17 Page 2 of the *Electric Circuit Conceptual Evaluation*

146   D.R. Sokoloff

Questions 8-12 refer to the circuit below in which four identical bulbs are connected to a battery. (The switch, S, is initially closed as shown in the diagram.)

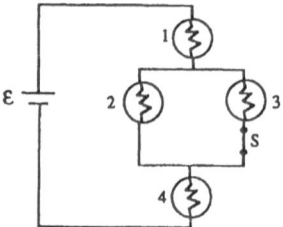

8. Which of the following correctly ranks the bulbs in brightness?
    A. All bulbs are just as bright
    B. 1 is brightest, 2 next brightest, 3 next brightest and 4 dimmest
    C. 1 is brightest. 2 is just as bright as 3 and both are dimmer than 1. 4 is dimmest.
    D. 1 is just as bright as 4. 2 is just as bright as 3 and both are dimmer than 1 and 4.
    E. 2 is just as bright as 3. 1 is just as bright as 4 and both are dimmer than 2 and 3.
    F. 1 is brightest, 4 is next brightest, 2 is just as bright as 3 and both are dimmer than 4
    J. None of these is correct.

9. Which of the following correctly ranks the currents flowing through the bulbs?
    A. All bulbs have the same current flowing through them
    B. The current through 1 is largest, 2 next largest, 3 next largest and 4 smallest
    C. The current through 1 is largest. 2 is just as large as 3 and both are smaller than 1. 4 is smallest.
    D. The current through 1 is just as large as 4. 2 is just as large as 3 and both are smaller than 1 and 4.
    E. The current through 2 is just as large as 3. 1 is just as large as 4 and both are smaller than 2 and 3.
    F. The current through 1 is largest, 4 is next largest, 2 is just as large as 3 and both are smaller than 4
    J. None of these is correct.

10. Which of the following correctly ranks the potential differences across the bulbs?
    A. All bulbs have the same potential difference across them
    B. The potential difference across 1 is largest, 2 next largest, 3 next largest and 4 smallest
    C. The potential difference across 1 is largest. 2 is just as large as 3 and both are smaller than 1. 4 is smallest.
    D. The potential difference across 1 is just as large as 4. 2 is just as large as 3 and both are smaller than 1 and 4
    E. The potential difference across 2 is just as large as 3. 1 is just as large as 4 and both are smaller than 2 and 3
    F. The potential difference across 1 is largest, 4 is next largest, 2 is just as large as 3 and both are smaller than 4
    J. None of these is correct.

11. What happens to the current through bulb 1 if the switch, S, is opened?
    A. It increases
    B. It remains the same
    C. It decreases
    J. Not enough information is given.

12. What happens to the current through bulb 2 if the switch, S, is opened?
    A. It increases
    B. It remains the same
    C. It decreases
    J. Not enough information is given.

©1992 University of Oregon
April 20, 1992

Figure 18 Page 3 of the *Electric Circuit Conceptual Evaluation*

7. Learning and Teaching Motion: MBL Approaches

Elena Sassi and Emilio Balzano

Universita´ di Napoli

Abstract. This paper deals with some MBL approaches to teaching/learning motion, mainly kinematics, in different school contexts. The key lines of the proposed pedagogical interventions, inspired by Open Environment Approaches, are described. The interventions refer to Secondary School Physics and Mathematics courses (14–16 year old students); University Introductory Physics Courses for physics majors (18–20 years); and activities for classes visiting a Laboratory for Science Education. The rationale is the same for all age groups, the depth of analysis being different. The contexts, contents and settings are briefly described, together with some examples of the learning activities. Positive global results indicate that open MBL approaches can introduce significant innovative changes in the teaching/learning process of the addressed content area.

7.1 Introduction

This paper refers to MBL supported approaches for learning and teaching motion in different school contexts. This work is part of a long-range research project aimed at restructuring basic physics education, carried out by an educational research group at Naples University [1]. The activities developed until now have addressed several sectors of educational research: from design/experimentation of software packages to teachers' training, to interventions in classes, to science popularization for the general public. As far as learning models are concerned the whole work refers to moderate constructivism and learner-centered strategies [2 through 9].

Here the pedagogical interventions on the study of motion, with particular emphasis on kinematics, with students in the age range 14–20 years are reported. The proposed approaches are the same for Secondary School students (14–17) and University ones (18–20), the difference being in the depth of level of investigation.

To frame this work, the main criteria, whose key words are Open Environment Approaches (OEA) and Integration between Physics and Mathematics (IPM), will be briefly commented on.

7.1.1 Open Environment Approaches

The Open Environment Approach (OEA) uses computer-supported learning environments which offer the maximum possible control to the learner, for instance through a friendly set of tools specially suitable for a specific content area. OEAs feature intrinsic flexibility and do not impose any preplanned educational path. Usually they are inspired by constructivist learning models and aim at creating learning situations in which the learners are encouraged to explore the content area at different levels, to express their own previsions and to compare them with results obtainable with the OEA system.

Initially OEAs were essentially software environments. In the past few years, educational research on transducers has produced powerful Microcomputer Based Laboratory (MBL) systems.

The intrinsic openness of these learner-centered learning environments (due to the absence of any pre-embedded pedagogical strategy) facilitates the acquisition of high-level cognitive skills suitable for different contexts and disciplines, such as constructive capabilities to transfer and correlate among different situations and contents; to self-evaluate and modify approaches; to propose and solve problems; to compare prevision with results; to use general purpose tools, etc.

More research is needed to prove in terms of hard data that learning activities based on OEAs facilitate these transfers of skills, which are not easy to measure; indications from most of our work are in this direction.

The new functionalities offered by MBL systems allow them to be considered powerful cognitive probes/tools. In fact they change and even subvert several aspects of the traditional learning of science (physics in particular). Namely, they allow students:

- to directly and operationally select sensors, variables, types of graphic representation, scales, etc.;

- to easily observe the time evolution of real and complex phenomena;

- to iterate the cognitive loop "prevision/MBL experiment/comparison";

- to focus on trend analysis rather than on the point-value usual in the traditional educational laboratory;

- to link, in many situations (e.g., motion, thermal processes, sound, light) perceptual insights, common-sense knowledge and acknowledged disciplinary structure.

Therefore, open approaches based on MBL analysis of the variations of variables, parameters, representations, models, etc., become for the first time possible and affordable in real-time. Students are empowered to grasp the internal structures of a physical situation, to distinguish between relevant and minor variables/aspects, and to clarify established correlations and create new ones. Our experimentation indicates that MBL tools, especially when used in open modalities and under the supervision of a good teacher, assuring the convergence toward a stable learning, are, from the above points of view, veritable cognitive probes which extend the areas of what is teachable and what is learnable in current schools.

7.1.2 Integration between Physics and Mathematics

Integration between Physics and Mathematics (IPM) is supported not only for cultural reasons but also because in the current science education, very often the students miss the links between the phenomena and their mathematical representations. This happens at different school levels and is among the causes of the current poor science literacy. Especially for young students the capabilities of MBL for supporting phenomenological approaches can also be used for connecting mathematical abstract topics with the modelling of experiments. This attitude of "rooting" an abstract formalism in the physical phenomena very often gives new clues for learning mathematics.

7.2 Rationale

The rationale of this research/experimentation can be subsumed into the following key lines. They define the general objectives, a guide for activities' design and also criteria to interpret results.

Addressing Common Learning Difficulties (CLD)

Although the physics of motion is taught in almost every introductory physics course at any school level, tests show that many students believe that moving bodies behave otherwise than what is predicted by acknowledged physics. There are many discrepancies and conflicts between the "common sense knowledge" and the theory and laws of motion as they are transmitted in a physics course. The teaching model of optimal accessibility, namely providing students with very good logical organization and complete exposition of the addressed content area, has shown its limits. Naive ways of looking at the physical phenomena have to be taken seriously into account if the teaching process is aimed at a coherent and lasting learning. Physics education research has studied the alternative conceptions quite extensively [10, 11, 12] and many other learning difficulties can be drawn from teachers' experience. To address CLD is therefore a crucial issue in planning a pedagogical intervention on the study of motion.

Learning situations inspired to open, constructivist approaches

We believe that learner-centered pedagogical strategies can facilitate the very complex process of knowledge acquisition. Computer technologies can contribute to creating learning environments aimed at giving the students the maximum possible control. For instance, the traditional educational laboratory can be substantially changed by MBL approaches; the affordable situations can be expanded quite dramatically with important consequences from a learning point of view. Therefore our proposal to teaching/learning motion is heavily based on MBL activities.

Impacting here and now the classroom practice through pedagogical interventions, derived from educational research and realizable under current school conditions and syllabus

Changes in traditional models of teaching demand deep mental changes and usually take a long time. Linear, sequential approaches need to give way to more open modalities. In the meantime the proposed pedagogical interventions need to preserve the efficacy of the teaching/learning processes while introducing innovative and motivating modalities as learner-centered learning situations, disciplinary integration, teachers' cooperative interaction, peer learning, etc. To support proposals for new teaching styles the teachers' enthusiasm is essential. Aiming at proposing educational research results into the current school situation, the proposed pedagogical interventions address very common content areas, so they have more chances to be accepted by teachers and the probability of conveying innovative ideas, methods and materials into classroom practice is increased.

Empowering students and teachers

Really innovative changes in the teaching/learning process can come from taking seriously into account students' interests and ideas and helping them to acquire high-level cognitive skills, suitable for different contexts and content areas. A deep change is demanded in order to consider the student as the very protagonist of the knowledge construction process and not terminal recipient of a transmission chain and to change the teachers' role from notions transmitters to supervisors of students' open learning paths which need to converge toward meaningful and lasting knowledge.

7.3 Contexts of the Pedagogical Interventions

The interventions here reported have been made in three different contexts.

*   Introductory Physics Courses for physics majors at Naples University: these students (18–20 years) are generally a motivated group which attend a Calculus based General Physics course, addressing Mechanics and Thermodynamics. One of us (E.S.) has been teaching this course for many years and also coordinates the computer supported activities.

    The MBL study of motion has been part of a pre-course "Link intervention," aimed at linking with the previous study of motion at the Secondary School [1]. All students (about 120) have been involved in interactive discussions based on MBL experiments (about 4 hours with a 30 student class) and in laboratory sessions (about 6 hours, in small group work). Furthermore, during the course, small groups, on a voluntary basis, have investigated specific experiments.

*   Secondary School Physics and Mathematics Courses attended by 14–16 year old students: Two Secondary Schools in Naples area, of the IPIA type (School for Industrial and Handicraft Technicians) were involved.[1]

    These students usually do not continue their studies at University level and have no particular interest in scientific education. Their schedule includes about eight

[1]The Italian Education System is a central one where curricula and time-schedules are nationally defined by laws. Education is compulsory up to Middle School (11–14); Secondary School (14–18) has three main streams: Licei, Istituti Tecnici and Istituti Professionali. In the first phase (1935–1991) of a National Plan for Informatic (PNI) in Secondary School, mathematics and physics of the first two years have been chosen as vehicle disciplines to introduce elements of computer science and computer supported learning.

periods per week devoted to Physics and Mathematics courses and long laboratory work for activities linked to their specific training. In the framework of the National Plan for Informatic (PNI) the mathematics and physics teachers have been trained in computer's uses; MS-DOS platforms are present in almost all Secondary Schools [13].

The intervention has been done in the framework of a current teachers' training project, about 40 hours for training and 20 hours for class experimentation. Ten teachers have been involved, few of whom have had previous training; some among the mathematics teachers were novices to laboratory activities and to computer use. Six classes, among their first and second year ones, were involved, the total number of students being about 120. This work has been described in more detail in [14].

• Classes from all types of Secondary Schools, visiting LES (Laboratorio per l'Educazione alla Scienza), an Institution for scientific education and science popularization recently active in Naples [15]:

At LES the single class offers several two-hour Didactic Activities (AD) on Motion, based on MBL experiments and held in specially designed learning settings (cfr. section 7.5). These AD have been designed on the basis of our research/experimentation and aim at involving the class in an innovative learning experience; they are facilitated by an LES person and the class teacher.

Many classes from different types of schools have been involved up to now; the feedback coming from this context is complementary to the ones from the two previous contexts. In fact the AD is a "once in a while" event for both the class and the teacher; when they are willing to continue similar approaches in their current work, support is provided through short series of AD, written materials and help. Because this work is in an initial stage, most of our comments refer to the previous two contexts; but there are indications that the MBL approaches are much appreciated by students and teachers at LES also.

7.4 Content and Specific Objectives of the Interventions

In this section we comment on the addressed content area, mainly kinematics, and some specific educational goals at which the work was aimed. The focus on kinematics does not automatically imply support for the view that an Introductory Physics Course (IPC) has to present mechanics and kinematics as the first subject[2]; it derives

---

[2]The problem of redesigning an IPC is a complex and controversial one; different points of view are being proposed and discussed (cfr. for instance the series of comments started in *American Journal of Physics* in January '88) and depend strongly on the intrinsic large variety of opinions on what physics is and also from one's own epistemological position. Among crucial issues are the risk of cognitive overload due to the addition of more and more contents in the syllabus, the difficulties in reaching consensus among physics educators on what could be the "minimal contents" valuable from a formative point of view, the role of modern physics (> 1900), etc.

from the situation of current curricula. Kinematics has been chosen because of several reasons, which can be summarized as follows.

- In the current curricula it has a rather important role.

- Many Common Learning Difficulties (CLD) have been reported, kinematics being among the content areas most studied by educational research.

  To empower the students in their learning process it is needed to help them overcome basic difficulties and become confident about studying physics and mathematics. Kinematics is usually among the first studied topics, so the addressing of kinematical CLD is an early intervention which can be beneficial for all science education. Furthermore, with MBL these difficulties can be addressed in real time together with the ones coming out in the local dynamics of the class from students' ideas and proposals, answers to questions, comments, etc.

- The study of kinematics has played an important historical role in the development of the Galileian method; comments from a methodological point of view can be easily presented together with the problem of the appropriate mathematical formalization.

- Reference to one's own perception is particularly easy when a student's walk is studied through MBL experiments. The analysis of body motion can be done relying on the intuitive awareness coming from psycho-motor experience; this naive physics knowledge can therefore be linked to the abstract description of one-dimensional kinematical graphs. In fact, the motion detector used "sees" practically only along a straight path and treats a moving human body as a small object reflecting the ultrasonic pulses. This feature adds an interesting pedagogical value to MBL study of walks: there are enough intrinsic irregularities, as the steps, for pointing out the difference between the experimental kinematical graphs and the idealized point-like objects one's presented in text books; at the same time a regular walk can produce such a smooth s(t) graph that its approximation with a linear function is a very good one. Body motion experiments can therefore be the phenomenological basis to start with in order to address basic kinematical concepts.

  In the following the main specific objectives, strategies and modalities of teaching/learning are described.

- Whenever possible a physical situation has been addressed by three points of view: a qualitative approach, often in terms of trends and order relations; a quantitative analysis, mainly for simple and idealized experiments, (e.g., low friction cart on ramp); and a semi-quantitative and approximate study of complex real motions (e.g., collisions and bouncing of carts against walls). The integration of these three types of analysis plays a key role in understanding physics. Furthermore, the capability of using different ways to look at a situation and to use them for knowledge construction seems to be an important aspect of the complex learning process.

- Multirepresentation and its usefulness has also been emphasized. The almost immediate change of scales and of variables allowed by MBL was used to clarify the meaning of the correlations between variables. The phase-space plot v(s) was introduced to Secondary School students as another powerful example of abstract space, and used to make visible crucial aspects of complex phenomena (cfr. section 5)

- The meaning of the modelling process was intensively addressed to clarify the relationship between physical concepts and mathematical representation. Modelling activities supported by spreadsheet (Lotus 123) work were frequently proposed to the Secondary School students also because they had been already exposed to the use of this general purpose tool.

- An early operative and very basic introduction to the use of linear fit, derivative and integral, was also proposed. These complex notions were only addressed in an intuitive mode (cfr. section 7.6.2) by using the software tools of the MBL system[3] to prepare the young students with a basis for future, more rigorous study of these topics.

7.5 Learning Settings and Materials

The learning setting for the small group MBL activities of University students has been a usual introductory physics laboratory equipped with several MBL motion detectors; three tutors for every 15 students were there to help the open approaches converge and to observe the students' strategies.

Special attention has been given to the semiotic of the learning setting in the Secondary School interventions, because of the young age of the students. It was chosen to evidence some messages and signs also by means of the logistic arrangement of the physical place; namely the structure of the science classroom usually proposes a code stressing the place/role of the teacher as the locus of control and authority [16]. Even when computers have been introduced in the classroom, almost always this type of sign is still there to convey codes supporting usual school practice.

The chosen learning setting was a combination of physics and computer laboratory and classroom, with no teacher's desk in front of students. One set of MBL apparatus was in a central position as the focal point so that a student walking in front of the motion detector could be easily observed by the whole group. Several MS-DOS computers were on tables around the room walls for small group spreadsheet work; the same tables were used as lab-benches when needed. This same setting has been used for the teachers' training. The choice to work with only one MBL system in central position, due in part to problems in arranging many MBL stations, facilitated the participation of the whole class (about 20 students) and became therefore a real choice. The two class teachers (Mathematics and Physics) and one of us were there;

---

[3]The ultrasonic motion detector has been develeped by TERC, 2067 Massachusetts Ave., Cambridge, MA 02140, USA. The data collection and analysis software, Mac Motion, has been developed by Center for Science and Mathematics Teaching, Tufts University, Medford, MA 02144, USA. We have also used a preliminary version for the MS-DOS platform, in the framework of a collaboration with CSMT.

the teacher/student ratio was higher than usual, our presence being essentially motivated by research purposes.

Each session was two 50-minute periods, from combining one period each of the physics and mathematics schedules. This co-presence of both teachers was a big effort which can usually be sustained only in part during usual class work. The effect was very positive both for students and teachers; after this intervention a stronger collaboration among the involved teachers is going on, and it is also spreading slowly among their colleagues.

The written materials to support classwork were produced during the teachers' training phase and were mostly open question worksheets and homework assignments, questionnaires, etc. To support the teachers in their use of open approaches a set of loosely structured examples, hints, suggestions and notes was developed. All these materials were updated in real-time during the class intervention to follow the feedback coming from the class dynamics.

As far as the LES learning setting is concerned, it is inspired to the same semiotic as the Secondary School one, with the difference that it has been designed from scratch; therefore audio visual and computer projection facilities have also been added. The written materials also focus on work which can be done at school after the activity at LES.

For all three contexts, whenever possible the structure of the proposed activities has been inspired to iteration of the cognitive loop "Prevision, MBL experiment, Comparison." Traditional assessment tools, with their advantages and problems, have been thoroughly adopted, as usual in Italian schools.

7.6 Activities Examples

In this section some examples of the realized activities are briefly commented on. Because it is not possible to go into detailed description of the different interventions, only the key ideas of some pedagogical approaches are sketched.

For the Secondary School students several of the most Common Learning Difficulties were addressed by studying the motions of walking students. Table 1 shows a non-exhaustive list of activities and related CLD. Even if the University students had already studied motion, they often have, with respect to phenomenological approaches, the same difficulties as Secondary School students; so these activities, even if in a shorter format, were very useful to them also.

As already briefly mentioned in section 7.4, the MBL experiments have also been used to address a physical situation, whenever possible, from the three points of view of qualitative study (often based on perceptive intuition, word description, trend and order relations analysis, and collective discussion on a solution's approach); quantitative analysis (with modelling and mathematical representation, when attainable within the students' knowledge); semi-quantitative and approximate analysis of complex phenomena (with attention to clarify the most relevant components).

Activity	Common Learning Difficulty
MBL study of many types of walks: from word description to trend analysis	Confusion between correspondence and function; s(t) and path
"Quasi linear" s(t) of very regular walk and its modelling with the linear function $y = mx + q$	Confusion between direct proportionality and complete linear function
"Quasi uniform motion" and slope in s(t) graph	Equivalent ideal uniform motion, sign of v, sign of slope
Correlation between s(t), v(t)	Inverting the motion as zero in v(t) graph
Two "quasi uniform" motions: overtaking in the same direction (passing in opposite direction) and solving a 2 linear equations system	Interpretation of s(t) graphs as if velocity was the same at overtaking (passing) point; meaning of the system's solution
From s(t) graph to v(t), a(t) graphs	v(t) as tangent to s(t), a(t) as tangent to v(t), geometrical representation of derivative
From v(t) graph to s(t) graph	Role of initial condition s(0)
Area under v(t), a(t) graphs	Dimensional analysis, geometrical representation of integral

Table 1   Activities and related Common Learning Difficulties

Three examples of approaches related to transversal themes are briefly described in the following.

7.6.1 Sudden Changes in Velocity

The theme of quick changes in velocity, with the correlated distinction between ideal sharp variations in mathematical functions and physical transitions, has been addressed going through a series of experiments: from inverting direction in a regular student's walk (Figure 1) to a uniform motion of a low-friction cart on a horizontal smooth floor (Figure 2), to a rolling ball on a ramp receiving a kick at the bottom and inverting its motion.

The proposed discussion about how to perform a "quasi uniform" walk with as sharp as possible a transition between the two constant (positive and negative) values of v(t) graph (Figures 1 and 2) allowed elicitation of interesting common ideas. Quite often, in the case of regular walk, training for faster reaction time was proposed, planning that an almost ideal transition, like the step function of textbooks, could be achieved with enough training. The link with very large acceleration when sudden velocity changes occur was therefore clarified and so the links between kinematics and dynamics, when the forces producing the due acceleration came into the game. Again here the perception of one's own motion was very useful as a starting point, the more idealized motion of the low-friction cart being a naturally following step.

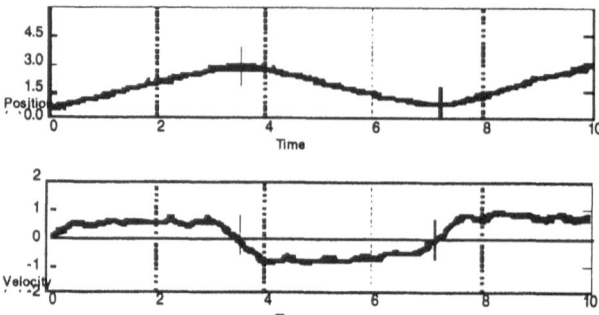

Figure 1   A student walking regularly in front of a motion detector inverts directions twice. The cursor positions point out the correlation between the positions when inverting the motion and the zeros in v(t). Both local variations in velocity due to the intrinsic irregularities of steps and the global linearity of "quasi uniform" walking are clearly shown.

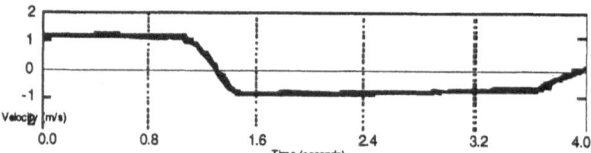

Figure 2   A spring-plunger, low-friction cart launched on a horizontal smooth floor bounces back after colliding against a rigid wall. Its motion is almost uniform and the smooth, not instantaneous, change of sign of its velocity is evident. The collision is not elastic, velocity is reduced to about 60%.

These activities also addressed the issue of joints between different trends in the kinematical graphs of motions with inversions of direction. The study moved from motions with constant speed, as the walk, to motions with constant acceleration, as the ball rolling on a ramp, to a motion with continuously changing velocity and acceleration, as a harmonic oscillator.

7.6.2 Early Operative Use of Mathematical Tools

The MBL software used offers the option of visualizing the geometrical tangent to any point of the current graph and the area under a portion of the graph. Both these tools were used to introduce the Secondary School students to a first intuitive idea of geometrical meaning of derivative and integral.

In quasiuniform motion produced by regular walk, while the tangent changes locally the whole s(t) is linear within a good approximation (Figure 3); therefore, the constant velocity of the equivalent ideal uniform motion can be easily calculated by finite difference for decreasing time intervals; the value of the instantaneous velocity as given by the tangent is intuitively linked to the sequence of velocity values.

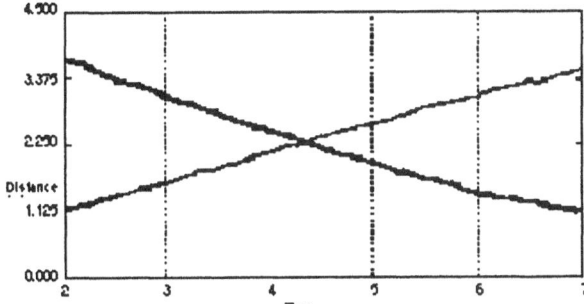

Figure 3 Position vs. time graphs of two students walking regularly and passing in opposite directions.

Activities based on v(t) graphs were used to introduce, in an intuitive way, the concept of integral. Figure 4 refers to a low-friction cart launched upward on a ramp. After a dimensional analysis of the area under the v(t) graph, the shaded triangular area under the positive part of v(t) is measured and compared with the distance Δs travelled by the cart in the same time interval Δt and with the numerical value given by the software tool Integral. The same approach is therefore used for a more complex v(t) (Figure 5) produced by regular walking, where the structure due to the steps is evident and the shaded area is estimated by approximate methods.

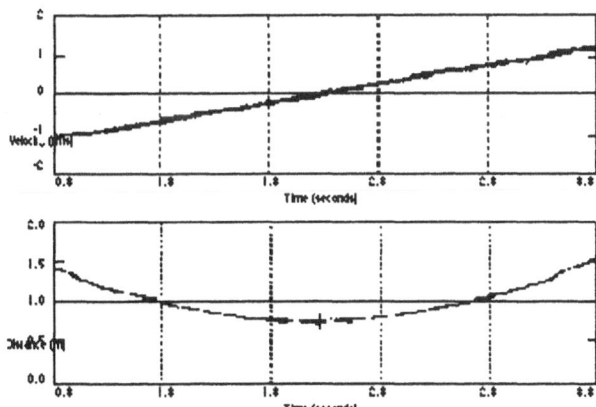

Figure 4 A low-friction cart, launched upward on a smooth ramp comes to a stop and rolls down. The value of shaded triangular area under the positive part of v(t) graph is calculated and compared with the distance travelled in the corresponding time interval, to introduce intuitively the integration process.

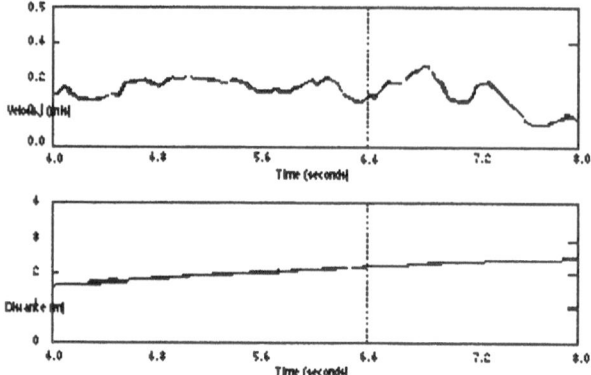

Figure 5 Quasi uniform motion of a student as described by s(t) and v(t) graphs. The integral of v(t), iconically shown by the shaded area with the intrinsic step irregularities, is compared with the travelled distance.

7.6.3 Phase-Space Plots

Multirepresentation and the meaning of different abstract spaces to describe a physical situation has been addressed since the beginning of the study of motion. Starting from the regular walk with direction's inversion time phase-space graph v(s) has been introduced (Figure 6) to reinforce the distinction between correspondence and function.

This plot was also used to address a type of description of motion which is often used "naturally" by young students: velocity with respect to position and not to time. In relation to this, a brief mention of historical approaches dealing with velocity versus position was introduced.

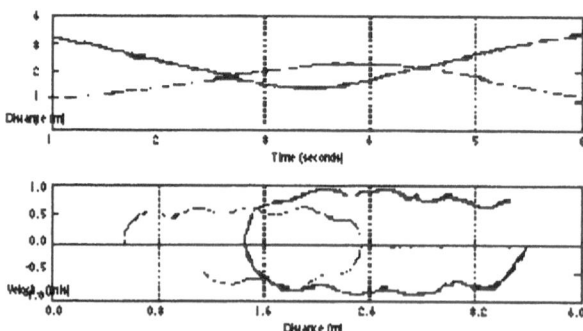

Figure 6 The phase-space plot of regular motion of two students moving toward and away from the motion detector is compared with the usual s(t) representation. Special attention is recalled on the direction the plot is built while the person is moving.

What a closed phase-space plot means, the direction it is built while the motion happens, and the meaning of the area under such a plot has also been addressed. Students learned easily how to use this new tool, for instance to estimate the percentage of kinetic energy loss in a collision between a low-friction cart with spring plunger moving on a smooth floor and colliding against a wall (Figure 7).

The phase-space plot was also used to make visible the representation of the collision duration, something which is usually not visible in other abstract representations. Plots like the one of Figure 7 clearly show the velocity decrease while the spring is being compressed during the collision, the spring's release and how the final velocity is less than the initial one.

The comparison between the phase-space plot of Figures 6 and 7 was also useful to point out the abstraction process needed to go from complex motions like human walking (irregularity of steps, sliding friction, etc.) to a more regular one (a low friction cart on a smooth floor) and to the ideal uniform motion of a point-like object (which very often is a starting point of the chapter on kinematics in text books). Through this type of approach, which appeals to everyday experience and builds on common sense knowledge, it is possible for young students to become familiar with a basic methodology in physics: starting from complex phenomena and arriving at ideal motions described by mathematically simple laws.

A more complex motion, namely multiple collisions of the same cart on a ramp against a wall at the bottom of the ramp, was qualitatively studied along the same lines (Figure 8). Again through the phase-space plot analysis (Figure 9) the visualization of the spring role and the hit duration was possible, together with a qualitative study of the pre-post collision asymmetry. Figure 10 shows the effect of collisions against two walls made with different material: rigid metal and semisoft plastic.

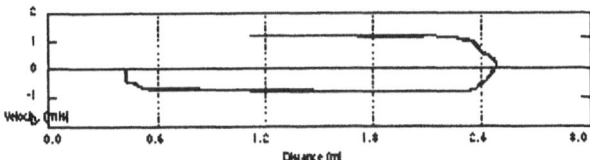

Figure 7   Phase-space graph of a spring-plunger, low-friction cart moving on a smooth floor and colliding against a wall. The velocity decrease and increase while the spring is respectively compressed and released are clearly visible, together with the decreased constant value of v(t) after the collision.

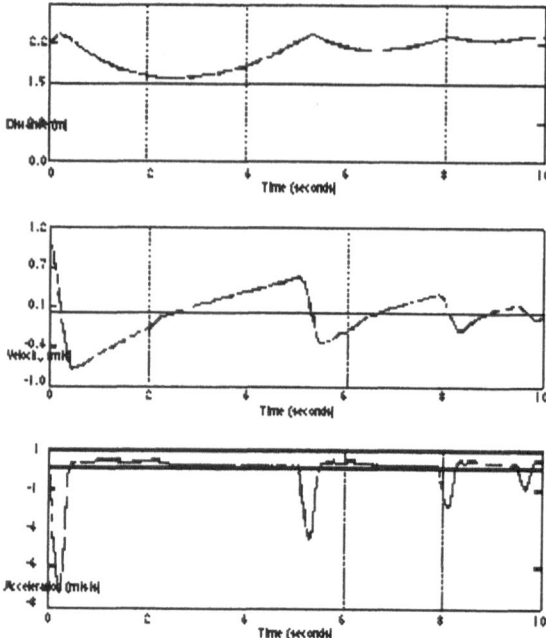

Figure 8   The same cart of Figure 7 moves on a ramp and undergoes several collisions against a wall at the bottom of the ramp. The correlation among distance at inversion, zero velocity and sudden acceleration during the hit is clearly visible.

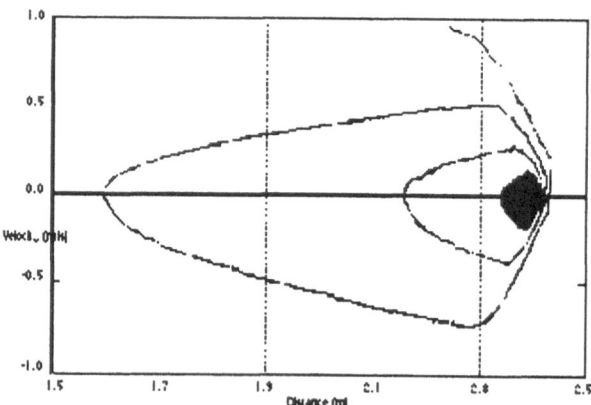

Figure 9   The same motion of Figure 8 is studied in terms of phase-space representation. The hit duration can be estimated to be of the order of a few tenths of a second; the ratio between velocity values pre/post collision allows us to estimate the anelasticity effect.

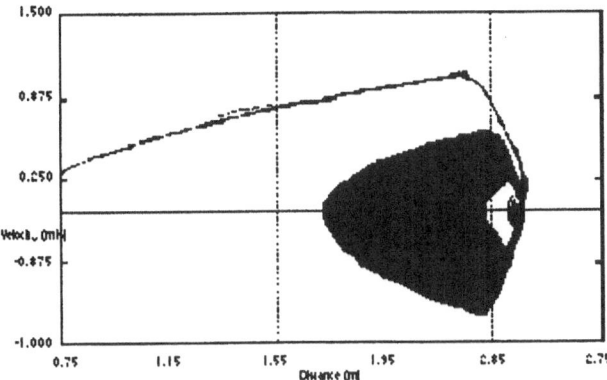

Figure 10 Repeated collisions of the same cart of Figure 7 against two different walls are compared. The more rigid metal wall (light line) absorbes less energy than the semisoft plastic one (dark line). The effect of a wall's elastic properties can be qualitatively studied.

7.7 Conclusions

Some positive global indications have come from all the interventions, independently of the specific context. They can be summarized as follows:

- MBL approaches have been very useful both for eliciting students' common-sense knowledge and reasoning strategies and for helping them to overcome learning difficulties.

- Secondary School students were helped by the rooting of mathematical abstraction in phenomenological approaches; in their reasoning patterns more links between Physics and Mathematics became evident. Generally, more focus on methodology resulted and the proposed mathematical topics were accepted more as meaningful tools than as a list of formulas to be memorized.

- The "prevision–experiment–comparison" loop was motivating both for students and teachers and more active and cooperative class dynamics were observed.

- The so-called "weak students" had new chances which allowed them to show some of their capabilities, which often were more related to intuition and global analysis than to precise recitation of notions and which usually do not enter the game in the current class practice; often an increase of self-esteem and better capabilities of expressing one's own ideas resulted.

- The so-called "best students" discovered that points of view other than the accurate study of the textbook can be as relevant as that one and were challenged to acquire more and different skills.

- For most of the students, about 90%, the acquisition of good operative reasoning capabilities resulted according to their teachers' evaluations.

- The Secondary School teachers much appreciated the cooperative design and production of written materials for students; this process increased their motivation and enthusiasm, which was transferred during the class work.

All these indications are coherent with results from our other work and suggest that research-supported MBL approaches can successfully be used in current curricula as powerful tools for big improvements in the quality of the study of motion. Today a variety of MBL systems with many powerful tools and functionalities can be used; a significant issue is to propose and test several pedagogical interventions aimed at providing Educational Added Value to these systems, in order to facilitate innovative changes in the teaching/learning dynamics in many content areas of scientific education.

References

1. Balzano, E., Guidoni, P., Moretti, M., Sassi, E., Sgueglia, G. (1992): Introductory university courses and open environment approaches: The computer as a multi-role mediator in teaching–learning physics, in *Intelligent learning environment and knowledge acquisition in physics*, Mandl, H. and Tiberghien, A. (eds.) NATO ASI Series F, Vol. 86. Berlin, Springer-Verlag, pp. 5–19.
2. Arcà, M., Guidoni, P., Mazzoli, P. (1984): Structures of understanding of the root of science education, *European Journal of Science Education*, Part I, Vol. 5, 1983, pp. 367–375; Part II, Vol. 6, pp. 311–319.
3. Cox, R. and Cumming, G. (1990): The role of exploration-based learning in the development of expertise, Proceedings IFIP WCEE 1990 (McDougall, A., Dowling, C., eds.), New York, North Holland, pp. 359–364.
4. diSessa, A. (1987): The third revolution in computers and education, *Journal of Research in Science Teaching*, 24, pp. 343–367.
5. diSessa, A. (1983): The phenomenology and time evolution of intuition. In: *Mental models*, Gentner, D. and Stevens, A. (eds.), Laurence Erlbaum Associates.
6. Dreyfus, A., Jungwirth, E., Eliovitch, R. (1990): Applying the "Cognitive Conflict" Strategy for conceptual change—Some implications, difficulties and problems, *Science Education*, 74 (5), pp. 555–569.
7. Forman, G. and Pufall, P.B., eds. (1988) *Constructivism in the computer age*, New Jersey, Lawrence Erlbaum Associates.
8. Mendelsohn, P. et al. (1990): *The Geneva Manifesto of intelligent learning environments*, TECFA Document 90-2, Faculté de Psycologie et des Sciences de l'Education, Université de Genève, Suisse.
9. Osborne, R. and Cosgrove, M. (1980): Lesson framework for changing children's ideas in learning in science, *The implications of childrens' science*, Heinemann, Portsmouth.
10. Driver, R. (ed.) (1989): *International Journal of Science Education*, special issue: Student Conceptions in Science Vol. 11(5).
11. McDermott, L. (1984): Research on conceptual understanding in Mechanics, *Physics Today*, July, pp. 24 ff.

12. Moretti, M. (1988): Difficulties in learning classical mechanics by physics majors. In: *Introductory physics, Course I* (in Italian). Degree thesis, Naples University, Department of Physics, unpublished.
13. Ferraris, Maria (1992): Information technology in the Italian school system: Some problems and perspectives." In: *Proceedings of TIE, European Conference about Information Technology in Education: a Critical Insight*, Universitat de Barcelona 1992, pp. 31–74.
14. Sassi, E. (1992): Basic physics education and computer supported open approaches, *Proceedings of TIE, European Conference about Information Technology in Education: a Critical Insight*, Universitat de Barcelona 1992, pp. 271–281.
15. Amodio, A., Balzano, E., Bobbio, S., Guidoni, P., Fusco, A., Moretti, M., Sassi, E., Silvestrini, V., Simone, P. (1992): *Futuro Remoto, Laboratorio per l'Educazione alla Scienza, Catalogo delle Attività*, Bolletino del LES n.0 Cuen (in Italian).
16. Groisman, A., Shapiro, B., Willisky, J., (1991): The potential of semiotics to inform understanding of events in science education, *International Journal of Science Education*, 13, pp. 217–226.

8. A Study of Pupils' Skills of Graphical Interpretation with Reference to the Use of Data-Logging Techniques

Laurence T. Rogers

Leicester University

Abstract. This paper considers the role of graphs in science education and examines the skills required by pupils in interpreting graphs to gain scientific understanding. Some common difficulties in graphical techniques are discussed and the opportunities for overcoming these and for amplifying the value of graphs through the use of the microcomputer are described.

8.1 The Claims Made for Graphs

> Graphing is one of the most interesting processes through which hidden patterns may be 'discovered.' Graphs can be made to reveal secret relationships. (Austin et al., 1991, p. 35)

> The skills involved in displaying data help pupils to organise their work and communicate their results and ideas to others. (Archenhold, 1988)

The use of graphics for presenting information permeates a wide spectrum of human activity, particularly in science, the media and the world of publishing. Scientists have long used graphs for conveying scientific information, and the ability to read and use information from graphs and charts has been considered essential in science education (Taylor and Swatton, 1990). The prominence and wide use of graphs may be attributed to the view that they provide a versatile language not only for presenting data but for analysing and communicating ideas.

It has been argued (Phillips, 1986) that the graph is a powerful memory aid and thinking aid which helps overcome the limit to the amount of unprocessed information we can memorise or manipulate. A graph effectively processes numerical data into a visual format which not only stores the information but allows it to be referenced as needed. It facilitates parallel processing by the user, enabling the comparison of sets of data which can be coded through movement, colour, brightness, size and shape of symbols. The view of graphs as a language is persuasive when one considers the analogy between items of data and graphs on one hand, and letters, words and sentences on the other: In the language of graphs, the features of the graph in terms of its shape convey ideas about physical change and invoke scientific understanding far beyond the scope of columns of numbers which comprise the raw data. These numbers have little significance until they are associated with each other or other data. Graphs provide this association and allow scientists to organise complex experiences into patterns which enrich their understanding (Swan, 1986). In the language of words, words have meaning which transcends the collections of alphabetical letters or phonics of which they comprise; the organisation of words into sentences is able to convey ideas; the exploration of ideas can endow understanding and insight.

Accepting these established claims for the benefits and value of graphs, it will be argued that the value is considerably amplified by the use of the microcomputer which can generate graphs with ease and versatility. In the first instance, microcomputers make graphical presentation available to a wide spectrum of users who lack the skill or time to construct graphs by conventional means, but who nevertheless can take advantage of the insights which graphs can offer. In the school science laboratory, data-logging software makes especially valuable contributions to the quality and scope of practical work. Such software manages the collection, display, storage and analysis of data obtained from physical sensors or data-loggers connected to the computer. In particular, the graphical features of the software enhance pupils' experience and use of the data in the following ways:

- The display of data is *immediate.*
- *Qualitative* display precedes *quantitative* analysis.
- Pupils may work with the data *interactively*, experimenting with scales and derived data, etc.
- *Analysis aids* support the process of interpreting graphs.
- Data may be *presented* in a choice of quality formats.
- A lower skill level is possible, improving access by lower ability pupils.

The significance of these features in helping pupils to acquire skill and overcome common difficulties will be discussed in the context of a survey of the skills associated with creating, manipulating and interpreting graphs.

8.2 Skills Required for Creating and Manipulating Graphs

When pupils are set the task of constructing a graph from their own collected data or from secondary data, they require at least four skills:

- organising the data
- choosing scales
- drawing and labelling axes
- plotting points

These can be taught through drill and practice, and surveys have shown that the majority of 15-year-old pupils can achieve success in these mechanical aspects (Kerslake, 1981, p. 120, and Archenhold, 1988, p. 38). The weakest aspect tends to be in pupils' understanding and use of scale markings. Errors frequently occur when a coordinate does not coincide with a scale marking or grid line, when more judgement is needed for placing a point. The effect of scales on the shape of a graph is a further aspect which reveals weak understanding. Unfortunately the time needed to plot a graph manually does not encourage the pupil to repeat the exercise if an unsuitable choice of scale is made. This is a clear case where the computer can come to the aid of the pupil and offer distinct advantages: Not only is the plotting accurate but, as a result of the speed of computer plotting, graphs may be treated dynamically and interactively; the graph parameters can be changed at will and the control exercised by the pupil takes effect promptly. Computer graphing shows pupils the effect of their decisions without delay, encouraging them to experiment with scales, etc., and generally explore their data. For example, the apparent rate of change of a variable depends upon the choice of scales. The zoom facility in software is a particularly friendly tool for adjusting the magnification and scales to provide a variety of different views of the data.

Of course, allowing the computer to take over the role of plotting the data has the effect of de-skilling the activity from the pupil's point of view, and this could be viewed as a loss. Further, when the computer is used with sensors and data-loggers, the manual skills involved in the collection, tabulation and plotting of data become redundant. However, there are substantial gains to be set against this loss:

- The graph is displayed while the data are being collected so that they are more easily associated with the phenomena they represent.

- Freed from the need to record the data item by item, pupils have more time to observe the phenomena in the experiment.

- The analysis of the graph may become an interactive process, encouraging pupils to test out their ideas about the data and seek their underlying meaning.

The success of the computer in empowering pupils to explore the data depends upon the design of the software giving them confident control over suitable analysing aids. The computer has great potential in shifting an emphasis in pupils' activity away from the gathering of data towards its interpretation.

8.3 Skills Required for Analysing and Interpreting Graphs

The work of Janvier has shown that, as previously noted, whereas most pupils may become proficient at the reading and plotting of graphs, the interpretation of graphs depends on the ability to understand global features such as intervals, maxima and minima, discontinuities and so on. Pupils are much less successful in these areas (Janvier, 1978). Janvier suggests that this could be due to the great emphasis placed by teachers on the 'accurate' skills: choosing scales so that the graph will fit the paper, generating and plotting points, joining them up with a smooth curve, reading off isolated values, etc. They appear to be emphasised to such an extent that the overall meaning of the graph and the significance of global graphical features are left unexplained in the pupils' minds. Swan suggests a teaching strategy to help correct the imbalance in this teaching through giving pupils practice in sketching graphs to match descriptions in words or pictures (Swan, 1986). It has been suggested also that the use of the computer can help shift the emphasis in pupils' activity towards the interpretation of the data. To consider this, the particular skills employed in interpreting graphs will now be surveyed.

There are several levels of sophistication in the process of interpreting graphs. At the simplest level, the graph shape may be viewed qualitatively, identifying trends and interesting features. Progressing to a quantitative treatment, information is obtained from the graph, reading values, performing simple calculations on coordinates and so on. Beyond this, the progression may involve attaching meaning, making generalisations and applying understanding derived from the graph. The following progression will be used in a discussion of interpretation skills:

- Viewing graph qualitatively
- Reading values
- Describing variables
- Relating variables
- Prediction
- Translation

8.3.1 Viewing the Graph

The most obvious feature of a graph is its shape. This can immediately convey information in a qualitative manner without concerning the observer in unnecessary, involved detail. The media and the press make prolific use of the graph as a device for communicating the 'feel' for a trend or a relationship. Likewise in science, without any recourse to numerical or quantitative consideration, pupils can gain a glimpse or a quick overview of what may be going on in an experiment; they can 'see' gradual or sudden changes, continuity or discontinuity and can select and give attention to particular interesting features. It is a valuable characteristic of data-logging software that the visual representation of the data is the first to be presented to pupils; the numerical attributes can follow when needed. Thus the traditional role of quantitative representation (tabulated results) being the pre-requisite of the qualitative image (graph) can be reversed, allowing pupils to explore the data qualitatively at first.

Such interpretation does not necessarily require axes to be calibrated, but clearly this imposes a limit to the precision of description. Descriptions in simple words can provide valuable stimulae to pupils' thinking and understanding, but the need for more precise description demands a more quantitative approach. Surveys have shown that children tend to opt for qualitative approaches to investigative practical work (Archenhold, 1988, p. 98) and are disinclined to adopt a more quantitative approach which would increase the complexity of the experiment. This reluctance has been identified in pupils even when they have demonstrated competence in constructing graphs in prescribed contexts (Strang, 1990, p. 21). Here, software can support the pupils by reporting initially in a qualitative mode, but then help to develop their ideas about the value of quantitative evidence through a variety of analysing and calculating aids.

8.3.2 Reading Values

Surveys have revealed that pupils tend to find difficulty in reading analogue scales correctly. The interpolation of values between scale markings and an understanding of subdivisions and decimal places are typical casualties. The problem appears to be compounded when two or more readings need to be compared and used to calculate quantities such as gradients and intervals. Using software, reading and calculating data from graphs becomes an almost trivial activity, merely requiring the pupil to manoeuvre a pair of cross hairs to read off values which are displayed in a panel on the screen. This type of facility may be used to gain automatic readouts of coordinates, time intervals, differences, ratios, gradients and areas.

In the traditional context of graphing skills it may seem an abuse to deny pupils opportunities to perform these tasks manually. However, by reducing these skills to a low level, higher order skills can flourish, empowering pupils to think more about the science of their experiment. This has an important impact on investigative and problem-solving approaches to practical science where the problem-solving process relies on service skills such as measurement not requiring thought; they should be effortless and as automatic as possible. The lack of automism interrupts the problem-solving process and can lead to errors (Underwood, 1990, p. 30).

Typically, the reliability of derived data generated by computer calculations is considerably higher than that which depends on pupils' own arithmetic. This makes the quality of the evidence available to pupils very high. Given that this enhancement makes the pupil better informed, we might hope that their judgements based on the information are also better.

Time dependent data are some of the most common in pupils' experience of school science. As a variable, time has a special quality; pupils have a natural 'feel' for it through a sequence of events or changes in other variables. Sometimes, so strong is the feeling of a schedule of events, pupils tend to interpret data in a 'pointwise' fashion rather than in terms of periods of elapsed time; they attach more significance to the time coordinates than the intervals between them (Leinhardt, 1990, p. 37). Software can compensate for this and help strengthen the notion of intervals of time by allowing pupils to choose an arbitrary origin for reading off time measurements. It can also exclude errors in reading the correct number of decimal places for the interval measurement, which often occurs when intervals are less than the interval between consecutive scale markings on the axes. This is similar to the difficulty which pupils experience in interpolating values between the scale markings. Both of these problems are alleviated when the software automatically adjusts the spacing of the scale markings according to the magnification and uses 'friendly' numbers for the scale markings so that interpolation only requires estimates of simple fractions of the scale interval.

Discrete data points on a graph can focus too much attention to the points such that terms like 'maxima' and 'minima' tend to be confused with the actual height on a graph rather than being identified with the steepness of the graph. Using the computer, data -logging typically involves much larger amounts of collected data plotted more densely on the graph, which helps to disguise the discreteness of the data and emphasise rate of change. Analysing aids make steepness (or gradient) easily computed, and display features can reinforce the concept of rate of change. For example, horizontal and vertical cursor lines may be locked on to the data so that as the pupil controls the movement of one cursor, the other cursor is constrained to follow the data values. The resulting relative movement of the two cursors gives a dynamic and visual indication of the rate at which one variable changes with respect to the other.

Alternatively, the confusion between slope and height might be an example of a linguistic problem in which pupils tend to interpret any words of magnitude casually as 'big' or 'small' without a clear context of the property being described. To make matters even more complex, this might also be compounded with conceptual difficulties associated with the variables. A common example is the confusion between velocity and acceleration which can cause the words to be used synonymously. Here software can assist by providing opportunities for working interactively with the data; through a range of display, analysing and calculating aids, pupils can readily manipulate their data to probe and test their understanding.

8.3.3 Describing Variables

To describe a variable, pupils need not only obtain information from a graph but also attach meaning to the information. It was noted above that qualitative descriptions may be based on the shape of a graph without any explicit reading of data values. To progress to a quantitative treatment, at least two items of information need to be taken from a graph and then compared in some way. Software analysing aids provide useful assistance in making the types of measurements which benefit such descriptions:

• Maximum, minimum and mean values

• Difference or ratio between two values

• Gradient for rate of change in a variable

The quantity of data contained in a graph naturally lends itself to the study of connections, patterns, and trends in the data. The visual aspect of the graph draws attention to these features more effectively than numbers in columns of tabulated values. Indeed, there is evidence that pupils are distracted from making generalised descriptions by obvious numerical patterns in tabulated results and instead describe patterns such as sequences of odd and even numbers, multiples and differences, etc. (Austin,

1991, p. 28). Numerical patterns are more difficult to spot when the data are not in serial order but for manual measurements serial data is common since most pupils are trained to increment or decrement the independent variable in an orderly manner. It is unfortunate that tabulated data encourages this type of stepwise analysis which misses the continuity in the behaviour of a variable. Even with a graph, when asked to draw a line through their data, pupils often attempt to join successive points with straight lines, again illustrating this stepwise perception. Software can help to move pupils' thinking towards a continuous view of the data by plotting a best-fit curve which emphasises an underlying trend represented by the shape of the curve.

However, even when pupils focus appropriate attention to the shape of the graph, this does not necessarily result in descriptions of a variable; some pupils instead describe the geometry of the graph line in terms of its shape, direction or curvature in isolation from the axes and from what those axes represent (Austin, 1991, p. 29). A similar geometrical viewpoint is revealed when pupils interpret a graph as the actual picture of a situation; for example, going up and down hills in distance-time graphs (Leinhardt, 1990, p. 39).

Much of this evidence suggests that the process of obtaining scientific descriptions of variables is fraught with distractions; pupils' descriptions are offered at a number of different levels of sophistication. One approach to deflecting attention from numbers, points and geometrical properties is to use software tools for manipulating the data, providing a variety of alternative views and presentations of the data. For example, the data may be smoothed to reduce the effect of 'noise' and emphasise the trend; short-term and longer-term changes may be distinguished; variations in the trend may be observed; the data may be overlaid and compared with a standard mathematical curve; pupils may experiment with matching 'trial fit' curves to their data; differences or 'residuals' between the fitted curve and the data may be plotted (Boohan, 1991, p. 10); scatter graphs indicate a correlation between two variables; or a first derivative (gradient) curve may be plotted. Through a variety of representations, pupils may be encouraged to think about the physical variable rather than the numbers and images which represent it. Software provides quick and convenient tools for all of these manipulations (Rogers, 1992, p. 7). In addition, data-logging in 'real time' helps to forge a link between the physical variable and its graphical representation since the latter appears on the screen simultaneously with the collection of data. The promptness of the display makes an interactive experience possible whereby pupils can alter conditions in an experiment and immediately observe a response (Barton, 1991).

8.3.4 Relating Variables

Identifying and describing a trend or pattern in the behaviour of a variable marks an important stage of sophistication in interpreting graphs because it indicates a perception which is generalized beyond the actual items of data presented on the graph. The majority of graphs generated by data-logging software typically show the time dependence of one or more variables, so the description of a pattern implicitly relates the physical variables to the time variable. Such descriptions are very frequently concerned with changes, rates of change, growth and decay, which are compound variables which relate a primary variable with time.

Describing the relationship between variables in a generalized manner is the key to developing scientific ideas from the graph, but pupils often find this difficult (Taylor, 1991, p. 15). Sometimes their difficulties are linguistic; they are unsure of the type of expression required; as previously indicated, they may find it easier to use geometrical or numerical descriptions rather than referring to the physical variables. Their terminology may lack precision; the use of vague terms such as "goes up" instead of "increases rapidly" might imply that they are perceiving variables separately and are not consciously relating them. Linguistic considerations may disguise pupils' understanding, but even withstanding this, it is a big step for pupils to see the graph as showing the relationship between two variables (Bell, 1987). Pupils need a teaching strategy which helps them ask appropriate questions and which nurtures and gives them practice in appropriate skills. They also need suitable tools for exploring graphs, tools which are readily provided in software.

The shape of a graph gives vital information about the relationship between the variables. A straight line or a curve have distinctive properties which provide insight and understanding of that relationship.

For a relationship represented by a straight line:

1. Changes in the variables occur at a constant rate.

2. For a given increment in one variable, the other variable always increases or decreases by the same amount. (For the case of a variable plotted against time, the increase or decrease in that variable in a given time interval is always the same.)

3. This is independent of the magnitude of either variable.

4. The ratio between the increases or decreases in either variable is constant.

5. When this ratio is not unity, one variable changes more rapidly than the other.

6. The gradient is the same at all places on the graph, i.e., it is constant.

7. When the gradient is negative, an increase in one variable is accompanied by a decrease in the other.

These descriptions are clearly equivalent to or follow from each other. Their significance is that they are individually testable using software tools: when a cursor is moved across the graph, changes in the variables may be read automatically and the rate of change calculated; measurements may be taken from any selected part of the graph; 'x' and 'y' cursors may be locked together showing easily the relative changes in two variables; the gradient at a cursor may be read automatically.

It is seen that pupils have a variety of ways of exploring the properties of a linear relationship. The same tools and methods may also be used to explore relationships represented by curves:

1. One variable changes more rapidly than the other.

2. The rate of change varies across different parts of the graph.

3. The degree of curvature shows how rapidly the rate varies.

4. Exponential behaviour may be identified when the rate of change varies in direct proportion to the vertical variable.

A further refinement in describing a relationship uses a curve-fitting facility: In a version called 'Trial Fit' (Rogers, 1992) the pupil can choose a general form of mathematical curve and experiment to find out the quality of the fit. The tool is designed so that the pupil is allowed to make a judgement of this quality. The three forms available have been chosen to identify the three most common relationships found in scientific data:

- *Linear*: used for identifying proportionality and extrapolating to find offsets and starting values.

- *Power Law Curves*: used for identifying parabolic, inverse and inverse square relationships.

- *Exponentials*: used for identifying exponential growth or decay.

As in previous examples, the speed of calculation and plotting of data through software provides an interactive tool for pupils. Different types of fit may be tried in rapid succession, and curves may be compared by overlaying so that pupils can look for similarities and differences.

8.3.5 Prediction

When a pupil has succeeded in elucidating a generalized relationship between two variables, it will not necessarily indicate that one physically influences the other, but it should make possible interpolation and extrapolation to predict new data values. Thus the successful interpretation can be put to the test by using it to predict new values. Software tools such as best fit curves support pupils in developing confidence in the concept that the description transcends the items of collected data and enables them to identify values between and beyond these items.

Similarly, pupils might predict data and descriptions of the graph shape for new compound variables. For example, predicting a velocity–time graph from a distance–time graph, or a power–voltage graph from a current–voltage graph. Software provides a variety of convenient calculating facilities for generating new data from the collected data allowing pupils to test their predictions.

8.3.6 Translation

A further stage of refinement of interpreting skill involves translating the relationship into an algebraic representation. Trends may be classified and described more precisely by association with a mathematical function, but unless the pattern is linear, the manual method of identifying the appropriate function usually depends upon transforming one variable and replotting it in the hope of obtaining a linear graph. This process can be convoluted, but software offers a number of alternative strategies:

1. The data may be linearised by a suitable function chosen by the pupil.

2. The trial fit technique indicated previously is useful for allowing pupils to interact with the data and use their judgement.

3. Finally, pupils may simulate a function, compare it with the graph of the data and alter the parameters of the function until the best match is obtained.

8.4 Software Design

Effectively designed software gives pupils a large measure of autonomy over the process of collecting and analysing data. The systems offered for control and the appearance of the screen have crucial roles in helping pupils select and exploit what features will be used. If the screen is cluttered with an excess of information, there is a risk that the pupil will be distracted from the purpose of the graph analysis. The 'tool' approach to software is successful when it allows the user to specify which features are to be enabled or hidden and this minimises the number of mandatory decisions when the program is used. However, it is important to give pupils control at a level they can cope with. This requires configuring default conditions so that they can get started easily and build their confidence.

8.5 Teaching and Learning Strategies

This paper has surveyed a range of pupils' difficulties in using graphical techniques and has argued that graphing software and data-logging software both offer significant benefits towards helping pupils overcome these difficulties. However, these software tools and the skills to use them are not enough on their own. Misconceptions cannot be removed by mere 'exposure' to the information contained in graphs. The graph is a thinking aid, but pupils need time to practise describing and using patterns and to engage in the necessary reflection upon their results and discussion with their peers and teacher (Phillips, 1986, p. 42; Taylor, 1991, p. 16). Fortunately, the computer is good at providing time because it performs its tasks so rapidly. It frees pupils to devote more attention to observation, reflection and discussion.

However, it is also necessary for teachers to provide a curricular framework which can challenge pupils with key questions which encourage the type of thinking which goes beyond the simple reading of information from the graph and description of a variable, and leads to an interpretation in terms of a generalised relationship between variables. Teachers also need to review the relative value they attach to different aspects of graphical technique. Pupils' success or failure depends on the incentives to meet targets set by the teacher. Targets implicitly indicate what is valued and important. If the emphasis is to be effectively shifted away from the routine collection and plotting of data towards the development and use of interpreting skills, the system of rewards and assessment of pupil performance needs to reflect this.

References

Archenhold, F. (ed.), 1988, Science at age 15—*A review of APU survey findings 1980–84*, London: HMSO.

Austin, R., Holding, B., Bell, J., and Daniels, S., 1991, *Patterns and relationships in school science*, London: SEAC.

Barton, R. and Rogers, L.T., 1991, The computer as an aid to practical science—studying motion with a computer, *Journal of Computer Assisted Learning* 7, 104–113.

Bell, A., Brekke, G. and Swan, M., 1987, Diagnostic teaching: graphical interpretation, *Mathematics Teaching* 119, 56–59.

Boohan, R. and Ogborn, J., 1991, *Making sense of data*; Inset Pack. Harlow: Longman.

Janvier, C., 1978, *The interpretation of complex Cartesian graphs representing situations*, Ph.D. thesis, University of Nottingham.

Kerslake, D., 1981, Graphs. In: Hart, K.M. (ed.) *Children's understanding of mathematics*: 11–16. Oxford: John Murray.

Leinhardt, G., Zaslavsky, O. and Stein, M.K., 1990, Functions, graphs and graphing: Tasks, learning and teaching, *Review of Educational Research* 60 (1).

Phillips, R.J., 1986, Computer graphics as a memory aid and a thinking aid, *Journal of Computer Assisted Learning* 2, 37–44.

Rogers, L.T., 1990, IT in science in the National Curriculum, *Journal of Computer Assisted Learning* 6, 246–254.

Rogers, L.T., 1992, *'Insight' measurement software—Teachers' Guide*, Cambridge: Longman Logotron.

Strang, J., 1990, *Measurement in school science*, London: SEAC.

Swan, M., 1986, *The Language of functions and graphs*. Shell Centre for Mathematical Education, Manchester: Joint Matriculation Board.

9. On Ways of Symbolizing: The Case of Laura and the Velocity Sign[1]

Ricardo Nemirovsky

TERC, Inc.

Abstract. This case study focuses on how a high school student, Laura, learned the meaning of the velocity sign. By moving a toy car she created many real-time graphs on a computer screen. The study strives to show that her learning was not just an acknowledgment of a rule, but a broad questioning and revision of her thinking about graphs and motion. Laura's process exemplifies what is involved in the learning of a way of symbolizing situations of physical change.

9.1 Introduction

The main focus of the Measuring and Modeling[2] project is to study how high school students who have not taken calculus courses learn about differentiation and integration notions when the problems are posed in the context of physical changes that students can control, predict, and measure. A part of this effort is the research reported here as a case study about Laura, a high school student, working with a computer-based motion detector. The motion detector enables one to produce graphs of a moving object in real time on the computer screen. Laura moved a toy car along a straight path in order to generate graphs. Most of the problems involved the translation between a given graph of position vs. time and its corresponding graph of velocity vs. time, or vice versa.

The case study focuses on how Laura worked out her ideas about the velocity sign. This paper explores more broadly, however, students' learning of ways of symbolizing kinesthetic actions and the roles that MBL (microcomputer based lab) tools can play in students' learning. Section 9.2 both describes the notion of symbol-use and ways of symbolizing and introduces central ideas for the analysis

---

[1] This chapter is also published as "On Ways of Symbolizing: The Case of Laura and the Velocity Sign" in *Journal of Mathematical Behavior*, Vol. 13, No. 4, pp. 389–422 and is reproduced with the permission of Ablex Publishing.

[2] The Measuring and Modeling project is supported by the NSF Grant MDR-8855644. Opinions expressed are those of the author and not necessarily those of the Foundation. The author wishes to thank the useful comments of John Clement, James Kaput, Steve Monk, Ann Rosebery, Andee Rubin, Cornelia Tierney, and Beth Warren.

of Laura's episodes. It contains examples of students' approaches (not part of the case study) in order to illustrate general issues on the learning of ways of symbolizing. Section 9.3 provides the background for the case study. Sections 9.4–9.6 include the description and analysis of Laura's learning. This report concludes with a discussion, Section 9.7.

9.2 Ways of Symbolizing

> ... the "law" side of things is not a bounded set of norms, rules, principles, values, or whatever from which jural responses to distilled events can be drawn, but part of a distinctive manner of imagining the real. (Geertz, 1983, p. 173.)

As a framework, I highlight the difference between symbol system and symbol-use. With "symbol system" I refer to the analysis of mathematical representations in terms of rules. For example, Cartesian graphs can be considered as a symbol system; that is, a rule-governed set of elements, such as points being determined by coordinate values in specific ways or scales demarcating units regularly. These rules can be syntactic or semantic. A semantic rule, for example, may establish how a measurement of a physical variable must be denoted on the graph. On the other hand with "symbol-use" I refer to the actual and concrete use of mathematical symbols by someone, for a purpose, and as part of a chain of meaningful events. Symbol-use may enact some rules of the symbol system that is being used, but the central point is that it cannot be reduced to those rules. Symbol-use is embedded in personal intentions, in specific histories, and in the qualities of a situation. Symbol-use encompasses a multitude of extra-symbolic components, without which it loses all its significance.

I was largely inspired on the distinction between symbol system and system-use by Bakhtin's differentiation between sentence and utterance. For Bakhtin, a sentence such as "I am very tired" can be subjected to a syntactical and semantic analysis through the application of formal rules describing the structure of the sentence, the role of each word in the sentence, and the manner in which the different word meanings interrelate to constitute the sentence meaning. On the other hand, the sentence is not an utterance. An utterance encompasses expressive intonations: who is the speaker, who is the addressee, what has been said before, what reactions the speaker is expecting, and so forth. For example, a person uttering "I am very tired" elicits completely different reactions if the communicative intention is "I am working too much" or "I am tired of you." The actual manifestation of these communicative intentions cannot be reduced to a formal analysis based on syntactical or semantic rules; one has to consider gestures, what had been said before, facial expressions, who is talking to whom, etc.

Situatedness is the fundamental aspect of an utterance. Given a sentence such as "I am very tired," we can understand, in Bakhtin's terms, "its possible role in an utterance"; that is, one can imagine situations in which someone could say it; but,

> The sentence as a unit of language, like the word, has no author. Like the word, it belongs to nobody, and only by functioning as a whole utterance does

it become an expression of the position of someone speaking individually in a concrete situation of speech communication. (Bakhtin, 1986, p. 83)

I define symbol-use as an utterance that involves mathematical symbols. Symbol-use has been studied in many fields, such as history of mathematics (Cajori, 1929), sociology of science (Lynch & Woolgar, 1990), and research in mathematics education (Carraher et al., 1988; Confrey, 1988; diSessa, 1992; Lave, 1988; Meira, 1992; Pimm, 1987; Saxe, 1982). Confrey (1988) has argued convincingly that the study of students' symbol-use may help to create new symbolic approaches that can be not only educationally fruitful but also offer new insights into the nature of mathematical ideas.

On the other hand, the field of research in mathematics education has a long tradition of striving to reduce symbol-use to symbol systems (just to mention two "classics:" Brown & Burton, 1978; Matz, 1982). The underlying assumption is that learning mathematics is learning new symbolic rules as well as abandoning "wrong" rules. Countless studies describe how students' mistakes relate to specific "alternative" rules. This is a progressive approach in comparison with another derived from the assumption that students' mistakes are random, senseless, or resulting from their inability to think properly. But what these studies leave unanswered is why learning is so complex and often difficult. One is tempted to think that in order to learn a new rule what is needed is more explicit teaching, stating clearly what the right rule is and practicing the right rule many times in many contexts. But when one gets immersed in concrete symbol-uses (and this paper is an example of such an immersion) we realize that "the rules" are just abstract descriptions of a very dynamic process that encompasses all the surrounding and historical elements of the situation. The rules are, using Bakhtin's words, "a fiction." A useful fiction for many purposes, but nonetheless the rules cannot be more than an inert abstraction of the lived utterances.

Before elaborating more on learning, I want to introduce the idea of this paper: ways of symbolizing. With "ways of symbolizing" I distinguish between various types of symbol-use. For the sake of clarity, let me introduce an example. There is a way of symbolizing that I call "interval analysis." To illustrate this, here is an example of a student who wanted to compare the speed of a car at different times with the position time graph illustrated in Graph 1.

She drew dotted lines on Graph 1, producing Graph 2.

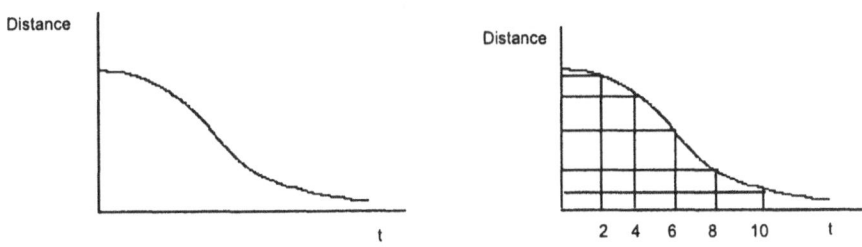

Graph 1   **Graph 2**

1 Student: Okay. Okay. It looks like at the beginning you're going slow and you
2   speed up a little, then you get a little more fast, then you go slower, and then
3   you go slower. [While talking the student acts out each interval moving the
4   car as in discrete movements between stops]

A particular process, interval analysis, is embedded in her symbol-use: the split-
ting of the horizontal axis in regular intervals, the focusing on the corresponding
"distances covered" and their qualitative comparison, etc.; it has certain commonali-
ties with other episodes of symbol-use with Cartesian graphs. However, let me be
clear, the point is not to reduce her action to a procedural rule such as: divide the axis
in regular intervals, trace vertical lines, etc. There is no fixed set of rules that exhaust
what actual symbol-users do. For example, the student decided to consider five intervals (How many intervals should be considered?); she wanted to quantify the length
of the time intervals but not the corresponding distances (When is the qualitative
comparison of lengths appropriate?); she said "you speed up a little" (How do you
know if it is "a little" or "a lot"?) and so on. One starts to make sense of why she did
what she did by not using the mere postulation of rules but by analyzing how her
interval analysis approach connected to her acting out of discrete motion—what she
is reacting to, what she is trying to do, who she is addressing and how she was using
the motion detector—in other words, by analyzing what makes her actions meaningful and how they are part of a personal history with a past and future.

The learning of a way of symbolizing, such as interval analysis, is not a matter
of being able to replicate a certain symbolic procedure. It involves developing a point
of view about the meanings of graphical shapes and about the situation represented
by the graph, from which a style of symbol-use emerges as a natural, fruitful,
and meaningful action in a specific context. Interval analysis configures a certain
style of using Cartesian graphs that varies from individual to individual and situation
to situation, but it is commonly adopted by practitioners in different fields of study.
Becoming a practitioner in a way of symbolizing involves what Bakhtin (1981)
calls appropriation:

> [The word in language] becomes "one's own" only when the speaker popu-
> lates it with his own intention, his own accent, when he appropriates the word,
> adapting it to his own semantic and expressive intention. Prior to this moment
> of appropriation, the word ... exists in other people's mouths, in other people's
> contexts, serving other people's intentions: it is from there that one must take
> the word, and make it one's own. And not all words for just anyone submit
> equally easily to this appropriation, to this seizure and transformation into
> private property: many words stubbornly resist, others remain alien, sound
> foreign in the mouth of the one who appropriated them and who now speaks
> them; they cannot be assimilated into his context and fall out of it; it is as if
> they put themselves in quotation marks against the will of the speaker. Lan-
> guage is not a neutral medium that passes freely and easily into the private
> property of the speaker's intentions; it is populated—overpopulated—with the
> intentions of others. Expropriating it, forcing it to submit to one's own inten-
> tions and accents, is a difficult and complicated process. (pp. 293–294)

Ways of symbolizing are not closed codes that one learns and, from that point on, one "has." For example, interval analysis opens up new avenues for thinking about the interplay between the discrete and the continuous or about the patterns in number sequences; furthermore, it is enlightened by other ways of symbolizing. Ways of symbolizing are always subject to growth and enrichment as well as to decline and impoverishment. To a large extent, this paper is an examination of the learning of a particular way of symbolizing as an opening up of new horizons of understanding. As happens with spatial horizons, one never reaches them; with each move forward, the horizon expands in all directions.

I want to invoke (yet) another idea of Bakhtin that seems to me central for the study of symbol-use; it is the notion of dialogism. In the formalist approaches to language the utterance is seen as an expression of a thought. The speaker, conceived as a carrier of thoughts, is an emitter of utterances whose meanings are his thoughts. For Bakhtin instead, the paradigm of how language is used centers not on the isolated thinker manifesting his thoughts, but on a dialogue in which the utterances react to each other and acquire meaning by mutual relationship and conflict. The utterance is voiced by an author whose words express a certain point of view that is recognizable, to himself and to others, only by reaction from other points of view. Dialogue is a constant and mutual shift of perspectives, and the multiplicity of perspectives is not only sequential but simultaneous. I sense this while I try to write this paragraph. I do not have a complete and unitary thought that I want to express. While I produce each utterance, a multitude of reactions come to my mind, some are reactions that I imagine from the readers, other are reactions to ideas that I held or met in the past. Before I finish typing an utterance, I feel this is not what I want to say. Finally I decide to focus on an example of a dialogism in symbol-use. It is a relief.

This example is analyzed in another paper (Nemirovsky, Tierney & Ogonowski, 1993) which includes several episodes of children's symbol-use. Jay is a six-year-old child who was interviewed during the summer before he entered first grade. Jay was given blocks to put in and take out from a bag. At one point in the interview Jay was thinking about the following sequence: start with four blocks in the bag, take out one, take out one, put in two, put in two. To show the sequence of changes to the interviewer, Jay wrote the following:

$$4-1 \quad 4-1 \quad +2 \quad +2$$

1 Interviewer: Explain to me what you just did [wrote].
2 Jay: I just did four minus one, then another four minus one. Then add two
3   more, and then another two more and that means you took away. [after
4   silent time pointing to the two 4-1's] All right. Four minus one equals
5   three So do another four minus one equals three [pointing to the second
6   4-1]. That gives six. Hmmm. That made it a little hard.

The expression has four pieces. Initially each piece meant a certain action of adding or subtracting a number of blocks (lines 2 to 4). For example, each 4-1 denoted the action of taking away one block. But as soon as Jay tried to calculate how many blocks "you took away," 4-1 "means" three, and three and three is six. At this point

Jay felt puzzled: "Hmmm. That made it a little hard." How could one take six blocks from four?

I include this example to illustrate symbol-use as a dynamic game among different simultaneous perspectives and as a constant meeting with the unexpected. At first the 4-1's meant taking away one block (lines 2 to 4) but later they meant three (lines 5 to 6). None of these alternative perspectives is "wrong," but each challenges the other in ways that one cannot anticipate. Symbol-use is not a redundant manifestation of what one already knows. Rather it is a conflictive attempt to deal with the multiplicity of angles from which objects and ideas can be seen.

9.3 The Case Study
9.3.1 The Velocity Sign

Use of the velocity sign to denote directionality of motion is a widespread practice. In motion along a linear path positive or negative velocity is commonly used (in textbooks, scientific practices, etc.) to discriminate direction. None of the students that I have interviewed used this symbolic criterion spontaneously. It was introduced by the MBL system. Some students made sense of the idea without apparent puzzlement. Other students, such as Laura, acquired fluency in using a velocity sign through an intense process of exploration and revision of ideas. This paper focuses on Laura's learning path, not because she necessarily represented what most high-school students would do in handling similar problems, but because she articulated some profound issues surrounding the understanding of the velocity sign. For some students the ideas that Laura struggled to make sense of are immediately accessible, but this is precisely why Laura's episodes are helpful to us: they are more explicit about what is involved in developing meaning for the velocity sign.

One can think of the velocity sign from two points of view: as a symbol system or as a symbol-use. As a symbol system the meaning of the velocity sign is a straightforward semantic rule: positive and negative velocity signs distinguish the two possible directions of motion in unidimensional motion. As a symbol-use, however, one becomes aware of how the determination of directionality and its relation to sign elicits broader and fundamental aspects surrounding the understanding of motion and signed mathematical objects. Focusing on symbol-use the velocity sign emerges as a particular way of symbolizing.

The MBL system displayed to Laura a representation of velocity with signed graphs, that is, graphs that can display positive and negative values of velocity. She had specific intuitions about signed graphs and to a large extent her learning path was a revision and reparation of what signed graphs meant to her and how they represented the motion of an object. Her learning was not just the acquisition of a semantic rule; it involved, among other aspects, a new understanding of how "more" and "less" are reflected in signed graphs, how less velocity may mean that the object moves faster, and how a change of direction has to include a "stopping" moment.

The velocity sign is a particular case of the rate-of-change sign. For example, flow rate is usually taken as positive for inflow and negative for outflow. The sign of the rate of change is used to differentiate increases from decreases. However, velocity sign for unidimensional motion probably elicits some specific connotations that are not shared by other rates. For example, there is a "natural" direction of increase for the volume of water in a container that is not present in horizontal motion; whereas "more" in the water-in-a-container situation is likely to correspond to more water being in it, there is nothing (aside from conventions) in the horizontal displacement of an object eliciting "more" as "more to the right." Similarly, this is also a difference between vertical and horizontal movement. For example, in vertical motion going up is likely to be seen as a "natural" increase.

9.3.2 Teaching Experiments

As part of the Measuring and Modeling project we designed and conducted a series of individual teaching experiments. Through them, each student worked with a specific environment exploring problems with either motion, air flow, or spatial contours. The use of microcomputer-based labs was intensive and critical in all the teaching experiments. The episodes reported here are part of a teaching experiment with Laura using the motion detector and a little toy car that she moved with her hand along a linear path.

The purpose of our teaching experiment methodology is to study the learning of the mathematics of change. We do not seek to produce models of "good teaching," because (1) the interview situation is different from the usual teaching context, and (2) there is immense individual variability. Often the same question, comment, or problem that is critical for one student will be obvious or irrelevant for another. We expect to inform better teaching but not through a literal or naive approach of looking at the interviews as showing good or bad ways of teaching. Instead, we strive to use the interviews to understand how the different elements of the situation (the MBL tools, the conversation with the interviewer, the problem-situations, etc.) contribute to the learning for particular students and how these students use what they know to deal with new situations. We expect that our own learning about these processes will enable us to illuminate new approaches to introducing calculus notions across educational levels. An example of this possible contribution is to achieve a better understanding of what it takes to learn a particular way of symbolizing or how the MBL tools may take part in students' making sense of the connection between symbols and physical behavior.

9.3.3 Laura

Laura was in the 11th grade during the teaching experiment (October 1991) at a private school in the Boston area. She was very active and spontaneous in the sessions. She had taken courses in Algebra, Geometry, Biology, and Chemistry, and was then taking Algebra II, Trigonometry, and Physics. She remembered that in 7th grade she had used the motion detector in school to measure the motion of cars that were assembled by the students themselves. Laura, who planned to study law after college, described herself as very active in class, having better grades in math than in science, and being perceived by teachers and classmates as "smart" in math but not in science. What she liked to do the most in mathematics was use computers. She enjoyed labs in school, where one can "experience things." Nevertheless the biology lab seemed boring to her and not very informative. In contrast, she loved the chemistry lab because "we did a lot of labs on our own." Laura thought that math and science "work together" and explained that some of the activities in geometry were equivalent to those one does in a science lab; however, "math is with numbers, this is why we do not do experiments."

Laura volunteered to participate in the teaching experiment; she was paid a student fee of $35. She worked with the motion detector and a sequence of problems that focused on qualitative graphs of distance and velocity vs. time. Typically, I asked her to figure out how the graph of distance or velocity would look, given the other one. Laura used the motion detector to test predictions and explore ideas. Each session lasted about 70 minutes. Laura was always very active and willing to think aloud. We used two video cameras, one for the computer screen and the other for the overall action. The audio was transcribed, and we kept all the drawings that Laura and I made on paper.

9.3.4 The Learning Path

This case study focuses exclusively on what happened during the teaching experiment. The episodes included in this paper took place during the first and second sessions. I selected these episodes because they encompass Laura's explorations of the velocity sign. Passage 1 was at the beginning of the first session. Passage 2 took the last 50 minutes of the first session. Passage 3 happened during the first half of the second session.

Each of the three passages (Sections 9.4–9.6) includes a Foreword that explains why it has been selected and its relevance to the whole case study, a Description, and an Analysis. There are always interpretative aspects in how one selects and presents the data, but explicit interpretations are included in the Analysis.

The motion detector measures the distance between the motion detector and the car and uses the position of the motion detector as the origin of coordinates. Therefore in this setting there is no real difference between distance and position of the car: its position is always its distance to the motion detector. For this reason Laura and I used both terms indistinctly. To show how Laura moved the toy car I will use a particular type of diagram, such as the following one:

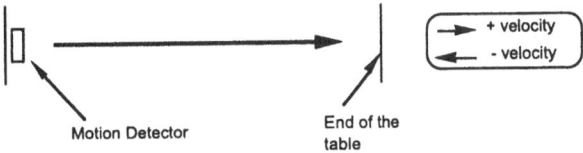

Figure 1

When Laura moved the toy car back and forth, each movement was shown with a different arrow. But all the movements were produced on the same straight line. An adhesive paper placed on the table marked a line on which the car was to be moved to keep it aligned with the motion detector. The arrows are displayed one on top of the other to indicate the order in which the movements took place. Figure 2 is an example.

Figure 2

Figure 2 shows that Laura moved the car from a to b and then to a, all on the same line. Laura was always standing in such a way that the motion detector was on her left. The car had a front and a back side. Whenever her language is ambiguous I indicate what she meant by "backwards" or "front ."

9.4 Passage 1: Graphing the Motion Shown in a Videotape

9.4.1 Foreword

This passage shows Laura's ability to distinguish consistently the speed, position, and directionality of a train moving along a track. It supports an analysis of the nature of Laura's "confusion" which appears later in Passage 2. Passage 1 also suggests how the graphical representation of directionality elicits more general issues about time and space.

9.4.2 Description

At the beginning of the first session I asked Laura to create a graph representing the motion of an electric train shown in a videotape. "Whatever you think, it's a good descrip-tion," I said. The video displayed the path of the train divided in units and the time, so pausing the tape she could take measurements of time and distance. The first video vi-gnette showed the train moving from right to left and slowing down. Laura constructed a graph of speed vs. time indicating a slowing down over time (see Graph 3).

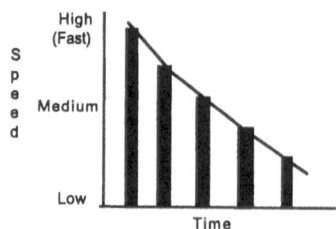

Graph 3

The second video vignette portrayed the same train moving in the opposite direction (left to right) and slowing down too. Laura drew the following graph (Graph 4):

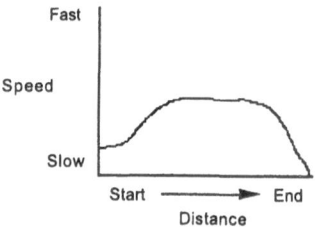

Graph 4

```
 1  Ricardo: You don't have time any more, you used to have time before [in
 2      Graph 3]. Do you have a reason for, for this change?
 3  Laura: Well I thought that it would be easier to show from the start to the end
 4      than the time on this one, 'cause the train was moving in the backwards
 5      piece, so I just didn't think of the time actually. But in the first one I
 6      thought of the time because it was going in a normal, in a normal fashion.
 7      The time didn't even faze me on this one because it was going backwards. I didn't
 8      think of the time in this one.
 9  Ricardo: This way you show in which direction it was moving?
10  Laura: Yes, it was moving that way [adding the arrow from Start to End in
11      Graph 4].
```

9.4.3 Analysis

The directionality of the first video vignette (from left to right) was not a salient feature to be expressed initially in Graph 3. But after watching the second video, directionality struck her as a major aspect to denote. Suddenly she perceived the previous direction as "normal fashion" (line 6), and the new one, an anomaly, had to be expressed. Directionality became relevant when she perceived it as a dichotomy. If she envisioned only one direction in the first video, it remained in the background of unremarkable default behavior.

As Laura tried to represent the new direction of motion she said, "The time didn't even faze me on this one" (line 7). Time has its own directionality, no matter how the train moves; time always goes in the same direction, hence a temporal graph was not helpful for her. The use of the sign of velocity to indicate directionality in a temporal graph was not used spontaneously by Laura (nor by any of the other students that I have interviewed). Since she wanted to show the "start" and the "end," distance emerged to her as more expressive than time.

In Graphs 3 and 4, Laura showed her ability to distinguish and coordinate speed, position, directionality, and time. Graph 3 displayed a decreasing speed over time; the position and the directionality of the train were not explicit. In Graph 4, directionality becomes explicit through the indication of the starting and ending positions. Through the two graphs Laura expressed her understanding that "fast" and "slow" may happen in both directions and that speed may vary across time or distance. It is also worth mentioning that Laura did not try a distance vs. time graph. The possibility that speed vs. time or distance is more expressive for students than a distance vs. time graph for horizontal motion has consistently emerged from several studies (diSessa et al., 1991; Tierney & Nemirovsky, 1991).

9.5 Passage 2: Directionality and Velocity Sign

9.5.1 Foreword

Passage 2 is divided into four vignettes; it is the main focus of analysis. During the first of these vignettes Laura encounters the computer's use of negative velocity. Each of the three following vignettes reflects a particular attempt to construct a meaningful interpretation of the velocity sign. Through this process Laura revises the connotations of the positive/negative distinction in a graphical context and its meaning of how the toy car moves.

9.5.2 Vignette 1: Encountering Negative Velocity

Description. I introduced Laura to the motion detector and the computer interface. Both of us produced and discussed graphs of distance vs. time as we moved the toy car back and forth. Laura interpreted the graphs in the following way:

12 Laura: The closer you go to the motion sensor the lower the graph goes. And
13   the farther away from it you go, the higher the graph measures, the farther
14   the distance, so the higher the graph goes.

At a certain point Laura produced on the computer a graph by moving the car as indicated in Figure 2. I asked her to copy the graph onto a paper sheet (see Graph 5).

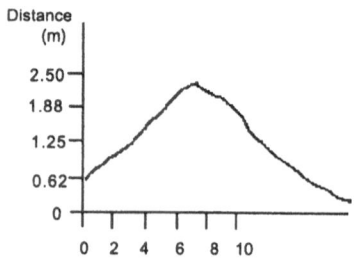

Graph 5

Then I asked Laura to sketch the corresponding graph of velocity vs. time. After a long silence she said:

15 Laura: I don't know where to start. I'm like, I have no idea where to start.
16 Ricardo: Would it help if you moved the car as you did it? And look at the
17   velocity?
18 Laura: Um, all right, wait a minute [preparing herself to act out], I started
19   here [at a, in Figure 2]. OK. I went up towards about the middle [b,
20   Figure 2] And then the car headed back down at a faster speed to about
21   there [a, Figure 2]

I tried to focus on the qualities that relate to the speed of motion:

22 Ricardo: How velocity changes. When is velocity high or low, when is the car going
23   faster, slower?
24 Laura: Oh, OK. Um, I think when you're coming down [from b to a, Figure 2]
25   it goes faster, when you go up [from a to b, Figure 2] it goes slower. So
26   it would start. Should I number it or? Should I number the speed?
27 Ricardo: Or just the shape.
28 Laura: Just the shape, OK. So whenever it goes up [from a to b, Figure 2] it
29   stays slower [Laura draws Graph 6]. And when it hits, um, all of a sudden
30   you have, it starts to get faster, like that [Graph 6]. Cause as it goes down
31   [from b to a, Figure 2] the car goes faster.

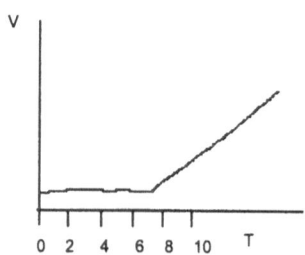

Graph 6

Later Laura used the software to get the computer graph of velocity vs. time. It had the following appearance:

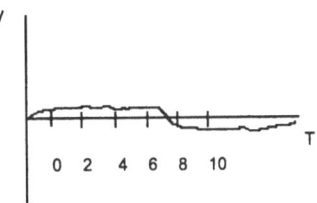

Graph 7   (on the computer screen)

32   Laura: So mine. Mine looks different. See, I don't know, because I didn't
33   number my velocity. I didn't. Negatives, that's why. Because I didn't know
34   there was negatives. So theirs was numbered different. I think the two
35   graphs are different, the way I saw mine and the way it did it.

Analysis. Laura mapped the height of the curve with how far the car is from the motion sensor (lines 12 to 14). From the very beginning the situatedness of her analysis is apparent. She does not talk about distance in general nor about objects moving "somewhere"; her talk is about being far from or close to the motion sensor and about how the distance is expressed in the height of a line appearing on the screen.

Even though Laura moved the car with more or less the same speed in her motion back and forth, she kept a clear subjective memory of going back toward the motion detector at a faster speed (see lines 20, 24 to 26, and 30). It is important that Laura was trying to figure out the velocity graph from her kinesthetic remembrance of how she moved the car and not from the distance in Graph 4 itself (lines 18 to 21). This way of thinking was prompted by my suggestion (lines 16 to 17); after Laura felt stuck (line 15), I suggested she think about the motion that she had actually executed in generating Graph 4. After that she used her memory of the actual motion instead of Graph 4 as a source of information for her prediction of velocity.

Laura's prediction (Graph 6) reflected a particular story, or account, of the motion that had taken place with Graph 5: go up slow, "hit the point," and come down fast (lines 28 to 31). Since the idea was to create a graph for velocity, the only qualities that seemed relevant to her were those connected to speed: first slow, and then, "all of a sudden," fast.

After seeing that the computer graph went down instead of higher when the car was in the condition of "fast," Laura recognized the "negatives" as the major difference between "the way I saw mine and the way it did it" (lines 32 to 33). Laura did not notice that the negative piece on the computer graph was more or less horizontal, she was primarily puzzled by the graph becoming negative.

9.5.3 Vignette 2: The First Attempt to Make Sense of the Velocity Sign

Description.

36 Ricardo: And what does it mean here [on Graph 7] negative?
37 Laura: That it went, as the car hit the point [b, in Figure 3] the speed went
38   lower.
39 Ricardo: The velocity is positive and then becomes negative. Why does the
40   velocity become negative?
41 Laura: Because as the car goes faster the velocity goes lower? I have the totally
42   opposite idea. I thought as the car went faster the velocity got higher. But as
43   the car goes faster the velocity goes lower.

I pointed to the time at which the velocity changed sign on the computer graph:

44 Ricardo: And here [on Graph 7] there was a time in which, according to this
45   graph, velocity was zero.
46 Laura: Mmm-hmm.
47 Ricardo: What does that mean?
48 Laura: Maybe that is when I hit that point, when I hit that... at some point I
49   started from going slow I started going faster, so around faster, it had to
50   change too. Like there had to be a change in motion.

Then Laura tried to make sense of why she had predicted a "wrong" velocity curve, instead of the one displayed by the computer (Graph 7). "I wasn't thinking of it in the right way," she concluded.

I posed to Laura a new but similar problem: to produce a distance graph like Graph 8 and to think about the corresponding velocity graph. Laura generated a distance curve like that in Graph 8, by moving the car according to Figure 3.

Figure 3

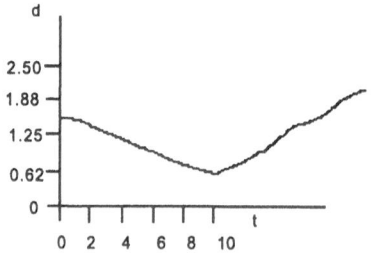

Graph 8

As Laura tried to figure out the new velocity graph she traced Graph 8 with her left hand while she moved the right hand to recreate the motion depicted in Figure 3:

51 Laura: When I started back here [a, Figure 3, just moving her hand] with
52   the car, I came forward... the car started out slower towards the beginning
53   [from a to b, Figure 3] and then as I came back [from b to a, Figure 3] it
54   got faster... Yeah, I think it was slower here [from a to b, Figure 3],
55   because like I said before when I first was doing it I wasn't rushing for a
56   time, but when I. You know, there wasn't much room left when I started to
57   go faster. And then as I started to go backwards it went faster. Yeah, I think
58   that's... I think. I'm trying to think of how I want to show it.

To resolve how to show velocity Laura analyzed again the graphs of the previous problem, but then she commented:

59 Laura: Doesn't a graph go like the opposite way though? When it goes down to
60   the negative isn't it going faster? When it goes down below isn't it going
61   faster? Is that the way it works? I haven't done this in a long time, that's
62   why.

Analysis. In this vignette Laura articulated her first attempt to make sense of the velocity sign (lines 36 to 43). She expressed her initial perception of negative as being lower: "speed went lower" (line 37). But Laura needed to account for her former reckoning that she went faster during the negative piece. She reacted to separating speed, how fast the car goes, from velocity, the curve displayed on the computer screen, so she said: "Because as the car went faster the velocity goes lower?" (line 41). However this conclusion was extremely counterintuitive to her: "I have the totally opposite idea."

One can see how her understanding of the velocity sign involved the coordination of two realms of experience: motion and signed graphs. I believe that this is a central point. She was not struggling with the motion of a car in itself, or the sign of a graph in itself, her problem was to make sense of their relationship. Laura had previous experiences in both domains, previous experiences that triggered specific expectations, such as if the velocity graph is negative, and therefore "below," the car should go slower. Through these vignettes Laura learned ways of relating the two aspects, and in doing so she had to revise what she knew about signed graphs and motion.

I asked Laura to interpret the velocity going through zero in Graph 7 (lines 44 to 50). Laura described the conditions in which that event happened: "when I hit the point," from going slower to going faster, and "there had to be a change in motion." The condition of zero velocity was elicited by the graphical representation. The case is not that the change of direction is lived as involving a stopping moment. Rather the change is a graphical necessity that in going from positive to negative the graph has to cross the zero line. In a non-signed graph (absolute value of the velocity, say) Laura could easily have argued that the graph did not "really" go to zero; but this is not an option in a signed graph. Then Laura developed the idea that the car goes through zero velocity in order to change speed.

Laura then thought about another velocity graph (lines 51 to 58). She took again the kinesthetic qualities of her motion as the guiding element for her velocity prediction. She constructed her hand motion in terms of this story: "slower towards the beginning and then as I came back it got faster" (lines 52 to 53). Laura strove to relive the subjective conditions of her actual motion ("I wasn't rushing for a time," "there wasn't much room left") in order to capture the changes of speed.

Then, when she was "trying to think of how I want to show it" (line 58), Laura re-encountered her puzzled understanding that the computer produced graphs by going lower for higher speeds (lines 59 to 62). On the one hand, through this vignette Laura articulated a way of making sense of velocity sign, but found it hard to believe. Her uneasiness was grounded in fundamental expectations about the relationship between up→more and down→less.

9.5.4 Vignette 3: Unsettling the Meaning of the Velocity Sign

Description. To explore more openly the issue of the sign I asked Laura to use the motion detector and move the car in any way she wanted while the velocity curve appeared on the computer screen, and to see when the graph was positive or negative.

First she moved the car according to Figure 4:

Figure 4

63   Laura: OK. All right, so. All right. So the closer you are. The closer you are
64   to it, to the motion sensor, the more negative it is. But the farther away you
65   go it goes up into the positive, the velocity.

Laura also tried to change the speed of the car by moving it according to Figure 5:

Figure 5

66   Laura: It seemed when I went faster towards the end [from b to a, Figure 5]
67   it went higher up [into the positive], but when I went faster towards the
68   beginning [from a to b, Figure 5] it kind of went down [into the negative].
69   That's what I don't understand.

Then Laura produced the motion depicted in Figure 6:

Figure 6

70 Laura: OK Well... Let's see. I started out here [from a to b, Figure 6] and
71   it was positive, and as I went back and faster [from b to a, Figure 6] it
72   went down to the negative it seemed like, but then I headed back [from a to b,
73   Figure 6] it went back to the positive. Why is it. It always seems to work
74   that way.
75 Ricardo: Which way?
76 Laura: That it goes. When I start out here [from a to b, Figure 6] it starts
77   out in the positive, and then as I go back it goes negative. And then I headed
78   back a little but it went back up to the positive. Let me try it again.

Laura moved the car according to Figure 7:

Figure 7

79 Laura: Mmm-hmm. It was positive through the whole way that I went [from a
80   to b, Figure 7], and then as I went far back [from b to a, Figure 7] it
81   went negative. I don't get it. It seems to be the closer I am, then it. It stays
82   positive when I'm at the closest range. But then it. Let me try it again.

Now Laura measured while she moved the car back and forth five times. The computer
graph looked like a wave.

83 Laura: Oh, I see. When. That's what it does. When I go back and forth it goes
84   [acting out] high positive, negative, positive, negative, positive, negative,
85   positive, negative, positive, negative, positive. That's how it went, so, does
85   that mean I was going fast when I was going backwards. Right? So I was
87   going faster when I went backwards. So the faster you go then it goes more
88   into the negative.

Later Laura acted out the car motion in such a way that I thought she had come to
see the velocity sign as unrelated to "how fast" the car moved. However, when in the
next moment I asked her to rethink the problem described in Graph 7 (that is, to
generate the corresponding velocity curve), she tried to figure out a sign meaning as
a combination of speed and position that would be compatible with her former observations. For example:

89 Laura: Far away but fast. Far away and fast is negative, and close up and
90   fast is positive.

But she could not find a way of putting all the pieces together. Looking back at Graph
6, she showed me that the sign of velocity did not change whether the car was close
to or far away from the motion detector:

91 Laura: I don't see a pattern. I mean even when I went far away it became
92   negative and as I came closer it stayed negative. But once you go far. It
93   doesn't make sense. It doesn't.

Analysis. In this vignette Laura was, for the first time, looking at a velocity graph while she was moving the car. Her initial reaction was to include the distance to the motion sensor as a factor for the determination of the sign. (See lines 63 to 65: "The closer you are to the motion sensor, the more negative it is. But the farther away you go it goes up into the positive.") One possible interpretation is that she did not sense a real difference in the speeds at which she was moving the car (see Figure 5) therefore speed only was not enough to account for the difference of sign. Laura's idea of involving the distance to the motion sensor in making sense of velocity sign was her attempt to add another factor so that she could account for a more complex experience.

Laura described accurately (in lines 66 to 69) how the graph behaved in correspondence to the car motion: "faster towards the end [away from the motion detector] it went higher up, but when I went faster towards the beginning it kind of went down." However, she stated, "That's what I don't understand." What does she mean? What is she not understanding?

I think that her language (in lines 66 to 69) gives us an important clue: more than positive or negative, it is a matter of graphical going up or down. Imagine for a moment that you visualize a vertical axis in which up is more of something and down is less of something, what could this "something" be? You are told that it is an axis for velocity, so first you try speed and it seems to go the other way around. You accept (reluctantly) that this instrument inverts the normal way: more (fast) is down and less (slow) is up. Then you try out other motions and it turns out that you find no clear correspondence between the graphical up or down and the car going fast or slow. Neither the normal mapping (more is up, less is down) nor the inverted criteria seems to hold. Something else has to be taking part. The distance to the motion sensor perhaps. This is a possible thread for Laura's experience.

The case is not that Laura was unable to identify the semantic rule: moving away from the motion detector is positive, and moving towards is negative. She produced a description of the "rule": "When I start out here [close to the motion detector] it starts out in the positive, and then as I go back it goes negative." (lines 76-78). But, she uttered: "Why is it?" (in line 73). This is a clear example of how making sense of the velocity sign cannot be reduced to the acknowledgment of a formal rule. The puzzlement that Laura was sensing was grounded in very fundamental expectations on how more/less is related to up/down, and has two aspects. One is parallel with the well-known enigma that, for example, -7 is less than -1 even though 7 connotes "more" than 1. The equivalent, in Laura's case, is that "fast," at least in some cases, is below "slow." The other aspect is this: if speed by itself does not determine how high the velocity graph is, what does?

In the next vignette we will see how Laura begins to construct a new meaning of velocity sign that relieves the tension accumulated through her struggle to make sense of the use of signed graphs for motion.

9.5.5 Vignette 4: Beginning of a New Understanding

Description.

94 Ricardo: Try to move it [the car] in such a way that it's only negative, without
95   becoming positive.

Laura moved the car fast and away from the motion detector. See Figure 8.

Figure 8

96 Laura: I can't. I can't. Maybe if I. I don't understand.
97 Ricardo: Or just positive, without becoming negative.

Now Laura moved the car slowly. See Figure 9.

Figure 9

98 Ricardo: OK, so it's always positive. Now just negative.
99 Laura: Now just negative [moves the car twice according to Figure 10].

Figure 10

100 Laura: I can't do negative, how come? You have to stay at a steady pace [slow]
101   for it to stay at a negative. But I don't understand. I can do the positive one,
102   but I can't keep it negative. That's fast [Laura does it again, from a to b,
103   Figure 10]. This is slow [from b to a, Figure 10].

Then she moved the car slowly according to Figure 11.

Figure 11

104 Laura: I don't understand.
105 Ricardo: Start from that side [from the extreme right of the table, far from the
106   motion detector].
107 Laura: Start from this side?
108 Ricardo: Yeah.

Laura moved the car according to Figure 12 observing on the computer screen a velocity graph that was only negative.

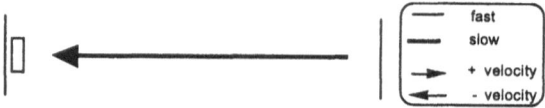

Figure 12

109 Laura: OK, so when you start far away and you come closer it's negative. But
110   when you start close up and you move backwards, it's positive. I know, I
111   figured it out myself! OK, so in this one [Graph 8], oh!, I understand now.
112   OK, so on this one [Graph 8].

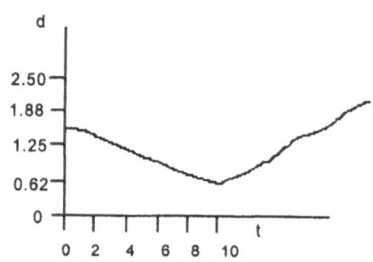

Graph 8

113 Laura: I came from the back where it started out negative [from 0 to 10
114   seconds]. And then I came forwards, right about here, [in the lowest extreme
115   of Graph 8] where it starts to go positive. And then it stayed positive all the
116   way, because I'm going front to back. [Laura draws Graph 9.] And it stayed
117   positive there. But this [at 10 seconds] is where I start going from. I
118   changed my direction.
119 Ricardo: And does it make sense that you go through zero when you change
120   direction?
121 Laura: Yes, because you have to. You have to like. Let me think...
122   because you're trying to stop and then you're going like that [Graph 9].
123 Ricardo: Let's produce this one [Graph 8]. Move the car to generate
124   something like this [Laura had produced Graph 8 before, but she had lost the
125   measurement with all the subsequent trials].

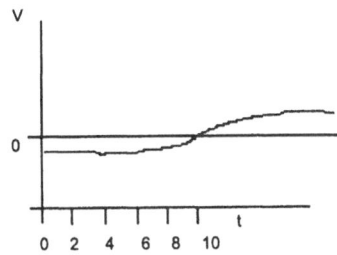

Graph 9

Now Laura moved according to Figure 13. The distance curve on the computer screen resembled Graph 8.

Figure 13

Then Laura asked for the computer velocity curve.

126 Laura: My graph [Graph 9] is similar [to the computer graph]. It's a similar
127   idea. I think I understand it now. OK, I get that.

Analysis. Laura started this vignette assuming her original construction of fast ↔ negative (low) and slow ↔ positive (high). Since for Laura the initial position of the car was not something upon which a decision was necessary, she started always close to the motion detector, and therefore she could do an only-positive graph but not an only-negative one. Focusing on the fast/slow distinction, Laura observed (lines 100 to 103) that now slow corresponded to negative (see Figure 10), which could be more reasonable for her (slow corresponding to below), but she noted, "I can't keep it negative."

When I asked Laura to start on the other side of the table (line 105), she reacted with some surprise: "Start from this side?" Through this interaction the starting point became something important to take into account. After Laura experienced the genesis of an only-negative graph she revealed the flash of her insight (lines 109 to 112). In her new account the starting points became the landmarks of her way of framing the velocity sign: "When you start far away and you come closer it's negative. But when you start close up and you move backwards, it's positive. I know, I figured it out myself."

I think that the problem Laura worked out in this vignette, that is, to produce an only-negative graph, is in itself not a "good" or "illuminating" one. At another point in her learning path she could have solved this problem by starting the car at different positions and concluding that it was negative because she went fast. Conversely, there may be many other problems that she might have experienced as the leading path to her insight. There are no "critical" problems providing a guarantee that if you solve them, you learn a way of symbolizing.

Reflecting on the nature of Laura's insight it is worthwhile not as a case in which suddenly everything was clear, or in which Laura reached a final understanding of signed graphs/velocity sign. Laura's insight was part of a continuum; it had a history and it was a turning point that opened up new horizons in her learning path. During this whole passage Laura accrued tension around the meaning of velocity sign. Through her insight this tension met relief, signaling the disclosure of a way of symbolizing velocity sign; it was not a mere recognition of a new semantic rule. The qualities of her new approach warrant analysis in lines 113 to 118. She pointed to Graph 8 and said: "I came from the back where it started out negative. And then I came forwards, right about here, where it starts to go positive. And then it stayed positive all the way, because I'm going front to back" (lines 113 to 116). Let me comment on this:

- Laura incorporated a new primary element to structure stories for the velocity of the car motion. Up to now her velocity stories had been sequences of "fast" and "slow." Now they were sequences of "coming forward" and "going backwards."

- For the first time Laura used the graph of distance in order to figure out the corresponding graph of velocity. Remember that throughout this passage Laura predicted the velocity graph only on the basis of her kinesthetic recall from her hand motion. This shift was possible because she was able to recognize directionality of motion from the distance graph. She did it from the very beginning of this passage (see lines 12 to 14).

- Laura's new perspective on directionality was a situated "towardness." "I came from the back": from the back of the table, toward the motion detector; "I'm going front to back": from the motion detector to the back of the table. Front and back were mutually defined by the position and orientation of the motion detector.

- The stopping moment and its surrounding slowing down acquired a new role. Her prediction in Graph 9 reflected this story (Figure 14):

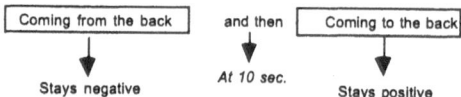

Figure 14

I asked her about "going through zero" (line 119). Let us compare her later answer (lines 121 to 122) with her former answer to the same question (lines 48 to 50). In lines 48 to 50 Laura understood going through zero as the marker of a "change in motion": from slow to fast; but now a new sense of necessity emerged for that stopping moment: "because you have to."

Laura felt reassured in her new way of using velocity sign by the experimental curve on the computer screen (lines 126 to 127). In sum, through this vignette Laura experienced a shift in her mode of understanding how signed graphs symbolize linear motion. A shift whose repercussions resonate over a wide domain of ideas.

9.6 Passage 3: Refining the Velocity Sign

9.6.1 Foreword

A general point I want to make is this: ways of symbolizing are not closed codes that at a certain point one gets and from then one "has." They open up new horizons of understanding that are always susceptible to growth, refinement, and enrichment, as well as to decline and impoverishment. This passage gives an example of refinement and enrichment. It marks the beginning of a new learning path through which Laura learns to coordinate the velocity sign with steadiness.

9.6.2 Description

At a certain point Laura moved the car in a manner that generated a graph on the computer screen like Graph 10.

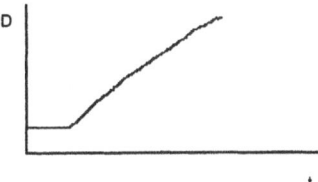

Graph 10 (on the computer screen)

I asked Laura to work with the graph that she had just obtained. First Laura copied the graph on paper. She drew Graph 11.

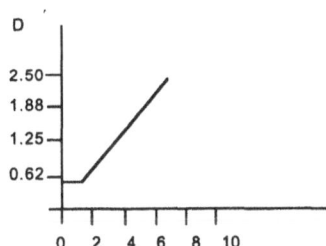

Graph 11

Then Laura started to think about the corresponding velocity graph.

128 Laura: [while drawing Graph 12] ...It would start up there and then slant this
129   way [down] because the farther away you get I think the closer it goes to the
130   negative, 'cause once you start coming back it goes down to the negative
131   ...Because, ahm, in a way I kind of think, I don't remember exactly but I
132   know when we started it the closer from the motion sensor the higher it
133   started in the positive and then as you went back it stayed in the positive but
134   as you come back forward it goes to the negative. So I think it would start
135   up here and then swoop down towards the negative.

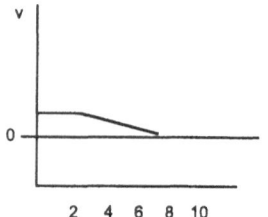

Graph 12

Later Laura asked for the computer velocity graph. After seeing it, she copied the computer graph over her prediction (see thick curve, Graph 13).

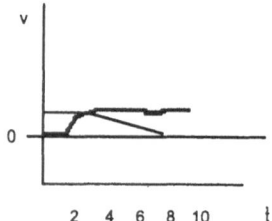

Graph 13

136 Laura: It starts on the zero and it goes up instead of down.
137 Ricardo: What do you think about the fact that the computer indicated zero
138   velocity here in the first part? [first two seconds]
139 Laura: Because we didn't, I didn't move, there was no motion during that
140   part, during that part right there, there was no motion at all, so there was no
141   velocity there.

Then Laura interpreted the computer curve:

142 Laura: Well, it starts out there at zero velocity 'cause the car wasn't moving so
143   it wasn't picking up motion and it goes up to the positive as you go away,
144   'cause like I said before, the velocity, when you go farther away it stays in the
145   positive... I was just going back down too soon... [the computer] stayed more
146   steady up here [10 seconds] and then it would start going down.

9.6.3 Analysis

With Graph 9 Laura brought forward directionality as the single most prominent feature of the car's velocity. She came to see the motion of the car as a sequence of directions or, more precisely, as a sequence of moving towards and away from the motion detector. Two components in her prediction can be traced (lines 128 to 135):

- "It would start up there." The initial velocity is "up there" because it has to start going from front to back, which is positive velocity (above zero).

- As the car moves "farther away ... the closer it goes to the negative." That is, as the car moves toward the back it has to get ready to become negative, therefore the velocity has to slant down. Laura rephrased this argument (lines 130 to 133). She thought that it had to stay positive but "swoop down towards the negative." In her prediction Laura projected the "towardness" of the directionality of the car motion onto the symbolic-graphical realm: the velocity graph had to slant "towards the negative."

Note how the two pieces "work" together: she had to start up in order to swoop down. If her prediction had started at zero velocity a gradual slant down toward the negative would not have been possible. This is an example of how, in assembling the several pieces of a curve, students have to solve problems of mutual relationship among the pieces in order to achieve the graphical connectedness of the curve. This source of anticipation is independent of students' perception of the physical behavior that is represented in the graph. Laura, for example, needed to start up in the positive in order to "go down" in the graph.

For Laura, the graphical space was one in which curves could go up and down, but now this vertical curving was qualified with a component of towardness: "swoop down towards the negative." Specifically, she described a "down" which is not just a down; it is characterized by its towardness, by its getting closer, but not into, the negative.

In analyzing the computer graph Laura highlighted what she found to be the most unexpected feature: "it goes up instead of down" (line 136). The beginning of the computer graph on zero made sense immediately to her in terms of the car's motion (lines 139 to 141). This necessarily created the need of going up, since "it goes up to the positive as you go away" (line 143). Laura reacted to the fact that the computer graph did not "swoop down" by repairing her own prediction: "I was just going back down too soon ... [the computer] stayed more steady up here and then it would start going down," (lines 145 to 146).

Toward the end of Passage 2, Laura learned to ignore how fast she moved the car in order to think about the velocity graph. She came to structure her account of the velocity graph in terms of directionality marked by towardness. Passage 3, on the other hand, initiated her process of reincorporating the speed of the car in her predictions of velocity graphs. Laura's words "stayed more steady up" announced what was to come. This first disclosing that "going towards the negative" did not necessarily mean to "slant down," moved Laura and me to explore the notion of steadiness and the meaning of horizontality in a velocity graph. But this is the beginning of another story.

9.7 Discussion

> Learning a language is learning how to do things with words. (commonly
> attributed to J.L. Austin)

Passage 1 is a clear manifestation of Laura's ability to understand and coordinate in sophisticated ways directionality, speed, and position. For example, she moved from speed vs. time to a speed vs. distance in order to show directionality; she also indicated how speed could change along distance and time. Her difficulties, and her learning, began when she met signed graphs and negative velocity using the motion detector in Passage 2. In Passage 2 Laura expressed ideas that could be understood, by a naive observer, as a "confusion" among directionality, speed, and position. For example: "When it goes down to the negative isn't it going faster?" (lines 60 to 61); "It stays positive when I'm at the closest range," (lines 81 to 82); "Far away and fast is negative," (lines 89 to 90). However, her struggle was not with the general distinction between speed, position and direction, but with the interpretation of the qualities of the toy-car motion with signed graphs. There were some basic connotations that Laura brought to her thinking with graphs, such as "up" was more and "down" was less, as well as negative being "below" positive, that she had to refine in order to develop a new approach to graphing motion in which the computer representation made sense. Initially, Laura read velocity graphs as a temperature graph in which "down" (positive or negative) is always "colder" (slower). Her learning path is not a story of learning to differentiate generic concepts, such as directionality and speed, but of learning a way of symbolizing, that is, of becoming fluent with a type of symbol-use.

Learning a way of symbolizing can be analogous to learning a second language. Laura did not begin to approach signed graphs of velocity from an empty state of mind. She had clear and far-reaching expectations about what those graphs meant. On the other hand, she encountered conflictive and senseless symbolic behaviors. She tried repeatedly to figure out ad hoc criteria, and she started to consider her own understanding more and more fragmented (e.g., "I don't get it. It seems to be the closer I am, then it. It stays positive when I'm at the closest range. But then it," lines 81 to 82). Later Laura began to recognize new patterns, but these were still in dissonance with more overarching understandings (e.g., "Why is it. It always seems to work that way," lines 73 to 74). This is a very important point. Laura said: "When I start out here [close to the motion detector] it starts out in the positive, and then as I go back it goes negative" (lines 76 to 77), but she could not, at that time, make sense of what that rule, "as I go back it goes negative," could possibly mean or how it might cohere with the other intentions and meanings that populated her thinking. Later, in Vignette 4, Laura's insight was not an acknowledgment of the "rule," but it marked her shift into a different mode of understanding within which she re-encountered fluency.

With Passage 3 I wanted to illustrate that the learning of a way of symbolizing is always open to new directions and to new connections with other ways of symbolizing. I believe that if I had ended with Passage 2, I would have given the impression that Laura's insight: "OK, so when you start far away and you come closer it's negative. But when you start close up and you move backwards, it's positive. I know, I figured it out myself" (lines 109 to 112) closed her understanding of velocity sign. Since the teaching experiment, I have spoken about Laura's ideas on velocity sign with a colleague who asked herself what should happen if the moving object were to go "across" the motion detector. I had not thought to ask this question to Laura, but it seems to me very likely that she would have said the velocity should change sign. After Passage 2, Laura always started to think of any new problem by discriminating directionality on the basis of the towardness of the motion; therefore, the motion through the detector would change from "going towards the motion detector" to "going away from the motion detector." Another colleague wondered what is special about linear motion that one has to stop in order to change direction, whereas in a plane one could change direction without ever stopping, as in Figure 15:

Figure 15

Perhaps in body motion one rarely really stops in order to change direction. I do not intend to elaborate on these ideas. Rather, I want to point out that ways of symbolizing are always subject to growth and that by learning how students learn them we may gain new insights into the nature of our own understanding.

Regarding the role of the MBL equipment through Laura's learning path I want to highlight two related aspects. The first one is that the motion detector allowed Laura to experiment with a symbolic behavior in terms of situated meaning. Without the motion detector, Laura could have produced signed graphs of velocity by following the explicit specification of new symbolic rules, so that she could plot data according to those rules. This difference is similar to a speaker trying to tell something in a second language conversing with another speaker of the second language, or a speaker assembling sentences in a second language by following the specifications of a grammar book. In the former what the speaker cares the most about is meaning; in trying to be meaningful the speaker is willing to use ungrammatical utterances or to complement gaps in verbal expression with gestures and idiosyncratic expressions. In short the speaker struggles with new ways of talking in the service of wanting to say something to someone else. Similarly, Laura used the motion detector to experiment with a symbolic behavior that was meaningful for others (for me, for the designers of the software, etc.), but that was dissonant with her own ways to meaning. Her hand motion, her expectations about graphical shapes, her kinesthetic sense, and her shifts of perspective, all became part of her wanting to mean in a new territory. This possibility of putting aside the grammar book, as it were, enabled by the access to the motion detector, seems to be a critical quality in Laura's learning path. The motion detector did not describe how-things-are, it offered Laura a symbolic behavior that she had to interpret according to multiple frameworks available to her.

It was this interpretative effort, in Passages 2 and 3, that moved her to revise her own understanding.

The second aspect is how the motion detector mediated our conversation. Thomas Kuhn, in a course that he offered at M.I.T., argued that an explanation is essentially a sequence of pointing acts. The central activity in explaining is the ability to point to something, say "look at this," (or "do this") and name it. Since I heard him articulate this idea, it has come back to me repeatedly as a powerful image. For example, in lines 105 to 106, I said: "Start from that side [from the extreme right of the table, far from the motion detector];" it was a pointing act that led Laura to an important insight, even though I did not add what is conventionally recognized as an explanation ("Because if you do this, then so and so ..."). I see the use of the motion detector in Laura's learning path as broadening, for both of us, the realm of the pointable. Almost any piece of Passages 2 and 3 can be used to exemplify this. For example in Passage 3: "What do you think abut the fact that the computer indicated zero velocity here in the first part?" (lines 136 to 137). This "here in the first part" meant for us both not just a little horizontal mark on the computer screen, I was also pointing to a particular movement that she had executed, that was not the movement she would have done following her prediction, that was below what she had expected, that had happened before she started to move away from the motion detector, and so forth. Perhaps rather than broadening I should say thickening the realm of the pointable. The experience with the motion detector mediated our conversation in a way that empowered us to mutually explain, in Kuhn's sense, our thinking, puzzlement, and suggestions with an extraordinary richness of detail and connectedness.

The case of Laura and the velocity sign helped me to understand that beyond the acknowledgment of simple rules a way of symbolizing is, using Geertz's words, "part of a distinctive manner of imagining the real," and that its learning is not a matter of showing how-it-works, but offering opportunities of symbol-use that help the learner to revise what she already knows and expects in order to make sense of what, for her, is a strange and puzzling symbolic behavior. This case study suggests, paraphrasing Austin, that learning graphing is learning how to do things with graphs.

References

Bakhtin, M.M. (1981) *The dialogic imagination: four essays.* Austin: University of Texas Press.

Bakhtin, M.M. (1986) *Speech genres and other late essays.* Austin: University of Texas Press.

Brown, J.S. & Burton, R.B. (1978) Diagnostic models for procedural bugs in basic mathematical skills. *Cognitive Science,* 2, 155–192.

Cajori, F. (1929) *A history of mathematical notations.* Chicago: The Open Court.

Carraher, T.N., Schliemann, A.D. & Carraher, D.W. (1988) Mathematical concepts in everyday life. In G.B. Saxe & M. Gearhart (Eds.), *Children mathematics* (pp. 71–88). San Francisco: Jossey Bass.

Confrey, J. (1988) *The concept of exponential functions. A student's perspective.* Paper presented at the conference Epistemological Foundations of Mathematical Experience, University of Georgia.

diSessa, A.A., Hammer, D., Sherin, B. & Kolpakowski, T. (1991) Inventing graphing: Meta-representational expertise in children. *Journal of Mathematical Behavior.* 10, 2, 117–160.

Geertz, C. (1983) *Local Knowledge.* New York: Basic Books.

Lave, J. (1988) *Cognition in practice: Mind, mathematics, and culture in everyday life.* New York: Cambridge University Press.

Lynch, M. & Woolgar, S. (Eds.) (1990) *Representation in scientific practice.* Cambridge, MA: The M.I.T. Press.

Matz, M. (1982) Toward a process model for high school algebra errors. In D. Sleeman & J. S. Brown (Eds.), *Intelligent tutoring systems.* New York: Academic Press.

Meira, L. (1992) The microevolution of mathematical representations in children's activity. In W. Geeslin & K Graham (Eds.), *Proceedings of the 16th annual meeting of the International Group for the Psychology of Mathematics Education,* 2, 96–104.

Nemirovsky, R., Tierney, C. & Ogonowski, M. (1993) *Children, additive change, and calculus.* TERC Working Paper Series, 2–93.

Pimm, D. (1987) *Speaking mathematically: Communication in mathematics classrooms.* London: Routledge Co.

Saxe, G.B. (1982) Developing forms of arithmetic operations among the Oksapmin of Papua New Guinea. *Developmental Psychology,* 18, 4, ~83–594.

Tierney, C. & Nemirovsky, R. (1991) Young children's spontaneous representations of changes in population and speed. In R.G. Underhill (Ed.), *Proceedings of the 13th Annual Meeting, North American Chapter of the International Group for the Psychology of Mathematics Education.*

10. Microcomputer-Based Laboratories in Inquiry-Based Science Education —An Implementation Perspective

Joke M. Voogt

University of Twente

10.1 Introduction

In the Experimental School Project, schools for secondary education cooperate with a research institute on the theme of computer use in education. The schools participating in the project (two in the eastern part of the Netherlands and one in the central part) are provided with computer and manpower facilities. The project started in 1987 and continued until 1993. Within the setting of this project several research studies were carried out.

One of these studies, the Computer-assisted Lab Work Project, concentrated particularly on students' inquiry skills. In this project use has been made of an MBL (Microcomputer-Based Laboratories) environment. The aim of the project was not only to integrate the computer in classroom instruction, but also to achieve innovative curriculum goals.

At the start of MBL high expectations about the potentials of MBL to support inquiry-based learning were stated, such as formulating hypotheses and predictions, analyzing results and acquiring some procedural skills like planning, testing and revising experiments. Linn and Songer (1991) show that inquiry-based learning doesn't occur automatically. Students need to be taught to observe precisely, and results of an experiment should be used for prediction of related experiments. After extensive use of MBL students were able to recognize inaccurately produced graphs, but only after instruction could they attribute the causes of inaccuracy. Striley (1987) found that students are mainly involved in the execution of their experiments. She proposes to focus the attention of the students to relevant features of their investigation and force students to draw conclusions about the underlying science principles.

Quite a lot of studies with MBL focus on one aspect of scientific inquiry: interpreting results displayed as graphs (e.g., Mokros & Tinker, 1987). Only a few studies highlighted other aspects of scientific inquiry.

The main part of this study deals with problems concerning the implementation of courseware (educational software and accompanying materials) for a junior secondary Heat and Temperature Curriculum.

A critical factor in implementation is the quality and practicality of an innovation (Fullan, 1991). Based on a review of the literature Van den Akker et al. (1992) argue that most courseware lacks quality. They state that most courseware is badly attuned to the curriculum, does not exploit the potential of computers for enhancing learning, is designed for use by individual students and not for classroom use, lacks support material for teachers and is hardly evaluated before publishing. The present study has tried to overcome these characteristics of poor quality of courseware in the design of the curriculum.

Van der Grift (1987) conceives three dimensions of the quality of an innovation: theoretical quality, empirical quality and practical quality. Theoretical quality has to do with the theoretical principles on which the innovation has been built. The aforementioned studies on MBL indicate that the potentials of MBL, particularly for interpreting graphs, are promising. So courseware taking into account the results and conclusions of these studies guarantee an initial theoretical quality. Empirical quality has to do with empirical evidence that the innovation has a positive impact on the students using the courseware. An innovation therefore needs to be thoroughly evaluated. Practical quality has to do with the workability of the innovation for the persons involved. Doyle and Ponder (1977–78) introduced the term 'practicality ethic' for the way practising teachers react to innovations. They pointed out three general dimensions—instrumentality, congruence and cost—of a change proposal, which determine whether an innovation is perceived as practical by teachers. Instrumentality refers to how clearly and specifically the courseware is presented. Congruence describes how well the innovation is aligned with the teacher's present teaching philosophy and practices. Cost is the teacher's estimate of the extra time and effort the innovation requires compared with the benefits the innovation is likely to yield.

Assessing the success of implementation relates primarily to studying the practical quality of an innovation. Empirical quality is perceived as a necessary condition for practical quality because it supports the possible benefits of the courseware for teachers.

10.1.1 Research Questions

Goodlad et al. (1979) proposed a typology for different manifestations of the curriculum: ideal, formal, perceived, operational and experiential. This typology has been used as a framework for the investigation of the implementation and effect of the courseware. In this report I will focus on the empirical and practical quality of the courseware. Empirical quality relates to the quality of the courseware for its effects on students (the experiential curriculum).

The following research questions have been posed for assessing empirical quality:

1. What is the impact of the courseware on students' achievement, including students' skills in interpreting graphs and controlling variables?

2. What is the impact of the courseware on students' motivation for an inquiry-based science curriculum?

Practical quality relates to teachers' execution and perception of the courseware. In this respect the characteristics of the courseware influencing the planning and coaching behaviour of teachers (considered critical in the operational curriculum) will be highlighted. Teachers' perception of the courseware is important to fully comprehend the operational curriculum. For assessing the practical quality of the courseware the following research questions have been posed:

3. Which characteristics of the courseware promote or inhibit the desired teaching and coaching behaviour?

4. For what reasons do teachers show or not show the desired teaching behaviour?

10.2 Courseware Characteristics

The courseware consisted of a software program containing a comprehensive set of laboratory exercises on Heat and Temperature, a student textbook and a teacher guide. In the project the computer has been used as a tool for collecting and displaying data during students' lab work. The courseware intended to improve students' inquiry skills, particularly interpreting graphs and drawing conclusions.

The software package, IP-COACH[1], which has been used in the project is an open-ended software program. IP-COACH can be considered as an MBL standard in the Netherlands. In close cooperation with the developers it has been made possible to implement a prefixed format for every laboratory exercise used in the curriculum. In this way teachers (or/and students) didn't have to bother themselves with, e.g., calibration and scaling but could direct their attention toward the physical experiment itself. Another important option in the software was the use of the mouse for making graphical representations of predictions of the experiment under different conditions to promote students' reflection as found by Linn and Songer (1988).

The student textbook included (in its last version) information and worksheets for executing the experiments as well as information and hints (suggestive questions) for building a theoretical framework and deriving physical rules as a result of the experiment. The student textbook assumed students to execute the experiments in groups of 3–4 students. Within each group fixed roles rotated among its members. During the execution of the curriculum students were more and more asked to make their own contributions in designing their experiment (e.g., formulating research questions) and reporting (e.g., deciding how data will be reported) about it. A concise manual of the software program, necessary for operating it, completed the student textbook.

It was supposed that most teachers participating in the project were not used to inquiry-based learning. Therefore the teacher guide paid special attention to recommendations for coaching students while learning by experience, next to information about technical aspects of the courseware.

---

[1] IP-COACH is developed by CMA, Nieuwe Achtergracht 170, 1018WV Amsterdam, The Netherlands.

From the formative evaluation it became clear that secondary science teachers tend not to pay attention to a separate teacher guide, but instead used the student textbook for their lesson preparation. Therefore in its last version the teacher guide has been integrated in the student textbook. In the teacher version of the student textbook didactic recommendations on critical points (which became clear during the formative evaluation) in the teaching process were put in the textbook at the opposite page in order to promote the desired coaching behaviour. The technical information for the teachers has been limited to necessary information for running the experiments.

10.3 Design of the Study

10.3.1 Participants in the Study

The students (N = 147) and teachers (N = 5) engaged in the research project came from six lower secondary science classes (two schools, three classes per school). Table 1 gives an overview of the numbers of students per class.

The courseware has been used by Havo (senior secondary education) and VWO (preuniversity education) students. The students in school I were mixed Havo/VWO students. The students in school II are following either the Havo or the VWO curriculum.

The physics curriculum in school I in lower secondary education is concentrated on a few topics. These topics are treated thoroughly. Normally a topic has been introduced through a lab, followed by a theoretical lesson with some applications of the topic. The physics curriculum in school II in lower secondary education has a concentric format: a lot of topics are treated in the eighth grade as well as in the ninth grade, relatively superficial in the eighth grade and somewhat more profoundly in the ninth grade. Theoretical lessons (eventually with in-class demonstrations) and applications are the main way of dealing with a topic. Sometimes labs are used as illustrations.

The average teaching experience of the teachers in the study is 11.6 years (s.d. 4.2). All teachers in the study, except for teacher E, have some experiences with the use of computers in their teaching. None of them, however, used IP-COACH in the students' lab. Teacher D used IP-COACH for in-class demonstrations.

	Total	School I			School II		
Teacher		A	B	C	D	E(1)	E(2)[2]
Boys	69	12	14	13	8	9	13
Girls	78	17	15	14	9	12	11

Table 1   Number of students (boys and girls) total and per class

---

[2] Teacher E participated with two classes in the project.

10.3.2 Instrumentation

Achievement. Before and after students worked with the courseware a 25-item test was administered to all students participating in the study. The test consisted of two versions (A and B) that were partly constructed as parallel-forms and partly contained the same items. The test included 20 multiple-choice items and 5 short-answer items. Some examples of test items are given in Figure 1. After omitting two items Cronbach's a for test A is 0.79 and for test B is 0.70. This is low but acceptable for a test of this length.

Questionnaire. Before the students worked with the courseware, data about their experiences with and expectations of investigating physical phenomena, doing lab

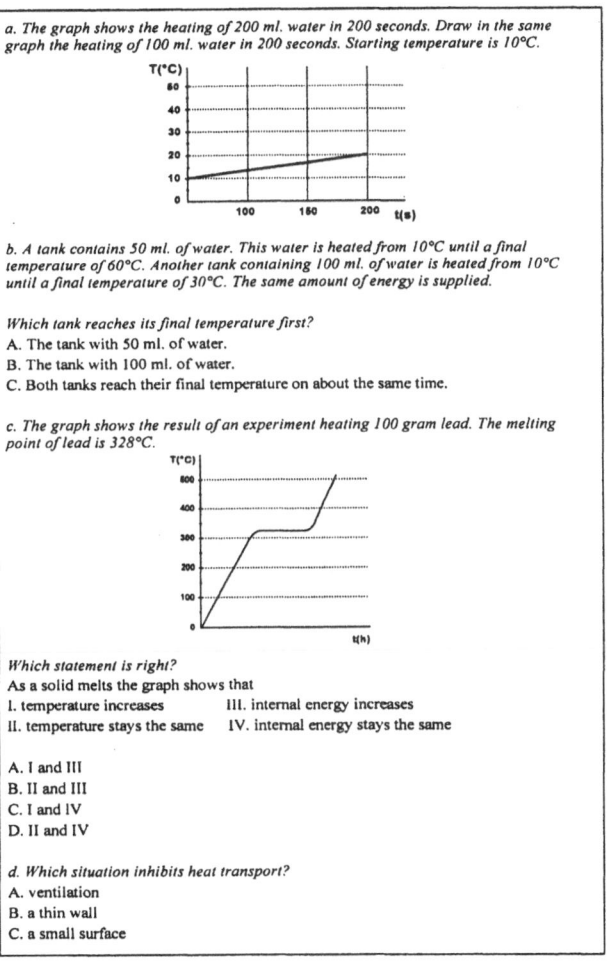

Figure 1 Test items—some examples

work and working with computers have been gathered in a 10-item questionnaire. Next to that information was collected about age and sex.

After working with the courseware the students participating in the study answered another questionnaire. This questionnaire partly contained the same items as the first questionnaire. It has been extended with items directed at evaluating the courseware. Examples of items from these questionnaires can be found in Figure 2.

Lesson Observation. All lessons (N = 80) with the courseware were both observed and audiotaped. The lesson observations were directed at gathering data about the actual lesson process, about the exact amount of time spent on each lesson, about student grouping, about the use by teachers of the didactic hints and suggestions, and about technical problems with the hardware and software during the execution of the lessons.

By audiotaping every lesson it became possible to follow the interventions teachers made when communicating with different groups of students. Written versions of the audiotapes were used for further analyses.

The lesson observations were primarily used supplementary to the data gathered by the audiotapes. In 11 lessons the observations had to be used as a replacement for

---

(Before, after)

I think physics is more fun when during physics lessons more time is devoted to investigating physical phenomena yourself

(strongly agree, agree, I don't know, disagree, strongly disagree)

(Before)

I like doing lab work

(strongly agree, agree, I don't know, disagree, strongly disagree)

(Before, after)

I think working with a computer during physics lessons makes physics more difficult

(strongly agree, agree, I don't know, disagree, strongly disagree)

(After)

I learned a lot about physics during the courseware lessons

(strongly agree, agree, disagree, strongly disagree)

(After)

In physical research computers are almost always used

(that is not at all true, that is not true, that is true, that is surely true)

(After)

I think the written material (textbook enclosing worksheets) belonging with the courseware is attractive

(strongly agree, agree, I don't know, disagree, strongly disagree)

Figure 2 Items from the questionnaire—example

bad or failing audiotapes. Lesson observation and audiotaping was carried out by research assistants.

Semi-Structured Interview. After the lessons had been carried out a semi-structured interview with each teacher took place. It addressed the teacher's perspective on and experience with the developed courseware. Next to some background information the topics of the interview were the goals of the courseware as grasped by the teachers, the fitting of the courseware within the existing physics curriculum, lesson preparation and execution and perceived benefits of the courseware.

10.3.3 Analysis

In order to assess the way teachers use the courseware in their lessons it was decided to use the curriculum profile method as proposed by Van den Akker (1988, 1991; see for an extended description Voogt, 1993) The constructed curriculum profile for the courseware consisted of five essential components: (1) *general skills* containing specific elements such as grouping, cooperation among students, planning of students, (2) *planning an investigation* with specific elements such as stimulating to formulate a research question, (3) *executing an investigation* including specific elements such as stimulating to predict a graph, (4) *building a theoretical framework* with elements such as relating results to physical concepts, and (5) *deriving physical rules* with elements such as relating a new rule to a former formulated rule. The curriculum profile has been developed for the courseware as a whole. Each teacher participating in the study obtained a user profile score that reflected the degree of implementation of the five components of the curriculum profile. Next to the total score, scores could be obtained per essential component. In this way a user profile score reveals the degree of implementation of the courseware in the operational curriculum: the desired teaching and coaching behaviour of teachers. The maximum assignment of scores per component is presented in Table 2.

Two trained independent raters assigned scores. To compute the interrater reliability Cohen's kappa (Cohen & Manion, 1980) was used. The interrater reliability of the six user profile scores varied between 0.85 and 0.98.

The semi-structured interviews were analyzed with Doyle's and Ponder's (1977–78) dimensions of 'practicality ethic' instrumentality, congruence and cost. An operationalization of these dimensions for evaluating the practicality ethic of

I	General skills	33	(18.3%)
IIA	Plan of investigation	22	(12.2%)
IIB	Execution of investigation	54	(30.0%)
IIIA	Drawing conclusions—building a theoretical framework	35	(19.4%)
IIIB	Drawing conclusions—deriving physical rules	36	(20.0%)
	Total	180	(100.0%)

Table 2   Maximum scores per component of the curriculum profile

courseware has been developed in a former study (Voogt, 1990). In this study the same operationalizations have been used for congruence. Instrumentality has been extended to all components of the courseware. 'In the long run time saving' as one of the operationalizations of costs (benefits) was omitted because this did not seem very appropriate, because the courseware opted for a longer, more profound approach of the topic, through an inquiry-based approach.

Students' results on tests and questionnaires have been analyzed with a t-test for all students in the study and per school.

10.4 Results

10.4.1 Use of the Courseware in Actual Classroom Practice: Opportunity to Learn

The time the teachers spent on the courseware as a whole turned out to be considerably higher than originally planned by the designers. The average time spent on the courseware is 566 minutes (sd = 30). On the introduction of the courseware, planned for two lessons (100 minutes), an average time of 120 minutes (sd = 29) has been spent. If we focus on the main part of the courseware (suggested time 125 minutes lectures and 250 minutes lab work) dealing with internal heat and temperature, the average time spent is 419 minutes (sd = 35). For the summary the average time spent is 30 minutes (sd = 19). From the data it can be concluded that the mean time spent on lab work in the main part of the curriculum is 247 minutes (sd = 55), which is about the planned time. The time spent on lectures, on the other hand, is higher than originally planned (mean 150, sd = 42). The rather large standard deviations show nonetheless the differences among the teachers about their distribution of time spent on the lessons. Particularly, teacher C has spent a lot of time on lab work in the main part of the curriculum (338 minutes). Teacher D devoted relatively much time to lectures (221 minutes, which is about the same as his time spent on lab work).

From the lesson sequence it can be inferred that teachers A (except for the introduction part), D and E (in both classes) followed more or less the proposed lesson sequence. Although teachers B and C followed the proposed sequence when beginning using the courseware, in the end they spent relatively little time on lectures. In these lectures B and C dealt with a rather large part of the curriculum. Nevertheless the data show that in all classes in the study the teachers paid considerable attention to the lab work, and that all planned labs were carried out by the students.

So a first general view on the operational curriculum shows us that students indeed had the opportunity to learn.

10.4.2 Assessing Empirical Quality

The Effect of the Courseware on Students' Achievement. Table 3 presents the mean scores and standard deviations on the pre- and posttest. The difference between pre- and posttest is significant as has been established with a paired t-test ($t = -17.83$, df = 146, p = 0.00). So the courseware has an overall positive effect on students' performance. When school I and school II are being compared, the results of school II appear to be significantly higher on pre- ($t = 6.31$, df = 124.6[3], pl = 0.00) and posttest ($t = 1.99$, df = 120.1, p = 0.05). The effect scores, however, are significantly higher for school I ($t = 3.32$, df = 145, p = 0.00). So the effects of the courseware are higher for school I than for school II.

Students' Skills: Constructing and Interpreting Graphs, Controlling Variables. Some items in the test particularly focused on students' skills in constructing, predicting (see, e.g., Figure 1a) and interpreting graphs (see, e.g., Figure 1c), and controlling variables. Students' results on these items are presented in Table 4. A paired t-test shows that on the posttest a significant gain of correct scores on these items could be established.

So with the courseware, students' skills in constructing and interpreting graphs and in controlling variables have been increased.

Students' Views on Traditional Physics. It has been asked what students think about physics as they know it from their traditional physics lessons. Table 5 presents the results for all students and per school.

| | Overall | School I | | | School II | | |
		A	B	C	D	E(I)	E(II)
mean before s.d.	11.49	11.52	9.59	9.52	11.82	12.76	13.75
	3.0	2.3	2.6	2.8	3.2	3.1	2.4
mean after s.d.	16.95	16.83	16.14	16.59	15.24	18.81	18.04
	3.1	3.4	2.5	2.4	3.6	2.8	2.5
effect score	5.46	5.31	6.55	7.07	3.41	5.05	4.29

Table 3   Mean scores, standard deviations and effect scores overall and per class

| | Percentage correct | | t-test | df |
	pretest	posttest		
Constructing a graph	37.0	89.8	-11.87***	146
Predicting a graph	33.3	58.5	- 4.55***	146
Interpreting a graph	15.0	59.9	-9.09***	146
Controlling variables	59.9	89.8	-6.86***	146

Table 4   Students' skills in graph construction and interpretation and on controlling variables (correct scores in percentages, t-test, p*** <0.01)

---

[3] Due to unequal variances a separate variance t-test has been used.

Above all physics is, according to the students, a subject where you use a lot of formulas. According to the students in school II physics is a theoretical subject. In this opinion they differ significantly from school I (t= -5.50, df = 142, p = 0.00). On the contrary, for the students in school I physics is doing a lot of lab work. The difference with school II is significant (t = 6.13, df = 144, p = 0.00). This reflects the different character of the physics curriculum in schools I and II. There is also a significant difference between the schools in the perception of difficulty of physics. Significantly more students in school II than in school I perceive physics as difficult (t = -2.18, df = 133, p < 0.05).

Students' Views on Inquiry-Based Physics. So there is a difference between the students of the two schools in their view of physics. What will their opinion be of an inquiry-based physics curriculum, as they experienced it when working with the courseware? Before actually working with the courseware and after having experience with it students were asked what they think of inquiry-based physics. Do they like it? Do they perceive it as difficult? Did they change their opinion after experiencing an inquiry-based physics curriculum? The results on these items are given in Table 6 for all students and per school.

Before as well as after experiencing the courseware students like an inquiry-based physics curriculum more than their traditional physics lessons. For school II the mean score afterwards, however, is significantly lower than before (t = -2.48, df = 53, p < 0.05). After experiencing an inquiry-based physics curriculum students view it as even less difficult than their traditional lessons. The differences are significant (overall significance: t = -5.53, df = 143, p = 0.00).

Students' Views on What They Learned. One of the questions was whether the computer, while being present very dominantly during the entire curriculum, was an obstacle to learn physics. So the students were asked whether they learned more about the computer or about physics in the curriculum. From the results (Table 7) it can be concluded that students agreed they learned about physics and tended to disagree that they learned about the computer. The differences are sig-

Physics...	Overall	School I	School II
is using a lot of math	76.7	78.0	75.0
is particularly theoretical	50.0	31.3	73.4
is using a lot of formulas	94.4	91.4	98.4
is doing a lot of lab work	58.2	78.0	32.8
you learn a lot of it by doing lab work	88.1	91.5	82.3
is a nice subject	61.2	67.5	52.5
is a difficult subject	54.1	46.2	64.9

Table 5   Students' opinions of physics (percentage of students that agree) overall and per school

Inquiring in physics is...	Overall		School I		School II	
	mean	s.d	mean	s.d	mean	s.d
more fun						
mean score before	1.74	0.81	1.84	0.87	1.61	0.73
mean score after	1.86	0.86	1.77	0.67	1.97	1.04
more difficult						
mean score before	3.18	0.98	3.63	0.97	3.23	1.09
mean score after	3.77	0.92	3.62	0.96	3.94	0.83

Table 6   Inquiry-based Physics Curriculum: more fun - more difficult (1 strongly agree – 3 don't know – 5 strongly disagree) mean scores and s.d. Overall and per class.

nificant for both schools (overall significance: $t = -5.32$, $df = 139$, $p = 0.00$). So the computer served as a tool in the curriculum.

10.4.3 Assessing Practical Quality

User Profile Scores. Table 8 shows the user profile score for each teacher. Teachers A and D have a relatively high user profile score. They showed about 50% of the desired interventions. The other teachers have a remarkably lower score. Teacher B and Teacher E (especially in one of his classes, E2) obtained the lowest scores.

I learned a lot	Overall		School I		School II	
	mean	s.d	mean	s.d	mean	s.d
Physics—mean score	2.26	0.71	2.16	0.70	2.38	0.72
Computers—mean score	2.87	0.84	2.51	0.83	2.90	0.82

Table 7   Computers or physics: about what did they learn (1 strongly agree – 4 strongly disagree) mean scores

Teacher	User profile score	
A	86	(47.8%)
B	42	(23.3%)
C	66	(36.7%)
D	93	(51.7%)
E1	70	(38.9%)
E2	48	(26.7%)

Table 8   User profile scores per teacher (scores related to threshold and maximum scores not related to threshold—absolute and in percentages)

In further analyses the degree of implementation per component has been considered. In Figure 3 the partition of each component per teacher has been compared by the ideal division of the distinguished components. Figure 3 presents the results for each teacher. It is striking that the results of both teachers A and D appear to be more or less similar to the ideal pattern. The other teachers tend to devote too much of their interventions to general skills (especially teacher B), while spending too few of their interventions to planning the investigation. The attention paid to the more theoretical part of the curriculum (building a theoretical framework and deriving physical rules as a whole) percentage of interventions are more or less similar to the ideal pattern,

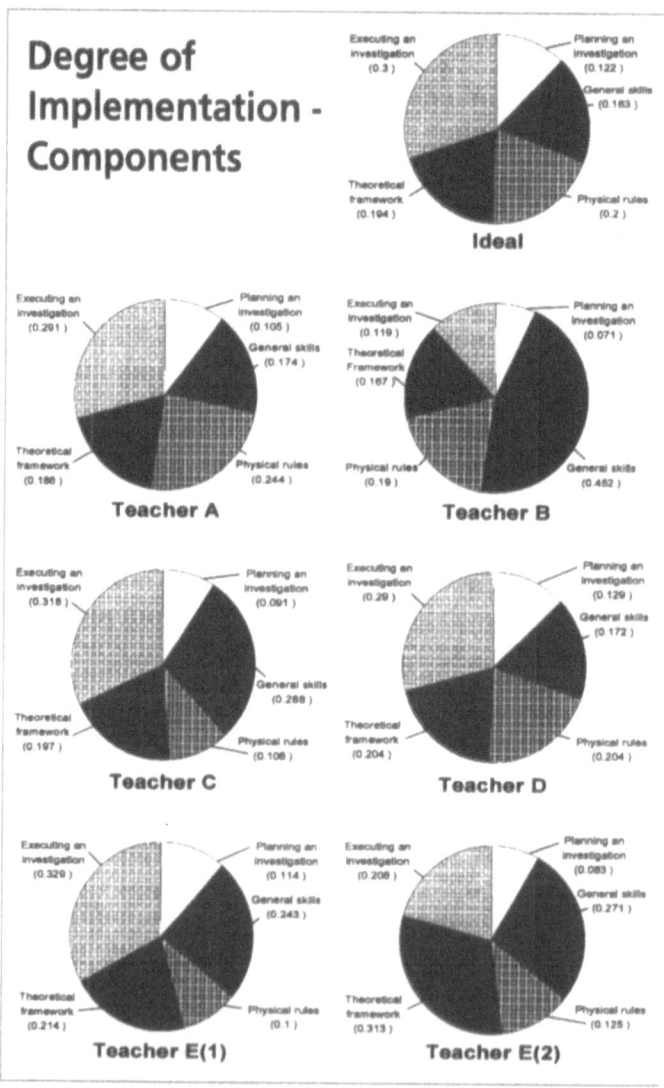

Figure 3   Degree of implementation per component

although the percentage of interventions directed towards deriving physical rules are somewhat low for teachers C and E.

Instrumentality, Congruence, Costs and Benefits. From the interview data teachers' perception of the courseware can be derived concerning the three dimensions of the practicality ethic as posed by Doyle and Ponder (1977–78). Table 9 presents a summary of teachers' perceptions on the distinguished elements.

Instrumentality applies to the materials itself: to what extent are the materials clear for lesson preparation and classroom use. Even in this version, where teacher guidelines are optically very close to students' texts and tasks, teachers tend to use the student textbook for their lesson preparation. Some teachers even say they behave as students when preparing a lab (A, B). Most teachers appear not to be very used to using a teacher guide. Only teachers A and D used the special suggestions as a guideline for their lessons. The suggestions didn't bother the other teachers though, except for teacher E who said that the suggestions did not fit in his approach. The software was for most teachers easy to use; they spent less than about a quarter of an hour to learn the program. Only teacher E, who claimed to be very awkward with computers, had much difficulty in learning the program.

Congruence refers to the extent the courseware fits into the classroom package and into the curriculum. For school I (teachers A, B and C) the topic of the courseware was not part of the current curriculum. Nevertheless, the fact the courseware opted for a profound approach with quite a lot of lab work made the teachers feel that it fitted into their curriculum and into the way they used to deal with a topic. Some remarks were made nonetheless. Teacher B and C wished for an approach where students had to deal with calculations on more complex applications of the topic. Teachers B and C also said tthey had problems with organizing their lectures. Both teachers said that one of the characteristics of their own labs is that students work independently. "I should be able to go away, when students do labs," says teacher B. For school II the subject was part of the curriculum, although the approach was quite different from the approach they were used to. This discrepancy appeared to be too big for teacher E. Teacher D, although not used to a coaching role to teaching, claimed to be very pleased with that role.

The costs and benefits for the teachers seemed to be somewhat diverse. All teachers said that the courseware was motivating for students. About the realization of educational objectives they were less clear. Most teachers agreed that students learned to work independently (except A). However they doubt whether students learned to use a systematic approach to planning and executing an experiment. Furthermore they valued some other objectives like the role of a computer in physics.

Technical Problems. From the lesson observations it became clear that school I had serious technical problems during the study, due to using probes they had assembled themselves. This caused especially problems for teachers B and C who experienced their lessons to be seriously disturbed by the technical problems. Teacher A said that, although he had problems in the beginning, in the end he had accepted the probes as

	A	B	C	D	E
Instrumentality					
• teacher guide	used as guiding points for accentuation in lessons	hardly used, partly because of time constraints; used notes of colleague	not used because of last year's experience; used his own notes; hardly uses teacher's guides	used; practical guidelines	not used; suggestions didn't fit into teacher's way of teaching
• student textbook	used like being a student	used like being a student; and recorded what to treat and what to omit during lectures	used; last year's notes served as a guide for students' difficulties and for the solution of problems	used; prepared student tasks at home; also made own flowchart from student activities	used to see what students had to do and should not forget
• software	easy to use	easy to use	knew program in advance	knew program in advance	difficult to learn and use it
Congruence					
• fit of content into current education	fits, because important to prepare students how to interpret graphs as a start for labs to come	fits, although would confront students with more applicatins if subject became part of curriculum	fits, not because of content but because of working atmosphere; more time necessary to confront students with applications if subject became part of curriculum	fits, replaces a part of our current curriculum; misses the explication of the concept of specific heat	doesn't fit basic knowledge of physics; better achieved with a theoretical approach (with demonstrations)
• fit of package into classroom practice	approach fits, although now labs take twice as much time compared with lectures	approach to labs fits, but in lectures too much subject matter had to be taught	approach to labs fits, but could not illustrate applications in lectures	doesn't fit, normally less coaching	doesn't fit, uses labs as an illustration
Cost					
• better realization of my educatonal objectives	not sure if students worked independent or systematic; doubt the effect of relation between graph and experiment; learned that the computer is just a tool	students worked independent; some groups worked systematic but others don't; they learned quite a lot about the subject	students worked independent but not systematic; learned that the computer is just a tool	students have a more realistic image of physics, worked in another way with graphs, worked independent and more or less systematic but did not internalize this way of working	students can recognize heating and cooling curves; students reflect too little; some groups worked systematic but others don't
• motivating for students	very motivating, students started working immediately	very motivating in the beginning; in the end somewhat less	motivating; students worked enthusiastically	motivating for students	the labs were motivating for students

Table 9   Teachers' perception on instrumentality, congruence and cost

they were and worked with them. Taking into account the poorly working probes, the fact that the technician in school I changed during the project was an extra handicap.

In school II there were only some minor technical problems, which could be easily remedied.

10.5 Discussion

In this study quality has been approached from an implementation perspective. Quality has been considered as one of the central factors in realizing educational change, such as the integration of MBL in the physics curriculum. This study focuses on the empirical and practical quality of the courseware. Empirical quality refers to the effects of the courseware on students' achievement and motivation. Assessing practical quality in the study meant studying the effects the courseware had on the behaviour and perception of the teachers in the study.

With respect to empirical quality it became clear that an overall effect of the courseware on students could be established. An increase in students' graphing skills and scientific reasoning skills (such as controlling variables) was also found. Similar results have been found in other projects (see the review by Nakhleh, 1994).

The outcomes suggest that students preferred inquiry-based physics to their traditional physics lessons, which above all in their view 'use a lot of formulas.' The students from both schools viewed inquiry-based physics as more attractive and less difficult than traditional physics. Although the approach to physics varied considerably between schools, only minor differences were found such as the very high expectations that the students in school II showed about the inquiry-based approach. Apparently the usual approach to physics did not influence the views of students on inquiry-based physics a lot. In both schools the students liked to use the computer in an inquiry-based physics curriculum and did not perceive computer use as difficult. Also other studies emphasize the importance of MBL for students' motivation (Nakhleh, 1994). Nevertheless, the results suggest that the students did not think that they had learned much about computers. This also implies that the computer had not masked the science content to be taught. Hence, the empirical quality of the courseware can be established.

Practical quality points to the characteristics of the courseware with respect to the workability for the teachers using the courseware in their lessons. These characteristics include two aspects:

a.   the material itself—the software, the student textbook and the teacher guide, and

b.   the inquiry-based approach applied in the software.

The characteristics of the materials include the pre-fixed format of the software, the design of the student textbook and the procedural specifications of the teacher guide. The inquiry-based approach to science education included the five components of the curriculum profile meant to provide the planning and coaching behaviour of the teachers: general skills, planning an investigation, executing an investigation, building a theoretical framework and deriving physical rules.

The pre-fixed format of the experiments within the software program made it easy to use according to most teachers (except for teacher E). It must be noticed, however, that all the teachers, except for teacher E, were quite familiar with computer programs, and that two of them had already used the software before.

For most teachers the student textbook was their guide in lesson preparation and execution. They regarded it as a clear guide. The student textbook contributed to the implementation of the curriculum at a general level: the execution of the proposed labs and lectures.

In order to influence the planning and coaching behaviour of the teachers involved in the study, procedural suggestions and hints in the teacher guide were supposed to support the teacher in his new instructional role. To promote the use of the suggestions and hints in the teacher guide, they were visually closely related to the student textbook.

Nevertheless, only two teachers reported using the hints and suggestions. It is interesting to note that these teachers (A and D) obtained the highest user profile scores on the curriculum profile. Besides that, the partition of the various components in the curriculum profile of teachers A and D was very similar to the ideal partition. Despite technical problems (for teacher A) and a weak congruence with classroom practice (for teacher D), these two teachers benefited from the teacher guide and showed an acceptable planning and coaching behaviour. Teacher A declared that his experiences in the introductory lessons prompted him to use the teacher guide in order to monitor progress in carrying out the courseware. Teacher D probably perceived the teacher guide as an operationalization of a coaching role which he would also need when using the new teaching method for physics, which the school had started to use in the eighth grade.

The planning and coaching behaviour of the other teachers hardly corresponded at all with the intention of the courseware. They tended to devote too much of their interventions to general skills such as cooperating and working independently, while spending too few of their interventions on planning the investigations, a central aspect of an inquiry-based learning approach. For teacher E it was clear that the courseware was not congruent with his classroom routines (Olson, 1988), which made it very difficult to structure daily classroom life during the execution of the courseware. His perceived clumsiness with computers was one aspect. However, a new didactic approach to physics and the use of computers made the innovation at stake very complex for this particular teacher. Apart from the technical problems which were a severe hindrance in the execution of the courseware for teachers B and C, it was probably the superficial congruence of the courseware with their normal lesson sequence which also caused problems. Teachers B and C tended to stress an independent working attitude as an important characteristic of their own labs for their students, while in the courseware, it was important that during labs students received guidance in order to learn by experience. Both teachers didn't know how to integrate labs and lectures, especially in the last part of the courseware. Teachers B and C underestimated the complexity of the innovation as far as it concerns its inquiry-based approach. Fullan (1991) states this as false clarity.

The importance of quality of materials (Fullan, 1991) is illustrated once again: easy-to-use software seems to be a necessary condition for the implementation of a complex innovation such as the courseware in this project, while technical problems threaten the implementation. Uncertainty about the support of competent technical support personnel seems an extra handicap for the implementation. Students' motivation when carrying out the labs supported the implementation of the courseware, because in some respects it reduced the costs (Doyle and Ponder, 1977–78) for the teachers.

References

Akker, J.J.H. van den: Ontwerp en implementatie van natuuronderwijs [Design and implementation of science education]. Lisse: Swets en Zeitlinger 1988

Akker, J.J.H. van den: Curriculumprofiel en Leerplanevaluatie [Curriculum profile and curriculum evaluation]. In: Knuppel, A. (ed.), *Twee conferenties van de werkgroep leerplanevaluatie SLO/SVO/Cito*. Enschede: SLO 1991

Akker, J.J.H. van den, Kenrsten, P., and Plomp, T.: The integration of computer use in education. *International Journal of Educational Research,* 17(1), 65–75 (1992)

Doyle, W. and Ponder, G.A.: The practicality ethic in teacher decision making. *Interchange* 8, 3,1–12 (1977–1988)

Fullan M.:*The new meaning of educational change*. London: Cassell Educational 1991

Goodlad, J.I., Klein, F. and Tye, K.A.: The domains of curriculum and their study. In: Goodlad, J.I. (ed.), *Curriculum inquiry*. New York: McGraw Hill 1979

Grift, W. van der: De rol van de schoolleider bij onderwijsvernieuwingen [The principal's role in educational change]. Den Haag: VUGA 1987

Linn, M.C. and Songer, N.B.: Teaching thermodynamics to middle school students: what are appropriate cognitive demands? *Journal of Research in Science Teaching*, vol 28(10), 885–918 (1991)

Mokros, J.R. and Tinker, R.F.: The impact of microcomputer-based labs on children's ability to interpret graphs, *Journal of Research in Science Teaching*, vol 24(4), 369–383 (1987)

Nakhleh, M.B.: A review of microcomputer-based labs: how have they affected science learning? *Journal of Computers in Mathematics and Science Teaching*, 13(4), 368–381 (1994)

Olson, J.: *Schoolworlds—microworlds, computers and the culture of the classroom*. Oxford: Pergamon Press 1988

Striley, J.S.: The computer as lab partner: classroom experience gleaned from one year microcomputer-based laboratory use. *Journal of Educational Technology Systems*, 15(3), 225–236 (1987)

Voogt, J.: Courseware evaluation by teachers—an implementation perspective. *Computers in Education*, 14, 4, 299–307 (1990)

Voogt, J.M.: *Courseware for an inquiry-based science curriculum. An implementation perspective*. Universiteit Twente, Enschede, dissertation, 1993

Part III

MBL and Learning

11. Computer Modelling for the Young—and Not So Young—Scientist

Leslie Beckett and Richard Boohan

University of London

Abstract. The laboratory is the natural home for the teaching of science where the real world can be formalised and controlled. Here, data gathering processes, now greatly enhanced by the use of the microcomputer, can reveal patterns and relationships demanding explanations and enabling models to be created which will lead to a greater understanding of phenomena. There is a need for a constant interaction between the roles which students adopt as experimentalists and as theorists. We argue here that computers have an important role in supporting theory-building by students. Exploring the consequences of theories is something which it is unrealistic to expect many students to be able to do without the help of the computer. The use of the computer as a modelling tool is thus a desirable means of complementing the careful measurement, observation and evidence gathering which characterise the Microcomputer-Based Laboratory (MBL). The effective integration of modelling within the context of MBL will be discussed.

We can develop explanations by creating models in a variety of ways. A model may be created as a set of quantitative relationships between the system variables, or a phenomena may be modelled through making visual representations of variables and indicating possible interactions between variables in a qualitative way. A third possibility we can explore is to model not through variables at all but through representations of objects and events. These three contrasting approaches to modelling will be discussed and illustrated by referring to examples of software which make this possible. These different approaches to modelling will be discussed in the historic order of the software development and references will be limited to examples of software which are familiar to the authors as developers and users. Readers viewing the discussion from another perspective may be aware of other examples of relevant software.

11.1 Dynamic Modelling

Consider first a typical example in physics of the oscillation of a mass restrained by springs and subject to viscous damping. This can be modelled by defining its variables, vis. position, velocity, acceleration, mass, spring constant, and damping factor and expressing relationships between these variables. The relationships formulate how the variables change in time as a consequence of the values of other variables and constants. The rules for the dynamic change of the system are thus the rules for computing the next value of each variable. The Dynamic Modelling System (DMS) (Ogborn 1984, 1986a) is such a modelling system. The program creates spaces which are available for defining the initial values of variables and for typing in a sequence of instructions which the system then converts to BASIC and inserts as part of its own program.

For the oscillator model these statements can be written as the following sequence:

```
F=-K*X-D*V
A=F/M
V = V + A*dT
X=X+V*dT
T = T + dT
```

where T is time, dT is a time interval, F is force on the mass M, and X, V and A are the displacement, velocity and acceleration during the oscillation under the restraint of a spring of spring constant K, moving with a damping factor D. The program fixes the initial values of the variables then iterates around these instructions which it places in a loop. The system can give tabular or graphical output of any two chosen variables or between any permitted functions of variables. Figure 1 shows the screen display of model statements and a graphical output comparing damped and undamped oscillations.

The screen can display one mode or be split between two, and different modes show initial values, model statements, table, graph, and a listing of models currently

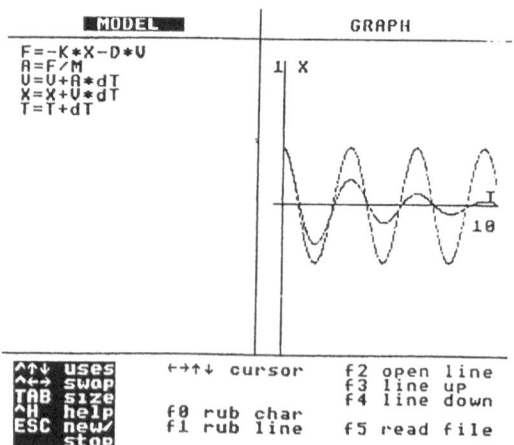

Figure 1 DMS model of oscillator

written in the disc. Limitations include allowing only one model at a time, and only one pair of variables plotted at any one time.

The worth of modelling software in science is most clearly seen, not in the display of the output from a refined and second-hand model but in the journey a student makes, by many stages, from an inadequate, unreal explanation to a model which most closely fits 'real world' data. The 'real world' may be accumulated statistics on population change which urge the student from simple linear growth models through exponential and logistic curves to possible oscillating and catastrophic changes in population. The 'real world' evidence is often the freshly logged data in an MBL which forces the student to adjust explanations to evidence.

The testing of explanations against real world evidence implies a need for close access to both a data display system and modelling system. This need has been recognized and met by the integration in the University of Amsterdam's 'I-P Coach' (Beurs and Ellermeijer 1992) of a modelling package based on DMS within a data collecting and display system.

One other benefit for scientific understanding which emerges from the use of this modelling tool is that in developing the modelling expressions students are forced to think deeply about the physical processes which lie behind the memorised mathematical expressions in physics. The student may know an elegant formula to express how the charge decays on a discharging capacitor but in giving modelling instructions to the computer it is necessary to understand what factors determine the small change in charge during a small interval of time. Such exercise should strengthen a student's understanding of science, but it will be obvious that the cognitive levels involved mean that this approach is only appropriate for older pupils in high schools and for students at college level.

Studies with grade 12 students (A level) (Robson and Wong 1985) have shown how experience with DMS has enhanced student understanding of concepts in physics. By giving instructions to the computer, students become subtly aware that understanding what goes on inside the computer is in fact understanding what goes on in the real world physical system being modelled. The availability of a machine which can mirror the intellectual processes previously internal to the teacher or learner means also that these processes are now external and available as topics of discussion.

11.2 Cell Modelling System

To overcome some of the limitations of DMS a Cell Modelling System (CMS) has been developed by Ogborn and Holland (Ogborn 1986a). This permits several graphs to be displayed at the same time as the model itself. Values are also stored so that new graphs can be drawn immediately.

Figure 2 shows a screen display in CMS considering how to model the possible rise in a departmental grant within a college or government. Computational cells are defined in much the same way as cells in a spreadsheet. Each cell has four slots, the first of which holds the name or symbol for a variable. The second slot of the cell

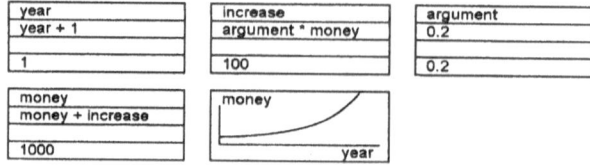

Figure 2   Exponential increase

indicates how the variable is calculated, and the initial or current value of the variable appears in the final slot. Any cell can be converted into a 'see' cell for graphics display of any two variables. Departments, particularly large ones, will argue that the increase should be proportional to the money they already receive. The size of the multiplying factor might well depend on the strength of the argument they present. Such a model would quickly lead to disastrous overspending as the money attracted rose exponentially. A more realistic model is likely to be produced if there is a cash limit imposed and if any increase depends on a multiplying factor f which decreases to zero as the money spent reaches the limit (f = I - money/limit). Figure 3 shows the

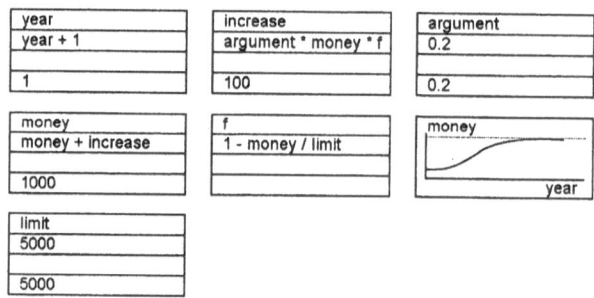

Figure 3   Logistic growth model

logistic growth of this money allocation towards the cash limit—a graph familiar in population studies when there are obvious limits of food or space.

The Cell Modelling System described above has the facility for designating cells for displaying data collected from the laboratory bench. So the discharge curves for a capacitor can be observed in real time on screen alongside a display being completed from a mathematical model. Sensors for rotation and motion now enable displacement/time graphs of damped oscillations to be compared with the graphs of proposed models. In such a case MBL provides the crucial immediate evidence which makes possible a study of factors which determine the damping forces, and refining of the theoretical model is stimulated by the readily available real world data.

11.3 Semi-Quantitative Modelling

Some models may be too complex for easy mathematical analysis—or we may be unable to use mathematical modelling because we do not have any way of knowing the quantitative relationships between some of the variables. Should we ignore these instances or is there not a case for attempting some kind of representation which although qualitative enables us to indicate how variables might affect each other (positively or negatively—strongly or weakly)?

We should remember that there is value and significance in many such qualitative explanations. Most explanations given by teachers are expressed without involving precise mathematical relations and pupils express their ideas in this way. "Winding a spring makes it stronger" might be the way a pupil would summarise an experience and in this way she would indicate the link between two variables in a situation and also the direction of that linkage as positive, i.e., more winding, greater 'strength' to resist the force of winding. "Eating more makes you less healthy" could be visualized as two variables 'eating' and 'health' linked by a 'minus' link indicating how health could decrease with overeating. Such knowledge about a system can be described as 'semi-quantitative'. It is the kind of knowledge which fills the business pages of newspapers which discuss 'trends', 'influences', even 'market forces'. By contrast, expressing knowledge in this way is not so common in science text books or examinations!

The project 'Tools for Exploratory Learning' has developed and tested with students in the age range 12–14 years a modelling program which uses visual representations of imagined variables and the connection between them without having to specify the form of the mathematical relationship. The design of the modelling program is in Miller, et al. (1990) and the results are discussed in Bliss, Ogborn, et al. (1992) and Bliss and Ogborn (1992). The modelling system is called IQON (Interacting Quantities Omitting Numbers). A variable is represented by a box and a marker line inside it represents the current value.

Consider a model of the quality of this NATO workshop expressed through IQON. If quality is high the happiness of the participants should increase as the week progresses and this may perhaps result in greater participation in the activities so that the quality of the workshop increases (Figure 4a).

This model is clearly optimistic. It contains positive feedback, so that if the happiness of the participants is increased by a small amount then after a while all the variables are driven to their positive limit (Fig 4b). A happy memory of a candle-lit canal cruise may set the feedback system operating! It does not matter whether the model is correct or complete; what matters is that it describes possible effects which will certainly arise in some cases.

As currently implemented all IQON variables are alike. An input from other variables modifies the rate of increase or decrease of a variable. Every variable has a central 'neutral' setting at which its output has no effect. Multiple inputs to a variable can be arranged, noting the various signs, to determine the rate of change, but some variables can be given greater weight than others (the 'plus' or 'minus' sign can be

Figure 4a Initial setting

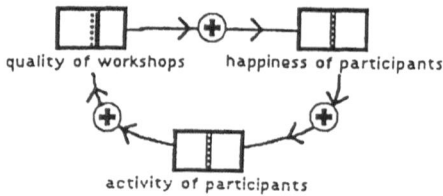

Figure 4b Positive feedback causes runaway

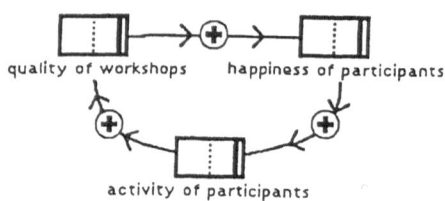

Figure 4   An IQON model for success of workshops

increased or decreased to indicate this). The response of each variable is made non-linear by using a 'squashing factor' which restricts its value to the range plus one to minus one and there is also internal damping. The idea has been to make the variables behave like some forms of artificial neurons (McClelland and Rumelhart 1987). The design features ensure that any system of inter-linked variables will have a smooth behaviour with no tendency for variables to go to infinity.

The figures 5 and 6 show two models in IQON created by pupils aged about 13. Nancy sees her fitness as depending on her general health, on the amount of sleep she gets and on her attitude (state of mind). And she manages a humorous dig at the teacher by suggesting that too much homework can ruin her health. Disease is perceived as something which undermines her fitness and also has a negative effect on her attitude to life. Whether Nancy is right with her model is not the issue here. The important thing is that she has produced a model which can be discussed and the surprises it produces may lead her to reconsider some of her ideas. But it is clear that the very exercise of creating a model had led her into a realisation of the complex interaction of several factors associated with her fitness.

A significant feature of Burgess' model of traffic congestion is that his 'variables' are less like an amount of something and more like objects. This was a common feature revealed in the school trials of IQON (Bliss and Ogborn 1992) in which many children had difficulty in creating 'amount like' variables, tending instead to create objects and events. This provides an appropriate cue for the discussion of a modelling system which is designed to use not variables but objects and events in which a world is created which younger children can relate to.

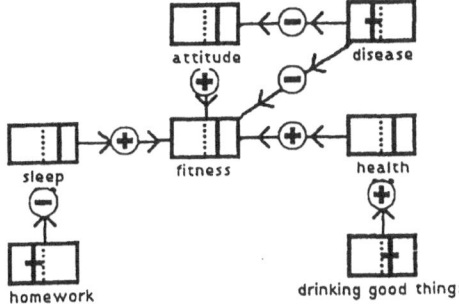

Figure 5 Nancy's IQON model for keeping fit

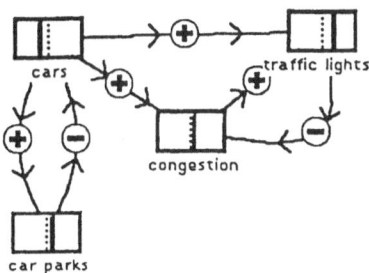

Figure 6 Burgess' model for traffic congestion

11.4 WorldMaker

In WorldMaker, models are made from objects and backgrounds (or things and places) which can be placed on a grid of cells using a set of plotting tools. Objects and backgrounds have rules which govern their behaviour when they are on the grid. Figure 7 shows a simple model (or world) which children have used when being introduced to the system. It consists of just one type of background ('wall') and one type of object ('bouncy ball'). When the ball is first put on the grid, nothing will happen to it. If, however, it is given a direction to point in, then it will move in that direction. It will continue to move in the same direction unless it hits a wall or another ball, in which case it will bounce. Figure 7 shows a solution to one simple problem: The question is "Where can you put the walls and the ball so that the ball will bounce all the way around the inside?" This world can allow younger children to investigate simple number patterns, while the same resources can be used by older children to model, for example, diffusion.

Another simple world is called 'glue' (Figure 8). This consists of two types of 'tube', each of which produces a different type of 'glue'. One type is free-flowing

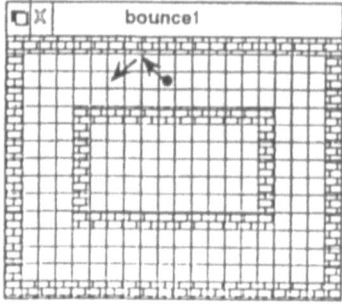

Figure 7 The 'bounce' world

and consists of objects which have a high probability of moving sideways. By using a smaller probability for this, the other type of 'glue' is made to be more viscous. The glue falls to the bottom of the screen where it forms a 'solid', the shape of this depending on the viscosity of the glue. So, having experimented with these objects, children may be asked to find out how to produce certain shapes using combinations of 'tubes'. Figure 8 shows an example, which was in fact produced by putting one of each type of tube next to each other at the top of the screen.

Playing with 'glue' generates much excitement amongst children, but there is a serious side to this too. When a tube is placed at the bottom of the screen, children immediately see that the model behaves like a volcano, with the 'glue' now acting like lava, creating a mound of solid and forcing its way up through a 'vent'. The shape of a volcano is determined by the viscosity of the lava.

These two examples give some idea about the kinds of worlds which can be constructed. The concept of WorldMaker derives from that of a cellular automaton (Toffoli and Margolis 1987), of which perhaps the most well-known is Conway's Game of Life. Using only backgrounds, WorldMaker would simply be a cellular automaton, but it has been extended to include the representation of objects which can retain their identity while moving around the grid. Many of the models which can be created with objects are probabilistic and draw on, for example, the statistical bead games described by Eigen and Winkler (1983). Marx (1981a, 1981b, 1984a, 1984b) has

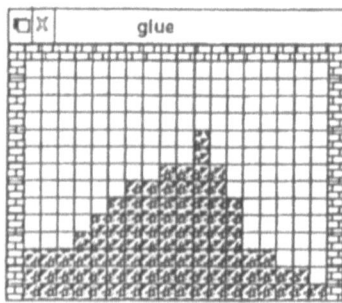

Figure 8 The 'glue' world

produced many examples of these kinds of models for microcomputers, but the rules are part of the program and are not accessible to the user. Recently, an interesting program, 'The Picture Simulator', has been produced (Camara, et al. 1991), which allows a limited range of rules to be defined for entities, but its interface was not designed to make the program accessible to children. The purpose of WorldMaker is to make the creation of rules simple enough to allow young children to construct their own models.

Even simple rules can lead to interesting and sophisticated behavior, so the software is not restricted to younger pupils. In developing the prototype version of the software, trials were carried out with pupils across a wide age range—from 9 year-olds in primary schools to 17 year-old students specialising in science.

11.4.1 Children Making Their Own Worlds

Worlds constructed with just a few simple rules can often show interesting and complex behaviours. One world which children have enjoyed exploring is 'rabbits'. Initially, the object 'rabbit' has no rules, so when it is placed on the grid it does nothing. Figure 9 shows how some simple rules can be constructed. A rule is represented by a picture showing a 'condition' and an 'action'. If the condition is satisfied, then the rule will 'fire'. With just the 'moves' rule, the rabbit moves randomly around the grid. With the 'breeds' rule, new rabbits are created—rapidly at first and then more slowly as the limits of space are reached. Rabbits can also be made to interact with other entities, for example, they can 'eat grass' or be 'eaten by a fox'. These other entities can also be given simple rules, such as making the grass grow, or making foxes breed and die.

Children can also explore the effects of changing the settings of the rules—each rule has a slider bar which controls the probability that the rule will fire if the condition of the rule is met. So, by changing the setting of 'breeds', for example, children can explore what happens if the rabbits are made to breed more slowly. Some changes to the model may simply cause the equilibrium position of the population to shift.

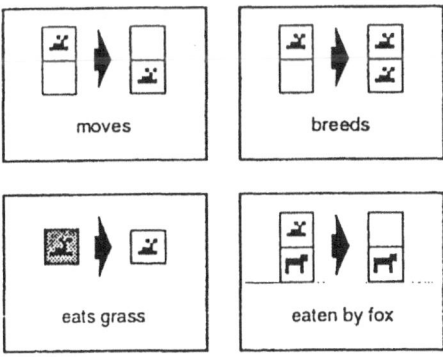

Figure 9   Some rules for 'rabbit'

Others can cause the populations to fluctuate. Often, changing the rules can lead to extinction. There are important lessons to be learned here about the balance of Nature. These examples of 'worlds' are all ones with which children are to some extent familiar. They have some intuitive understanding of the kinds of behaviour that the models may show. The point here is not so much that pupils should gain a deeper understanding of the phenomena, but that they should understand that these phenomena can be modelled. They are learning about the relationship between the real world and a simplified model (or theory) about the real world which has consequences. The value of the model can be judged by the extent to which its consequences are reflected in real world behaviour. Knowing that a model is not the same thing as the real world is essential for pupils when they turn their attention to exploring and modelling unfamiliar phenomena encountered in the laboratory.

An example of a phenomenon which can be explored in the laboratory and easily modelled is radioactive decay. This can be done using just one simple rule—in fact, the simplest rule possible of one object changing into another (Figure 10). With a large number of A's on the screen, a graph can be used to follow the decay of A to B. Changing the probability that the rule will fire will change the rate of decay. The model could be extended by using another rule in which B changes slowly into C. Or an equilibrium situation could be modelled by adding the corresponding rule for B (B → A). Changing the probability of the forward and backward changes would alter the position of the equilibrium.

Figure 10 A rule for modelling radioactive decay

More complex equilibria can be easily modelled. One such equilibrium often studied is the formation of ethyl ethanoate and water from ethanol and ethanoic acid. Two rules are sufficient to model this system, A+B→C+D and C+D→A+B, with additional rules to make the particles move about. Students often find it difficult to predict what happens to the position of such equilibria when one of the factors is changed, for example, if more water is added. These effects can be studied in the laboratory and related to changes in the behavior of the model. One difficulty for students is that these kinds of phenomena need to be thought about at both the macroscopic and at the molecular level. A system such as CMS may be used to collect and process experimental data and to make comparisons with a model of the macroscopic level. The advantage of using WorldMaker is that it is modelling the equilibrium at the molecular level, so encouraging students to interpret experimental data and the equilibrium laws in terms of the behaviour of molecules.

Very simple rules can lead to surprisingly complex behaviour. Figure 11 shows a simple rule in which A changes to B into more A. This might be a model of grass spreading through a field, or of an infection spreading through a population of people. It is an autocatalytic change—it starts slowly at first, but as more A's are created so more B's are changed into A, and the change becomes faster and faster until the limitations of space are met.

Figure 11 An autocatalytic rule

If the corresponding rule for B is added (BA→BB), then unlike the equilibrium model, this model will never tend to a mixture of A and B but will always end up as either all A's or all B's. This is because the more objects of one kind there are, the more likely they will increase their numbers in the next generation (positive feedback). If, by chance, after the first generation there are more A's than B's, then this early advantage will tend to increase in further generations. Each time the model is run, it is not possible to predict whether it will be A or B that will dominate, but it is possible to predict with certainty that eventually, one or the other must do so.

The rules governing growth are critical if 'evolution' is to take place. In the equilibrium model, even when one object has a much greater advantage, the other object does not die out but remains in the population. In the positive feedback model, the problem is that with only a small number initially, an object even with a great advantage is overwhelmed by the large number of the other objects and dies out. For evolution to take place, an object with a relatively small advantage, and in relatively small numbers, must be able to survive, multiply and eventually replace the other object.

In developing tasks for classroom use, pupils using WorldMaker have been able to come up with many ideas that they would like to try. The problem has not been to encourage children to explore, but to ensure that their exploration is productive. In general, the most successful kinds of task are those in which they are given a limited range of resources and a clear problem to solve with those resources. A set of 'learning tasks' have been developed which progressively introduce new features of the software as they are needed. These learning tasks are grouped into three stages:

- Initially, pupils are given backgrounds and objects which have already been created. They make models by choosing how to place backgrounds and objects on the grid.

- The next stage is to modify the behaviour of backgrounds and objects by changing the 'slider bars' of their rules.

- Finally, pupils create entirely new types of backgrounds and objects by building new sets of rules.

Many other kinds of models have been developed. There are a wide range of examples related to science—bacterial growth and decay, epidemics, predator-prey systems, desertification, forest fires, diffusion, crystallisation, nuclear chain reactions, rates of chemical reaction and chemical equilibrium. But there are plenty of opportunities elsewhere—in mathematics, models can be used to study number patterns and probability, and in geography, models which have been developed include rainfall and coastal erosion.

11.5 A Journey Through Modelling

We have discussed a number of different modelling systems, as it happens, in the historical order of their development. But we propose that children learning about modelling should make this journey in the other direction. They should start by modelling situations directly by defining the behaviours of the objects in the system. Later they can begin to think about the world in terms of variables and the ways in which these variables can affect one another. Eventually, their models can express these relationships mathematically. Of course, there is a reason why these journeys run in opposite directions. The development of increasingly powerful computers is making modelling accessible to younger and younger children.

References

Beurs, C. de, and Ellermeijer, A.L.: *Computer applications in physics*, paper presented to NATO Advanced Research Workshop, Amsterdam 1993. Chapter 12 in this book.

Bliss, J. and Ogborn, J.: Tools for exploratory learning: End of award report 1992

Bliss, J., Ogborn, J., Boohan, R., Briggs, J., Brosnan, T., Brough, D., Mellar, H., Miller, R., Nash, C., Rodgers, C. and Sakonidis, H. (1992) 'Reasoning supported by computational tools', *Computers in education*, Vol. 18, No. 1–3, pp. 1–9, reprinted in Kibby, M.R. and Hartley, J.R. (eds.) (1992) *Computer assisted learning*, Pergamon Press 1992

Boohan, R., Ogborn, J. and Wright, S.: *WorldMaker: Software and teachers' guide*, ESM, Cambridge 1994

Camara, A.S., Ferreira, F.C., Nobre, E. and Fialho, J.E.: *The picture simulator*, Uninova, Portugal: Newsoft Group 1991

Eigen, M. and Winkler, R.: *Laws of the game: how the principles of nature govern chance* (translated by Robert and Rita Kimber). Harmondsworth, England: Penguin Books 1983

Marx, G.: Some simulations of science I, *Physics Education*, Vol. 16, pp. 152–158 (1981a)

Marx, G.: Some simulations of science II, *Physics Education*, Vol. 16, pp. 212–217 (1981b)

Marx, G.: Simulation games in science education, *European Journal of Science Education*, 6, 31–5 (1984a)

Marx, G.: *Games nature plays*, Roland Eotvos University, Budapest (1984b)

Miller, R.S., Ogborn, J.M., Turner, J., Briggs, J.H. and Brough, D.R.: Towards a tool to support semi-quantitative modelling, *Proceedings of International Conference on Advanced Research on Computers in Education*, July 1990, Gakashuin University, Tokyo (1990)

Ogborn, J. and Holland, D.: *Cellular modelling system*, Microcomputer software, London: Longmans (1987)

Ogborn, J.: *Dynamic modelling system*, Microcomputer software, London: Longmans (1984)

Robson, K. and Wong, D.: *The dynamic modelling system*, School Science Review No. 236, p. 682 (1985)

Toffoli, T. and Margolis, N.: *Cellular automata machines: a new environment for modelling*, Cambridge, Massachusetts: MIT Press (1987)

12. Computer Applications in Physics: The Integration of Information Technology in the Physics Curriculum

C. de Beurs and A.L. Ellermeijer

University of Amsterdam

Abstract. There is a growing interest in using MBL and modelling tools to support research activities in the classroom. One of the major challenges is to create the conditions for pupils to 'do' science; to give them the opportunity to formulate (at least partly) their own theories, to build and validate their own models. Possible results of these activities are the excitement of scientific discovery, the feeling that 'understanding the complex world' is a puzzle which can be solved, leading to a better understanding of the scientific approach.

To do this, the availability of user friendly tools is important, but not sufficient. Pupils need laboratory techniques and modelling skills, and need to have some basic knowledge of system properties (hardware). We cannot just presume the existence of such techniques, skills and knowledge. They have to be taught!

In this paper we report on a project which aims to integrate the subject 'computer applications in scientific research' into the traditional physics curriculum of secondary schools. In section 12.1 we review the history of the project. In section 12.2 we discuss the aims and contents of the course. In section 12.3 we discuss the implementation, and in section 12.4 we consider the first results of classroom experiments.

12.1 Information Technology in the Physics Curriculum of the Upper Level in Secondary Schools of the Netherlands

New Information Technology (IT) is rapidly changing our world. In the Netherlands there is a broad consensus about the desirability of paying attention to these developments in general secondary school. In 1985 the national government, in cooperation with three computer-enterprises, started a project for all the 2000 secondary schools (pupils of ages between 12 and 18) with the aims:

- To provide every school with 11 IBM-compatible computers (IBM, Philips or Tulip).

- To develop and offer extensive in-service courses for teachers, from general awareness to subject related courses.

- To develop courseware (educational software with instructional materials for pupils and teachers).

This NIVO-project finished in 1988 and was followed by a second project, more directed to implementation, the PRINT-project.

In 1986 the national government formulated a so-called "Informatics Stimulation Policy." According to this policy the original ideas of introducing a new and separate school subject 'informatics' were rejected. Instead an integration of skills and concepts from information theory in existing subjects (mathematics, physics, economics and orientation on society) was advocated.

Since 1982 the Physics Education Department of the University of Amsterdam has been engaged with research concerning microcomputer-based laboratories (MBL). In 1986 we started a project to integrate aspects of information technology (IT) in the physics curriculum of the upper level in Dutch secondary schools. This project, 'Information Technology Applied in Physics', known as ITN-project, formed a part of the NIVO-Project mentioned above.

In the first stage of the project we had to find out which goals and what topics would be relevant when integrating information technology into the physics curriculum. Our next task was to elaborate the contents and find suitable contexts. Finally, we designed several alternative blueprints for possible courses. Each blueprint contained a detailed description of the subject and some matching contexts for a course of 20 lessons. These blueprints were designed to serve as a starting point for the development of material for use in the classroom.

An important policy (1987) which influenced our project-activities was the revision of the upper level physics examination-programme. Changing the physics curriculum was necessary for two main reasons:

- The need for shifts in the teaching approach because of new didactic insights [1]:

  - More effective and motivating learning by treating subjects in real-life contexts.

  - More active learning by starting from concrete situations and by stimulating classroom discussions, practical work and students' personal investigations.

- The need for updating the physics curriculum by including new subjects from biophysics, quantum physics, electronics and modern research methods, computational physics.

As a result of this policy the ITN-project received the responsibility for all activities necessary for a proper introduction of the new subject into the physics curriculum. Besides obtaining a consensus about objectives, contents and contexts and designing blueprints for courses, we also had to develop and distribute classroom materials (learning packages, teachers handbooks, hardware and software). This led us to develop material for teachers' courses and to organize in-service training throughout the whole country.

Based on a research programme [2] and an extensive consultation of experts in the fields of informatics, control technology, computational physics and education, we finally opted for the development of two courses:

- A course 'Process Automation' for pupils of the fourth class of HAVO and VWO (age 16). (VWO is of a pre-university level, while HAVO is of a more general level). In this course (about 20 lessons) a treatment, based on practice, of Information Technology from the context of 'measurement—and control—systems in daily life applications' was chosen.

- A course 'Computer Applications in Physics' for class 5/6 VWO (ages 17, 18). The main subject of this course is the application of Information Technology in modern scientific research (on-line measurement, data processing and modelling) and the use of IT-applications in the school laboratory.

In the next paragraphs we will concentrate on the second course.

12.2 Computer Applications in Physics: Course Aims and Contents

There is a strong interaction between physics and information technology. On one hand physics contributes to the development of micro-electronics and computer science and on the other hand the products of information technology provide powerful tools for experimentation, data analysis and calculation. Research on particle physics, for example, would be impossible without the applications of IT in almost every stage. In the course 'Computer Applications in Physics', the second side of this interaction is reflected. The general goals of this course were adopted from both physics and informatics:

- Pupils should have an 'active' knowledge and understanding of modern research methods.

- Pupils should have an 'active' knowledge and understanding of basic concepts in the field of information processing (acquisition, processing and distribution).

We use the word 'active' in the description of these goals, because we intend to emphasize problem-solving skills and not merely the transfer of descriptive knowledge on the subject.

In the new physics curriculum all pupils of pre-university level (ages 17, 18) have to carry out their own research project as a part of the examination. 'Computer Applications in Physics' is also intended to support the development of skills to solve research problems with the help of hardware and software tools [3, 4, 5]. To attain these goals we developed a course, containing the following modules:

1. **Data acquisition** (how to use the computer as a tool for data-acquisition)

    It is crucial that pupils gain insight into the operation of measurement systems:

    - Knowledge of the function of the system components (sensors, amplifiers, filters, AD-converter) and their influence on the quality of the measured signals.

    This knowledge is meant to enable pupils to design measurement systems, using standard building blocks, for their own research projects. It is not intended to treat the system components on the level of electronics. It is sufficient that they understand the function of the subsystems relating to the main system goal. In addition, they have to understand how electrical signals are able to carry and transport data and how subsystems (e.g. amplifier, AD-converter) influence the representation of data by means of signals. The understanding of measurement systems on this level will help pupils with the interpretation of the graphic results and the estimation of measurement errors. It enables them to distinguish between signal changes caused by the physical quantity being measured and changes (e.g., noise, signal distortion, quantisation) caused by specific properties of system components.

2. **Data analysis** (how to use software tools for analyzing data)

   It is crucial that pupils learn which software tools are suitable for specific tasks and how they can effectively be applied when investigating physical data:

   - display data in a more exploratory way, for instance by using different kinds of graphic representations.

   - analyzing graphics by scanning, determining the slope or area under the graph, curve-fitting.

   - modifying data by using zooming—or filtering—techniques.

   - processing graphed data by replacing a graph by its derivative or integrated version.

   - processing numerical data by performing calculations on data columns (spreadsheet operations).

3. **Modelling** (how to build and use computer models in scientific research)

   It is crucial that pupils understand how computer models can be used as powerful tools to predict the consequences of sometimes complex theoretical assumptions. Computer models offer flexible opportunities to 'play' with ideas. In the classroom the use of modelling tools improves the ability of pupils to solve realistic physics problems, without tripping over inappropriately difficult mathematical obstacles.

   Using models in research projects in the classroom presupposes at least basic knowledge of the numerical approach. The ability to formulate—unambiguously—physics theory in terms of differential equations and initial conditions, using the programming language of the modelling tool is required. In addition some understanding of approximation errors is needed.

   In this module we concentrate on these kinds of knowledge and abilities.

4. **Research projects** (how to solve research problems with the help of hardware and software tools)

   Research activities in the classroom demand that pupils have at least some 'active' knowledge of the scientific method. They need investigative skills to solve research problems. Many of these skills will be developed (and improved) in doing the job. In this section, however, we discuss the scientific approach by an elaborate example. We show how MBL and modelling tools can be used in different research stages:

- In the stage of concept building and formulating hypotheses, by doing test experiments. The answers to 'what if' questions are easily investigated with the help of the MBL equipment. In some cases modelling tools can be used to investigate whether existing theory is sufficient in describing the phenomena properly.

- Modelling tools can be used in the stage of building theoretical models for the phenomena under investigation. Computer simulations offer flexible possibilities to investigate the explanatory power of the model for different conditions and the validity of built-in hypotheses.

- Direct comparison of measurement-graphs from the MBL-environment with the theoretically-predicted graphs from the model makes it possible to test hypotheses in a very informative way.

For use in preparation for their own research activities, we also offer some 'easy to do' exercises in testing hypotheses. In these cases pupils receive a diskette with measurement data and a simplified numerical model. They are instructed to complete the model with their own hypothesis and test this hypothesis by comparing measurement results with the results from model calculations.

12.3 Successful Implementation through a Coordinated Approach

The development of a course 'Computer Applications in Physics', including the production of learning packages, was just one of the objectives of the ITN-project.

The complete implementation of the new subject in the physics examination programme appeared to be no small task. We were responsible for all the necessary activities such as information supply, the production and distribution of classroom materials, and the development and execution of in-service training.

With financial support from the government we were able to distribute equipment for microcomputer based laboratory experiments among the schools. We developed the building blocks (interface-board, sensors, amplifiers, filters) for data-acquisition systems. We also developed the open software environment (IP-Coach) with versatile tools for measuring, analyzing and modelling. At present every HAVO/VWO school in the Netherlands has at its disposal at least a minimum of 'Coach equipment' (hardware and software) to support MBL activities in the classroom.

We developed instructional material for teachers and organized in-service training (a 45-hour course) throughout the whole country. Financed by the government, in total more then 40% of all physics teachers of upper level secondary schools (at least one teacher at each HAVO/VWO school) followed a refresher course.

In 1993/1994 the subject 'Computer Application in Physics' will be examined for the first time in the school exams (research projects) and a year later in the national exams. In the meantime most authors of physics schoolbooks have already integrated the subject into their methods. The ITN-material undoubtedly led the way the subject is treated in the schoolbooks. The ITN-learning package is also used at many schools along with the regular schoolbook.

We owe the success of the implementation, in such a relatively short period, to the fact that all phases of the innovation were coordinated from one institute. During the in-service training teachers became familiar with the Coach environment. They worked with learning packages from the ITN-project and were shown elaborate examples of lessons that have proven to be useful in the classroom. This coordinated approach enabled a consistent transfer of subject oriented and didactic views.

12.4 Classroom Experiences and Some Conclusions

The ITN-materials were carefully tested on so-called 'frontschools'. These frontschools played an important role in the implementation of the new physics examination programme. Before the programme was implemented on a national scale the frontschools (5 HAVO schools and 5 VWO schools) had to do a try out.

The first classroom results are promising. As expected, the material stimulates an investigative attitude. Research activities are more challenging (and longer) because less time has to be spent at laborious data processing activities. The results of measurement and calculations are available quickly and presented in a very informative way.

The experiences indicate that pupils have surprisingly few problems with the use of software tools (Coach-environment) for measuring and analyzing data. It is sufficient that they have a functional picture of the possibilities of the programme. It also appears that knowledge of the measurement-hardware (on a system level) helps pupils with the correct interpretation of graphs and with the finding of solutions for measurement problems.

As expected, pupils did have tremendous difficulties with building models for physics phenomena. At the end of the course most of the pupils could not build models for new phenomena independently. At best they could 'read' existing models, change models and build analogue models.

Independent model building demands at least:

1. The ability to work with the modelling tools at the level of the user interface.
2. Understanding of the numerical approach and how the computer calculates model oxidation.
3. Some knowledge of the syntax and semantics of the programming language.
4. Understanding of the physics theory needed to explain and describe the physics phenomena.
5. The ability to translate physics knowledge unambiguously into terms of a working model.

The sequence of this list is not arbitrary. Apart from item 4, classroom experience indicates that the list gives a hierarchy of difficulty. What's more, activity 5 will surely fail if something goes wrong with the other four items.

The number of lessons (5 to 6) that is available within this course for teaching modelling is surely not enough to reach the goal in item 5; at the end of the course pupils will still need help with the forming of models.

If we are aiming at independent research activities in the classroom, more time and attention are needed for the numerical approach and model-building. Perhaps we have to do this at the cost of the traditional analytic approach; spending less time on the mathematics of calculating solutions and more time on the physics of forming models for phenomena.

Future research will have to find an answer to what can be reached and at which costs.

References

1. Driver, R. and Bell, B., Students' thinking and learning of science: a constructive view, SSR, 1986.
2. Beurs, C. de, Ellermeijer, A.L. and Heijeler R., Integratie van informatica-onderdelen in het natuurkunde onderwijs op bovenbouw HAVO en VWO (Integration of informatics in physics education of the upper level HAVO and VWO), Tijdschrift voor Didactiek der ß-wetenschappen, nr. 2, Utrecht 1990.
3. Rogers, L.T., The computer-assisted laboratory, Physics Education 22, pp. 219–224, 1987.
4. Thornton, R.K., Tools for scientific thinking—microcomputer-based laboratories for physics teaching, Physics Education pp. 230–238, 1987.
5. Tinker, R.F., Computer-based tools: Rhyme and reason, In: Redish F. and Risley J.F., Computers in Physics Instruction: Proceedings, pp. 159–168, North Carolina 1988.

13. Global Lab: From Classroom Labs to Real-World Research Labs

Stephen Bannasch and Boris Berenfeld

TERC, Inc.

13.1 Introduction

The Global Laboratory Project combines several innovations in science education to create a community of students, teachers, scientists, engineers and educators investigating local and global environmental change. A key goal of the project is to extend the investigative scope of a science class from the classroom laboratory to research problems in the real world. To do this, the project focuses on a topic of interest to many students: the environment. We use a curricular model of project-based science along with telecomputing and innovative instrumentation to facilitate teacher and student entry into real-world research. We have expanded microcomputer-based laboratories (MBL) to include remote data collection capabilities along with low-cost adaptations of state-of-the-art instrumentation. In this fertile mix of methods and resources, students share data, reports, and analyses about investigations that matter to them, and in doing so, learn how to use the tools and techniques of scientific research.

Over the past ten years, the concept of MBL has proven to enhance science education. MBL enables students to learn basic principles through an interactive hands-on approach rather than memorizing texts and repeating traditional lab experiments. However, unless the laboratory addresses real-world problems that concern students, many students will remain unmotivated and passive learners in the science classroom. Many may even avoid science because they fail to see its relevance to their lives.

Though MBL is an increasingly popular educational tool, its applications too often focus on abstract concepts whose value is more apparent to professional scientists than to high school students. Since students frequently find science's theoretical underpinnings less than relevant to their lives, MBL, though successful in universities, is not as effective in high schools. MBL will have a much greater impact on the high school level if it is used for activities whose value and utility are quite clear. MBL must be made as relevant to high school students as it is to graduate students, and therefore its pedagogical applications must be refined and broadened. There are many possibilities for creating new MBL materials and tools that combine scientific and technological advances with a pedagogical approach that puts the student investigator first.

It is critically important that students investigate what is actually happening in the world around them, first in order to know the physical truth of the situation, and secondly in order to investigate options to remedy discerned problems. Extending a laboratory in this manner turns the repetition of a boring procedure into the means by which important questions are identified and addressed. For a student who eats fish, discovering the concentration of PCBs in the fish caught from a local river may be more important than correctly answering a similar question on a test.

13.2 Three Strategic Objectives

At TERC, we have developed a threefold strategy to maximize the educational value of MBL: first, MBL studies must focus on issues salient to students' lives; second, MBL studies should be project oriented; and third, telecommunications must play a significant role in the curriculum. Let us consider each step individually.

13.2.1 Student Relevance

For the educational process to be most effective, students must want to learn, and students are motivated when they appreciate the relevance of the curriculum. What is relevant to students' daily experiences that, when scholastically examined, can also reveal good science? Most things. Consider, for example, an artifact of most students' lives — the ubiquitous slice of pizza. A simple question such as "Why does the smell of a pizza increase when it is heated?" can stimulate curiosity more effectively than an abstract textbook on molecular behavior.

One field that is both scientifically fecund and interesting to high school students is environmental studies. There may not be many kinematics clubs on the high school level, but there are many environmental clubs. Arguably, students are more concerned with the well-being of the environment than are adults. Whether teenagers' concerns stem from a fear for the planet they will inherit or plain horse sense, environmental studies can draw upon this relevance and root science education in a real-world setting with real-world concerns.

13.2.2 Project Orientation

MBL-curriculum should be project-oriented rather than just sets of academic exercises with often questionable relevance to both students' lives and each other. Project-based curricula offer a hands-on approach to science education. They provide well-defined objectives and compel students to understand if not develop strategies and methodologies for achieving those objectives. They also encourage collaboration and cooperation between students.

13.2.3 Telecollaboration

Finally, telecommunications should play a vital role in MBL applications. Telecommunications can greatly enhance the scope of MBL by enabling geographically disparate schools to communicate and collaborate with each other. In such a scenario, the resources of any one MBL are augmented by those of its collaborators. Collaborative MBL projects enable the development and implementation of regional if not global strategies and effectively educate students about the rigors and procedures of real world science.

13.3 Delivering on the Promise

Are project-based, collaborative, real-world environmental studies possible with MBL? Yes, and a case in point is the Global Laboratory Project implemented by TERC and funded by the National Science Foundation. Now in its third year, Global Lab is a dynamic network of 100 schools from 18 countries scattered around the world. Following a specially designed curriculum, schools conduct identical environmental studies on their respective locales and exchange their data via computer and a dedicated telecommunications network. Alone, each school is an MBL, but when united, they literally comprise in both theory and practice a true global science community capable of gathering original and scientifically useful data.

13.3.1 Environmental Snapshots

An example of this capability is a set of Global Lab activities called Snapshots. In the beginning of the Global Lab year, each school chose an accessible plot of land called the study site to examine environmentally. Students then performed initial environmental assessments of their study sites. As in any collaborative scientific endeavor, they conducted their analyses according to uniform procedures so that findings would be comparable even when the study sites themselves were quite dissimilar. By the middle of the school year, the Global Lab schools had become experienced in following standardized procedures and exchanging their findings via computer.

In order to further refine students' data-collecting skills, we asked them to perform a precise set of procedures that would be synchronized in time. On one pre-arranged day, and often at the very same hour, the Global Lab schools, using low-cost, high-tech tools supplied by TERC, conducted an environmental "snapshot" of their study sites. Within days, they had collated their data into standardized templates and forwarded them to TERC where we assembled the data into a project-wide database.

Shortly after the first Snapshot, which occurred in December, 1992, the schools performed a second, this time on an indoor environment. Then in March, 1993, we asked them to conduct another snapshot on their study sites that was identical to their first. Doing so enabled students to not only compare the environmental parameters of their study sites at a precise moment in time, but examine the changes these parameters undergo over the course of several months. As with the first two Snapshots, schools forwarded their data in a uniform format for inclusion into the burgeoning Global Lab database. This impressive set of data is eloquent testimony to the abilities of schools to implement MBL for collaborative, project-based environmental studies.

13.4 The Success of the Global Lab

The reasons for Global Lab's success lie in its core. Its relevance stems from its focus on local environmental issues, its structure is project oriented, and the glue bonding everything together is telecommunications.

13.4.1 Real World Studies

The project stimulates relevance by taking students out of the tedium of textbook studies and lab exercises into the reality of their own worlds. Real-world environmental studies allow students to study what is immediate to them—their homes, their neighborhoods, and their communities. These studies range from measuring the thickness of the ozone layer, to testing food and soil for pesticide residues, to investigating the quality of the indoor air students breathe.

In another departure from conventional classroom curricula, we asked students to select what they wanted to study. They chose their study sites, thus ensuring that their objects of study were of concern and interest to them. We found that empowering students to choose what to study increased both their motivation and curiosity.

13.4.2 Real Research

We took pains to communicate to our students that they would conduct real scientific research. Upon completion of the Global Lab year, the students would know the key parameters of their study sites better than anyone else, including scientists. For the duration of the project, they would be true pioneers, and even their teachers and textbooks would not know the answers to the questions we posed. They could take justifiable pride in knowing that one column of data in the Global Lab database was their very own. It was a tangible and useful result of their hard work.

13.4.3 Project-Based

Conventional MBL studies are often compartmentalized. Students tackle one task one week and another on the following week, and quite often these tasks have only tenuous connections to each other. The project-based design of Global Lab, on the other hand, enables students to build toward identifiable goals over the school year. Its structure unifies for students the project's many tasks into a common purpose. Their education becomes a succession of interrelated exercises that build upon each other. Students know and understand what they are working toward and can enjoy a well-deserved sense of accomplishment upon completing their enterprise.

13.4.4 Telecollaboration in the Global Village

When computers first arrived, they were perceived primarily as number crunchers, and one of their most powerful features, telecommunications, was overlooked. This remains true today in many MBLs. Computer-based communications is what makes Global Lab possible. Structurally and metaphorically, Global Lab is designed as a real world networked science laboratory. Each school, or MBL, is a fully functional node on the network, and all interaction between schools is computer-based. Over dedicated teleconferences on Econet, students can converse with their colleagues from around the world and post messages on bulletin boards. We have seen posted everything from greetings and ruminations to brainstorming and reasonably sophisticated analyses. In schools from New York to Moscow, from the deserts of Qatar to the Arctic Circle, our students know that they belong to an interactive global village populated by teenagers such as themselves. When they collaborate, students can enjoy productive ongoing working relationships with their colleagues just like scientists.

Exchanging Data. Another advantage to MBL telecommunications is the ability to exchange data. This is the capability that finally delivers on the promise of student-based science networks. Displaying a very impressive success rate, nearly all of our Global Lab schools, regardless of their country or culture, translated their data into standardized units, usually metric, placed their data into a uniform format, and telecommunicated them onto the network. In the process, schools overcame considerable challenges, with hardly the least being processing large data sets of over one hundred individual measurements.

Collaborative Analysis. Global Lab schools have demonstrated admirably that they can follow the regimen of professional scientists. They can identify what to study, describe it, monitor it, and evaluate its characteristics. They can gather time-sensitive data according to controlled procedures, collate this data into a strict format, and telecommunicate their data to their far-off peers. However, the one step that has confounded most of our schools is the final one of their Global Lab research—analyzing the rich body of information they so fastidiously gathered. Finding meaning in a large body of numbers can tax even professional scientists, let alone high school students, but we believe that if students can at least visually depict their findings in graphs, they can identify basic principles such as baselines and trends. Again, we see MBL as the solution to this problem.

With MBLs, students can use simple spreadsheets to chart rows and columns of numbers in various ways. One imposing obstacle to this approach, however, is the diverse hardware platforms used at our schools. In response, TERC is developing a unique software package called *Alice* that will provide uniform spreadsheet and graphing capabilities in key operating environments such as Windows and Macintosh. With its gentle learning curve, *Alice* will enable students to explore their numbers by displaying them in a variety of charts. Such a strategy can help students to better understand their findings, teach them research basics, and add impact to their work.

13.5 Examples

13.5.1 The Example of Moscow School #520

It should be noted that more advanced schools did find analysis to be a worthwhile challenge. For example, School #520 in Moscow examined the project's database and discovered through trend-line analysis that some schools were reporting erroneous results because they had deviated from the project's stringent data collection procedures. The Moscow students posted to all schools a very intelligent and insightful call for more precise standardization so that all data are accurate and thus useful. We can only speculate on what else these sharp students would have detected in the data had they had a powerful tool such as *Alice*.

13.5.2 Measuring Indoor Air Quality

Some classes expanded their work beyond the curricular program of the project. For example a class in San Antonio, Texas, measured CO_2 levels in their classroom and found very high levels of over 2000 ppm. CO_2 levels outdoors are approximately 350 ppm. Though not toxic, these indoor levels were high enough to cause headaches and fatigue. From background material we supplied, the students knew that high CO_2 levels result from inadequate ventilation and can correlate with high levels of other indoor pollutants. They conducted a CO_2 level survey of their entire school and found high levels in all classrooms except for several shops. These shops had large garage doors that were occasionally open to the outdoors.

To measure CO_2, the class used a technique that the Global Lab project adapted from methods deployed by industry. It featured a gas sampling tube and an inexpensive air pump designed at TERC. The sampling tube is made of glass and is filled with a chemical mixture that changes color in response to the presence of a particular gas, in this case, CO_2. The color change progresses down the tube as more of the gas is pumped through. Hundreds of different gases can be measured with this technique. The tubes are single-use tools and cost between $2 and $4 a measurement. The development of an inexpensive, continuous electronic CO_2 monitor for this application would be desirable.

After reports reached the school administration, environmental inspectors were sent to the classrooms. Both the students and the inspectors were surprised to discover that they were using the same gas sampling tubes to measure CO_2. The inspectors got the same readings as the students even though they were using air pumps costing $300 compared to TERC's $10 pump.

Once the readings were confirmed, maintenance personnel repaired the ventilation systems and the students made follow-up measurements. They were happy to discover that the CO_2 readings were now acceptable.

The steps taken by these students are profound and deserve to be listed.

1. The students discovered a problem by measuring their local environment with a low-cost variation of a high-tech instrument.

2. The students surveyed nearby classrooms and discovered an even larger problem.

3. The school administration sent in environmental inspectors to check on the air quality and the inspectors confirmed the findings of the students.

4. Resources were expended by the school district and the ventilation system was fixed.

5. The students were able to confirm the proper operation of the ventilation system with their own independent measurements of air quality.

It is difficult to overstate the importance of this experience for these students. Not only was the initial research they performed confirmed by experts, the quality of life was improved because of the actions taken as a direct result of student research.

Obviously this experience cannot be shared by every class measuring indoor air quality. Many classrooms will not have such high levels of CO_2. However, we believe that a rich set of measurement tools and instruments along with a project-based curricular focus set the stage for relevant student research.

In order to extend the measurement tools available to student-scientists measuring indoor air quality, we have developed an inexpensive hot-wire anemometer to measure air flow in ventilation systems between 10 and 300 meters-per-minute (25 to 1000 feet-per-minute).

13.5.3 Developing a School-based Network for Monitoring Stratospheric Ozone

Another example of student research is the work schools around the world have done along with engineers at TERC and collaborating scientists in the development, testing, and calibration of the TERC Total Column Ozonometer (TCO). (See Figure 1.) This instrument is a 2-channel UV spectroradiometer that measures narrow bands of UV-B and UV-A light. With appropriate calibration, the instrument can measure the thickness of the ozone layer.

Over 30 TCOs have been distributed to schools and projects in the US, Poland, Australia, Italy, and the Czech Republic. A TERC TCO has even been used in Antarctica, where its measurements of ozone were compared to those taken with the standard device, a Brewer spectrophotometer. The TCO's measurements were within 5% of the standard readings. We were particularly impressed because the TCO was operating outside of its design parameters.

The TERC TCO costs about $600 to build and calibrate. This is a great deal of money for a school; however, the chance to participate in this important area of scientific research has proven to be interesting to a number of schools. Standard devices for measuring ozone cost approximately 50 times what the TCO costs. With a research commitment there are many opportunities for creating low-cost advanced instrumentation for student use which can give professional and scientifically valid data.

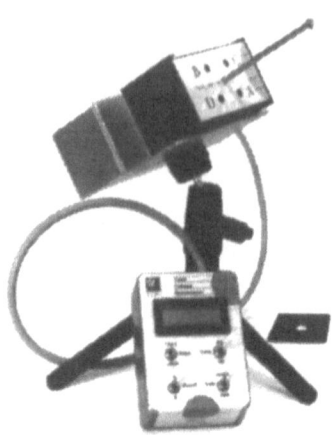

Figure 1   TERC Total Column Ozonometer (TCO)

Many schools participated in the development of the instrument by collecting data and performing intercomparisons with standard ozone measuring instruments at scientific institutions around the world. Each school was supplied with preliminary operating instructions and a large selection of background material relating to ozone measurement and interference filter-based instruments. The students and teachers are collecting important data that are being used in the design of the next generation of TCOs.

The amount of ozone measured is described as a "thickness," usually expressed in milli-atmosphere-centimeters called Dobson Units (DU). A typical level is 300 DU. The total quantity of ozone overhead at standard temperature and pressure is a thickness of 3 mm. That much ozone totally absorbs the short wavelength solar UV (UV-C, with wavelengths less than 280 nm) which is extremely damaging to biological systems. The ability of ozone to absorb light decreases rapidly with increasing wavelength (in this near-UV part of the spectrum). Thus, UV-B (280 to 320 nm) is only partially absorbed, and the portion which reaches the ground can cause serious biological and material damage. UV light with wavelengths longer than 320 nm (UV-A) is hardly attenuated at all by ozone, but is relatively benign in its biological effects.

The consequence of the wavelength dependence of ozone absorption is that the amount of UV in the critical range of 300 nm to 310 nm that reaches the earth's surface depends heavily on the amount of ozone overhead. This fact is both the reason we care how much ozone is above us, and the reason we are able to use a simple technique to measure that amount.

13.6 An MBL for the Field

We are developing an easy-to-use data-logging and control computer we call the TERC DataLogger (TDL). Prototypes of this datalogger have been paired with weather and climate instrumentation developed in Moscow at the Institute of New Technologies (INT). The pair form a monitoring station for schools to investigate their local weather and climate. This is an extension of the traditional MBL from the lab to the field. One difficulty students have with this type of data collection is that the data are viewed in their entirety only after a lengthy collection period. This delay means that the data are more abstract and removed from the students' experience than are data collected with traditional MBLs, which are displayed as the experiment is in process.

In order to help overcome this problem, one teacher had his high-school students drive by the operating datalogger at night and shine their headlights into the light sensor (which was mounted sideways). The students recorded the time when they did this and later were able to pinpoint on graphs and data tables when they had driven by the school. The point is that such teaching methods lessen the abstraction of data collection that occurs unattended outside the lab.

The TDL uses a powerful but inexpensive microcontroller along with memory, clock, LCD display, a wide range of interface circuitry, and a battery to create a portable computer that can monitor and log data from many types of sensors and probes (see Figure 2). What makes the TDL unique is the software interface and architecture that allow it to be easily controlled and interrogated from any computer with a modem.

One of our design goals for the TDL is that it be easy to use on the many different types of computers used in schools and laboratories in the United States and around the world. TERC pioneered the development of MBL software for PC and Apple II computers. Our software, which makes full use of the computers' capabilities, is renowned for being both powerful and easy to use. For several reasons, however, this type of custom software development would be inappropriate at this point for the TDL. The cost of developing custom programs for different computers would be high, and designing the flexibility needed to support the many capabilities of the TDL would be difficult. This model also requires all improvements and extensions to be implemented by a small programming team at TERC.

We have a different idea. The TDL is accessed and controlled in the same way as a computer network or bulletin board. A telecommunications program is used to establish a connection, and the TDL responds with a full set of easy-to-use, text-oriented menus. The menus describe the type of data collection services available. Using the menus, the time and period of collection are specified and a collection is started. After collection, the data are displayed on the computer in calibrated physical units and saved or copied to a spreadsheet for analysis and graphing. Each probe can be calibrated and the calibration saved in battery-backed memory on the TDL.

If we were implementing this interface in the micro-controller's native machine language, the programming task would be horrendous. Instead we have embedded a high-level language specifically designed for data collection and control of small computers. The TDL has the language FORTH built in. Over 90% of the code we have written is in FORTH; the rest is in machine language for speed. FORTH has the advantages of an interpreted language in that code can be tested interactively and it approaches the speed of a compiled language.

With FORTH built in, a much larger group of people can now write programs for the TDL. We realize that not many people may take advantage of this capability. Writing small programs is not difficult if you have some experience with programming languages.

The TDL comes in an inexpensive ($15) waterproof plastic case. Wires for probes can be run through holes drilled in the case and sealed with silicone rubber. If needed, extra cases are available.

The TDL includes a small LCD display and several buttons so that it may be operated in the field without a separate computer.

The Weather and Climate monitoring stations developed jointly by TERC and INT use the TDL as the datalogging computer. The TERC/INT stations record such data as wind speed, wind direction, solar insulation (watts/sq. meter), barometric pressure, outside temperature at two levels above the ground, soil temperature and soil humidity. The probes for soil moisture and barometric humidity are experimental; all the equipment needs testing and feedback by Global Lab classrooms.

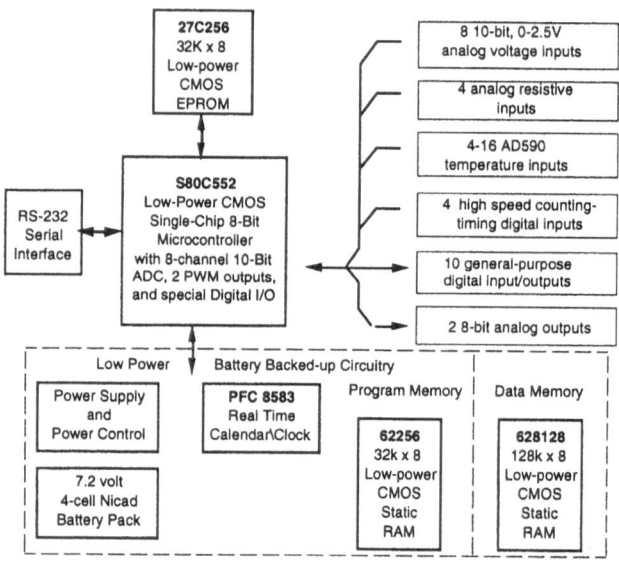

Figure 2 TDL Block Diagram

13.6.1 Remote Probes and Sensors

We are very interested in exploring the use of inexpensive radio-based telemetry and control. If dataloggers such as the TDL could be connected through a radio link to a desktop computer in the science classroom, the difficult and labor intensive process of downloading data would be eliminated. The advantage of making long-term data collection less abstract is even greater. Even if a class project involves a month-long collection of data, the availability of current data at all times will keep students more interested.

If the costs can be sufficiently reduced, the possibility exists of clusters with ten or even a hundred individual sensors. Each cluster would report to the central data-consolidation computer back in the science classroom.

13.7 Conclusion

There are rich opportunities for instrumentation development and adaptation awaiting educational innovators. By making visible and concrete environmental measurements in the air and water around us, as well as in the atmosphere above, students can participate in areas of science investigation that are interesting to them and possibly important to current scientific research. However, the difficulties in such development are large. The design of probes for measuring artfully crafted models in the physics classroom is less challenging than creating instruments that measure phenomena in the real world. In addition, the costs are almost always higher than schools would like. However, the payoff in student interest in doing science and the possibility of students discovering important data relevant to their lives and/or the larger scientific community is great.

Computers are becoming essential science education tools and the time has come to broaden their use. Projects like Global Lab indicate that students are indeed capable and willing to engage in hands-on, real-world, collaborative science curricula. We easily can envision a day where 1,000 or 10,000 schools are networked together to conduct environmental experiments so accurate that the scientific community would rely on their data. From such synergies, the scientists of tomorrow will emerge.

14. Dynamic Physical Representation of Real Experiments

Dieter Heuer

Universität Würzburg

Abstract. With adequate software, computer-supported experiments can provide important experiences for the student. The learner can test, question and even correct his/her own physical concepts, while the teacher may only provide some assistance. By coding experimental results using a suitable but simple notation, connections to existing ideas can easily be established. For example, vectors, arrows and rays are examples of notations that can be comprehended more directly and easily than graphs. Instead of "reading" a graph the student has to comprehend an ideogrammatically represented process. This process can, by a sequence of pictures, be directly related to the inner picture of the course of the experiment. In contrast to the textbook, which only offers one single static representation of the course of the experiment, the computer can show the transformation into a dynamic physical representation in parallel to the actual progress of the experiment.

This supports the crucial direct comprehension of physical concepts and ideas. Such a system paves the way for new exciting didactic possibilities for teaching. This paper illustrates some of these possibilities by using the open PACMA-System[1] to analyse different types of motions and capacitor charging and discharging. These examples show an additional learning step facilitated by the system: not only are the qualitative connections better comprehended by the dynamic physical representation, but this representation also simplifies the necessary transition to working with the graph as a more precise instrument for analysis by simultaneously providing both representations.

14.1 Developing Adequate Physical Concepts

How can we help students to learn physics more effectively, to develop adequate conceptions, and to overcome inadequate conceptions? I think there are two main directions in which we can proceed.

---

[1] PACMA hardware and software is obtainable from Microsystems, Dr. H. Parsche, Neufahrner Str. 21, D-85748 Garching, Germany

The first possibility is to find out about misconceptions and existing learning difficulties. Then, effective educational concepts which are to overcome these barriers have to be developed. Both steps are quite labour-intensive as well as time-intensive.

Although there are numerous publications dealing with this matter, scientific analysis of this problem has just begun. As yet, educational schemas to overcome specific learning obstacles are still rare. This is a long and thorny path, but it is worth the effort since fundamental results are likely to be achieved. As for the immediate and medium-term future, it must be realized that this way will only be of partial use for students.

Taking into account the still relatively poor understanding of learning concepts and the rapid progress of hardware and software development with its ever new opportunities to evolve better and more effective teaching concepts, other solutions have to be considered.

14.2 A Generic Approach to Make Physics Learning More Successful: Qualitative Understanding and Active Working with One's Own Ideas

The other approach to make physics learning more successful is more generic. The general insight we gained up to now into the learning of physics should be taken into consideration more than we have done in the past when shaping the learning process. This includes that the learner has more opportunity to

• use qualitative understanding for the explanation of the course of physical events.

It is equally important to

• actively test ideas in physics on a broad spectrum of concrete situations and problems. If necessary the learner should at the same time extend and correct his/her own ideas.

There are several, very different methods to approach these goals, all of which have only been tried and tested insufficiently as yet. A. V. Heuvelen [1], for instance, has suggested going through formal solution-steps in problem solving. By beginning with written descriptions the learner should always proceed to a visual, then to a physical and finally to a mathematical representation without omitting any steps. This schema can help students to evolve the solution of exercises step by step. Each of the solution steps ought to be an important help for the next step.

14.3 Computer-Supported Experiments as a Basis for Experience

Experiments provide a completely different starting-point to achieve the goals of qualitative understanding and active working in physics education. Traditional lab sessions are, in general, not suitable for this since the focus is on measuring and testing a formula. Only with computer-supported experiments, where taking data and graphing is automatised, the student has sufficient time to compare his/her ideas about the experiment with the resulting graph (D. Heuer [2]). In the area of mechanics, especially kinematics, R. Thornton [3] has shown how effective such lab experiments can be when they are embedded in a suitable learning strategy.

To be able to realize versatile computer-supported experiments in various areas of physics we have created an open experimenting system at the department of physics education didactics of the University of Würzburg. We call it PACMA: Physics ACtive Measuring, Modelling, Analysing and Animation [4]. Our most important goal was to achieve openness as well as easy handling of the system. The experimenter should be able to make changes easily which influence the evaluation as well as the representation. Furthermore, he/she should also be able to adapt the system easily to different problems. The basic conception of PACMA was first realized on the C-64-computer (D. Heuer [5]). German teachers have been using this version over several years. Later on the system was available for the AMIGA in a widely extended form. Now the transfer onto the PC-platform under Windows 3.1 is completed. Numerous new conceptions for experiments with the open experiment and modelling system as well as first reports on experiences in teaching with this system have been published (see W. Reusch, D. Heuer [6]). A special advantage of PACMA lies in the possibility to not only record the course of an experiment and analyse it by representing relevant graphs but, in addition, to model the course of events on-line, i.e., in real time.

14.4 Advantages of Different Notation Systems Compared to Traditional Graphs

The use of computer-supported experiments as proposed in the past has always been concerned with the illustration of experimental results by graphing functional relations between physical quantities. For experts who already have adequate concepts of physics this is a very efficient method, but it always requires one transformation: the graph has to be "read," i.e., its characteristic results (these are often qualitative) have to be comprehended and connected to the inner picture of the experiment and its progress. If this inner processing is successful, interrelations can be inferred correctly during a mental reconstruction of the experimental situation. Ideally, the qualitative form of the graphic representation can be reconstructed without necessarily remembering the concrete picture of the graph that has been produced in the actual experiment.

However, for a non-expert there are two main difficulties involved in following the several steps of processing:

- Students must "read" the graphs, i.e., they must deduce characteristic results from the graphs with regard to the given experimental situation. As R. Thornton's research [3] shows, this can be taught and, given enough practice, done successfully.

Nevertheless it is always necessary to be very concentrated and afterwards to think back to the initial situation.

- The results which are found must then be related to existing ideas. This is a further difficulty for students (Th. Wilhelm, D. Heuer [4]).

By coding the characteristic results of an experiment in a suitable notation system, connections to existing ideas can be made more easily. These notations and the corresponding visualisations help the student to recall the results later. The notation systems already have been developed with the evolution of physics. Vectors, arrows and rays are examples which, depending on the context, are representing various physical quantities and thus can be used in very different content areas. Their frequent use, especially in introductory textbooks, obviously shows that statements can be classified very concisely with their help. However, as yet learners use them relatively rarely for problem solving. It is obviously necessary to create a physical representation for situations in physics which can be varied in many ways, thereby supporting qualitative deducing and argumentation. But it seems just more important to acquire and elaborate such concepts in connection with experiments than with simulations. This is why the learner is really challenged to transfer the shown representation to reality.

14.5 Dynamic Physical Representation of Experiments

Dynamic representations can be supported by computers. In connection with experiments this area has not been looked at so far. When a representation of the experiment (which, of course, is the immediate level representation rather than the traditionally used, hard-to-understand graph representation) can be shown on screen during the course of the actual experiment, the student will benefit in two ways:

- The "reading" of the graph can initially be omitted. Instead, only a very simplified representation of the process has to be comprehended.
- This picture sequence of the process can directly be related to the inner picture which the student has of the experiment and its events. Often this sequence can be condensed to a single picture together with a statement about the change of a relevant property.

In comparison to a single picture in a textbook, the computer clearly has the advantage that it can show and illustrate the course of the experiment and, with it, the dynamic changes of the physical representation of the experimental situation. In principle, this method of showing actions and events in physics through a suitable representation can be used in simulations as well as in conjunction with real experiments. For simulations examples can sometimes be found where especially motions are represented by acceleration and resulting force-vectors, e.g., in the software Interactive Physics. However, what is missing are systematic methods to design these notations in connection to real experiments in order to get the concepts across.

14.6 Examples of Physical Representations of Real Experiments

The general thoughts which I've set out so far should become clearer with some examples. The examples given here are all using very different possibilities of representation to show the versatility of the PACMA environment.

14.6.1 Motion of a Glider which Is Exposed to a Constant Force, e.g., on an Incline

The motion of a glider on an air track which is recorded through a precision travers-ing wheel (resolution ca. 1/10 mm). The motion of the schematically represented glider is shown true to scale on the screen. The momentary velocity v (inferred form $\Delta s/\Delta t$) is represented as a vector. The change in v is clearly visible since the old velocity v_o-vector from the last time-interval and the gain in velocity Δv-vector are shown on the same screen (Figure 1). All vectors are moving along with the glider. In order to indicate the continuously running time a symbolic clock appears on the screen as well.

The goal is to explain how v and Δv are correlated with the force acting on the glider by showing v and Dv as vectors. The course of the experiment shows always the same Dv, therefore v increases constantly over time. In the elastic collision of the glider at the end of the lane, Dv is suddenly very large and in the opposite direction;

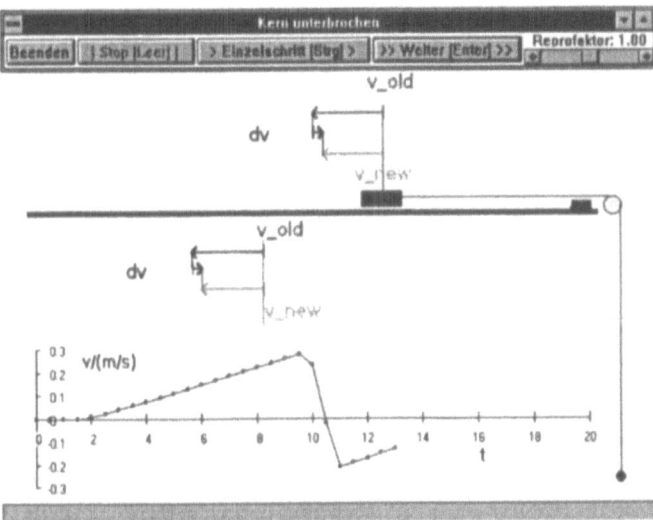

Figure 1   A glider moves along an air track under influence of a constant force. Shown is the present velocity v, the "old" velocity v_o and the change of velocity Δv. In addition the graph v(t) is shown in the lower window. During instruction this graph, like all the following graphs, is only shown later on in a second discus-sion phase (s. section 14.7).

thus v reverses its direction. After the collision Dv has again the initial direction. Now, however, the length of v is constantly decreasing until the direction is reversed and v increases again. Students can next vary the time-interval Δt, the force F and mass m of the glider.

Other results which can be shown. The value of Δv depends on the force F but when Δt is doubled or halved, Δv changes accordingly (Δv ≈ Δt). There are two possibilities to explain this:

- The effect of the force F on Δv is increasing proportionally over the time Δt: not F but the quantity FΔt determines Δv. Including the mass m the assumption is: FΔt ~ m · Δv.

- The force F only determines Δv with the same Δt. Since Δv ~ Δt only the quantity Δv/Δt is a measure for the effect of the force on the motion.

When these interrelations are to be shown, the physical representation has to be extended; then in addition the F Δt and the change of the momentum have to be included.

14.6.2 Oscillation of a Spring Pendulum

Corresponding to the constantly accelerated motion, here the position of the oscillator, the displacement of the spring, again v, v_0 and Δv are shown, and in addition the acceleration a. Compared to the experiment with a = const., an important new result

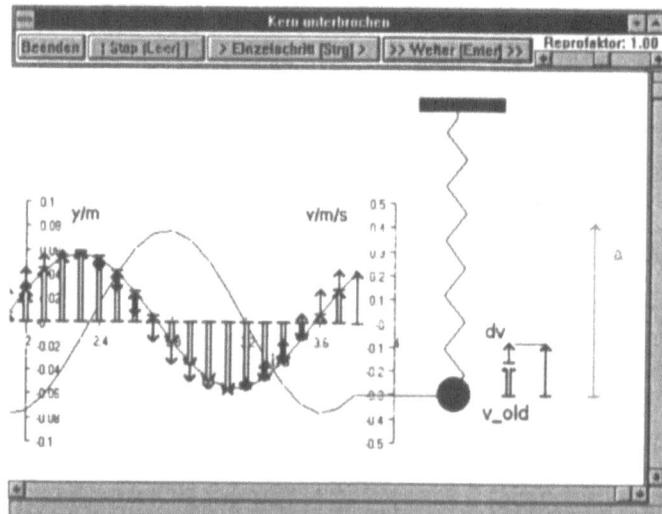

Figure 2   Momentary picture of the dynamic graphical representation of an oscillating spring-mass-system. In addition to the quantities from Figure 1, the acceleration a is shown here. The length of the a-vector is proportional to the corresponding displacement of the spring. The graph shows v(t).

can be presented: Δv and, more clearly, a are always pointing to the equilibrium position. The value of Δv and a increases according to the displacement of the spring (a ~ -x) and thereby according to the restoring force (Figure 2).

In an extended representation it is furthermore possible to measure the force F of the spring on the oscillator with an electronic dynamometer and to represent F_D together with the gravitational force F_G of the oscillator and the resulting force $F_R = F_D + F_G$. In addition the vector of the acceleration is drawn; thus the relation between the forces acting on the mass and acceleration is visualised.

14.6.3 One-Dimensional Collisions

As with the motions described above, here it is again possible to visualise relevant quantities through the velocity vectors v and their changes Δv in order to comprehend and describe collisions. In a collision of two identical masses (Figure 3) the transfer of velocity from one glider to the other can be observed very clearly when the collision takes place slowly. This can be realized through very soft springs. When the masses are different, the weighted velocities and thereby the momenta have to be shown. Thus the transfer of momentum during the collision can be followed directly in the physical representation.

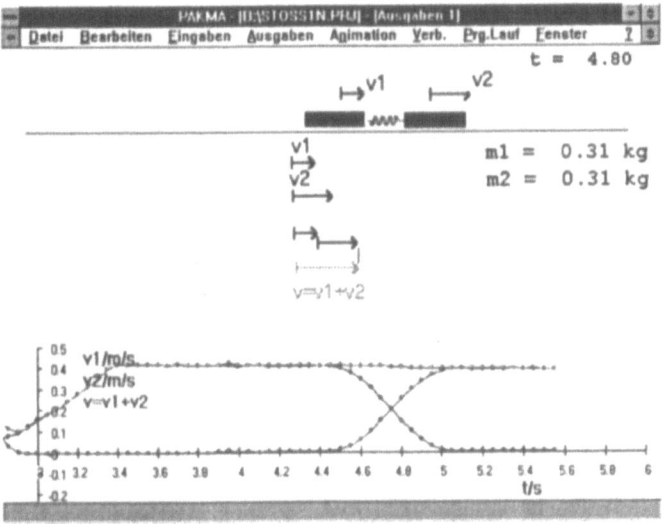

Figure 3   Momentary picture of the dynamic representation of collisions with equal gliders. The velocity vectors v_1, v_2 are drawn at the two bodies 1 and 2; in addition they are drawn from a fixed point. Thus their mutual change during the collision can be compared better. The collision takes place through very soft springs to allow students to follow the course of the experiment easily.

14.6.4 Two-Dimensional Motion

Guided two-dimensional motions can very easily be realized with a computer-mouse. Thereby, not only the position x(t) of the mouse, but also its velocity v(t) and acceleration a(t) can be shown. Since the already drawn vectors are not, in general, covered up by the vectors which are to be drawn newly, the former will not be erased. Consequently the construction-algorithm from v to a can be seen and understood very clearly when the "old" and the "new" velocity vectors are drawn beginning at the same starting point, P. Krahmer, D. Heuer [7]. (See Figure 4.)

14.6.5 Motion of the Centre of Gravity and Internal Motion of Connected Oscillators

Beats, as they appear with connected oscillators, can mathematically be explained easily as a superposition of two sine-oscillations with different frequency. The relation of the course of motion to the symmetrical and the anti-symmetrical oscillation can be comprehended directly only with difficulties. When, in addition to the displacement x_1 and x_2 of the two oscillators, the displacement x_{cg} of the centre of gravity and the displacement x_{in} of the inner motion are shown on the screen (Figure 5), the round sinusoidal course of x_{cg} and x_{in} and the somewhat different frequencies can be seen from the changes over time.

Not only the construction of x_{cg} and x_{in} from x_1 and x_2 can be derived directly from the representation (Figure 5). From the initial dispositions x_{10} and x_{20} of the oscillators the size of the corresponding x_{cg} and x_{ino} can be derived. When the initial velocity at the starting point equals zero, the parts of the motion of the centre of gravity and the inner motion are determined. With help of the separate representation of the motion of x_{cg} and x_{in} we want to show that, at each moment in time, the motion of the two connected oscillators can be seen as superposition of two independently

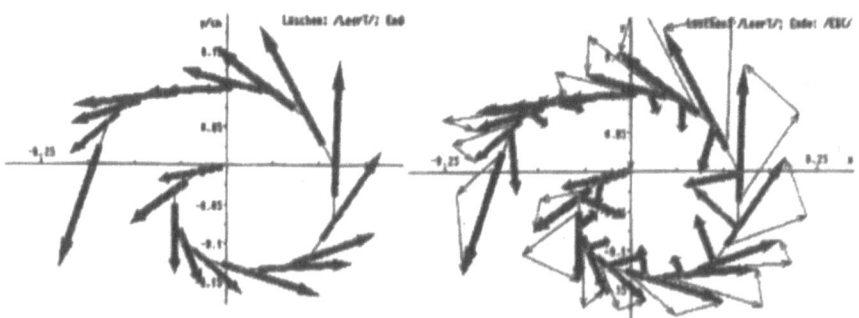

Figure 4   In addition to the motion of a computer-mouse the velocity- and acceleration-vectors are shown. Furthermore it is made obvious how the acceleration vector is constructed.

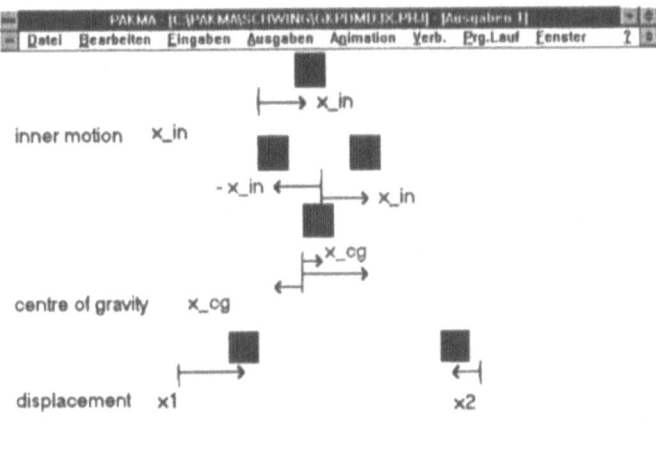

Figure 5 In the lowest section the actual displacements of the coupled pendula are drawn. Above, the position x_{cg} of the centre of gravity is shown. In the next section the quantity x_{in} is constructed using the position x_{cg} and the displacements x_1, x_2 which are here drawn from the common equilibrium position. Finally, in the uppermost section, the inner motion x_{in} is drawn as it is seen by an observer who moves along with x_{cg}.

oscillating systems which are representing the motion of the centre of gravity and the inner motion.

The creation of a beat and its course over time can be conveyed very graphically with the well-known pointer representation. Here the projection of a uniformly rotating pointer results in a simple harmonic motion. All you have to do is to put together two independent pointer motions. The left hand side of Figure 6 shows the analysis: from x_1 and x_2 are derived x_s and x_{in} by means of addition and subtraction. On the right-hand side we see the synthesis: in the lower part of the screen the pointers are shown. From their projection we get x_s and x_{in}. These pointers are added above and the projection is compared to x_1.

14.6.6 Charging and Discharging of a Capacitor

When potential differences and currents are represented by vectors, the charging of a capacitor can be characterised by the schema in Figure 7. With increasing potential difference at the capacitor U_C the difference between working voltage U_B and U_C is decreasing and according to $U_B - U_C = i \cdot R$ the charging current i is decreasing as well. With this schema the function of U_C as the source of the potential difference becomes clear. Furthermore it becomes obvious that the changes are happening more and more slowly, the closer U_C and U_B get. In order to show the value of the current

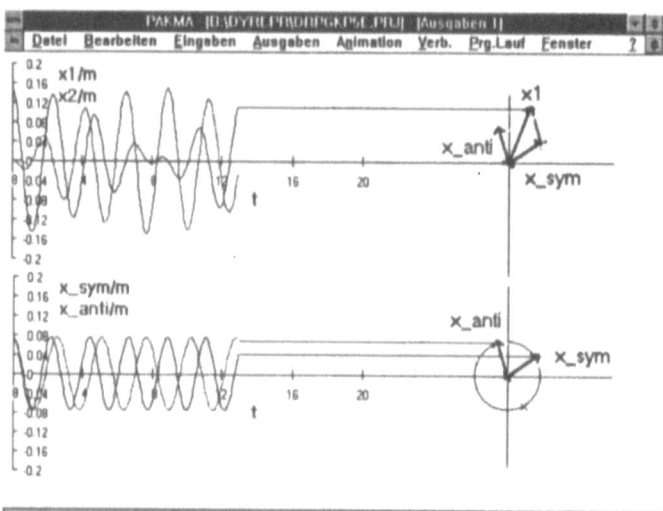

Figure 6   The beat in the upper left corner is below divided into the two fundamental
oscillations with different frequency. These two graphs result from the two rotating pointers on the right side. The addition of the two pointers (only the addition
is shown above) and respectively the subtraction again create the beat in the
projection on the vertical axis.

i an arrow with fixed length has been chosen. Its width is changing proportionally
with i. In order to support the idea of a current the direction of the arrow can be
chosen to be the same as the direction of U_B and U_C.

Our goal is to explain the charging and discharging procedure with an expressive
structural model and to use this model for qualitative predictions with varying
U_B, R and C.

Extensions. Introducing the quantity R into the potential difference-vector diagram.
When the distance between the U_B- and U_C-vectors is chosen, the current i is proportional to the slope of the connecting line between U_B and U_C. The charge on the
capacitor can be shown through a rectangular area, beside the U_C-vector, with the
sides U_C and C. ($Q = C \cdot U_C$). In a 3-dimensional representation the energy-contribu-
tions to the charging of the capacitor can, according to $U_C \cdot DQ$, be represented as a
volume. Its total value can be inferred directly from the graphical representation.

Another example for using dynamic physics-representation for electricity is the
branching circuits. Again we can represent the currents through arrows with variable
width (see R. Girwidz [8]).

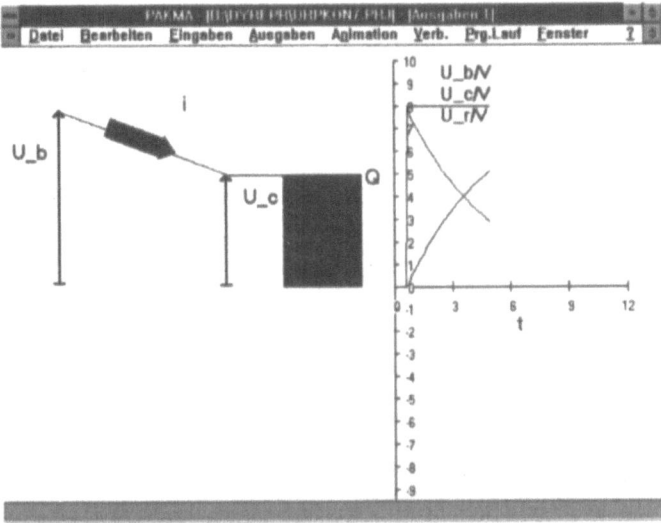

Figure 7   Scheme of charging and discharging of a capacitor. The working voltage U_B and the capacitor's voltage U_C are shown as vectors, the current i as an arrow with fixed length, its width changing proportional to i. The rectangle represents the charge Q on the capacitor.

14.7. Didactic Goals of Dynamic Physical Representations

The examples show how qualitative results and statements can be conveyed directly through a suitable visualisation using dynamic physical representations. The goal is to show the learner on screen suitable physical representations in parallel to the course of the experiment in order to clarify underlying important basic concepts. With the physical course of the experiment and the representation being parallel and through special variations of the experiment, students should become able to build up their own physical concepts. With the help of these concepts the physical course of the experiment should be satisfactorily comprehended. Only after the student has acquired such concepts in a basic form and has comprehended the interrelations, graphs should be used in order to make the physical interdependencies more precise, which need further processing for interpretation. In this transitional phase the simultaneous availability of both representations is absolutely desirable. The reason for this is that going back to the known representation is helpful for dealing with the more abstract graphical representation.

14.8. Realizing and Using the Dynamic Physical Representation

The dynamic physical representation has been realized in PACMA in two very different ways. The earlier Amiga-PACMA provides single commands for the drawing of vectors, helping lines, filled areas, etc. In the PC-PACMA version under Windows 3.1, which is available since 1995, a separate graphics-editor is provided for the creation of the dynamic physical representation. In this editor the graphic elements are called up, arranged and their connections to the physical output quantities are defined. These output-quantities are determined in the core program which presents an open system with which the user realizes his/her ideas of measuring, analysing, modelling and representing. The dynamic representations, of course, are also designed to be constructed and changed as easily as possible. Since they are especially used for getting to know basic experimental procedures, students will work with ready-to-use representation programs, which can be built easily and changed by teachers. Now in 1995 the conception of the dynamic representation is being examined in a school-test, to find out how far the proposed dynamic physical representations support the learning of physical concepts in dynamics.

Acknowledgements

I am grateful to the programmers of the PACMA, most of them advanced students at the University of Würzburg: for the PC-PACMA version under Windows 3.1: Stefan Hild, Kings College, London (main part of PACMA), Michael Schmidt (Animation part), Stefan Hahn (Measuring routines) for the earlier Amiga-Version: Martin Schröder (Animation part), Holger Graefe, Martin Müller-Sommer, Martin Vaeth. They all worked really hard to realize as many conceptual details as possible. I also thank the numerous teachers using PACMA for physics teaching for providing feedback with their impressions to me and for offering many constructive suggestions. The development of the PC-PACMA was supported by grant from IBM Deutschland.

References

1. Heuvelen, Alan Van: Learning to think like a physicist: A review of research-based instructional strategies, Am. J. Phys. 59, 891 ff. (1991).
2. Heuer, Dieter: Changing misconceptions through MBL—a concept for LAB-sessions. Chapter 14 of this book.
3. Thornton, Ronald K., and Sokoloff, David R.: Learning motion concepts using real-time microcomputer-based laboratory tools, Am. J. Phy. 58, 858 (1990).
4. Heuer, Dieter: Offene Programmierumgebung zum Messen, Analysieren und Modellieren: Ein Werkzeug, physikalische Kompetenz zu fördern, Physik in der Schule 30, Heft 10 (1992).

15. Changing Misconceptions Through MBL—A Concept for Lab-Sessions

Dieter Heuer

Universität Würzburg

Abstract. Students' difficulties in solving qualitative problems are well known. Obviously students aren't sufficiently required to use qualitative argumentation in problem solving. Lab experiments offer various possibilities to enhance the understanding of physics if we are willing to radically change the goals of these experiments. Instead of experiments emphasizing measurements we need experiments emphasizing conceptual understanding. By using the computer for the recording and analysis of the measured data the student can then concentrate on applying his/her physical ideas to predict and interpret the course of the experiment and in case of unexpected events, to test his/her own conclusions and concepts. For the example of the gravitational pendulum ten problems are proposed which require the student to work intensively on parts of these problems which include everything from the observation and analysis of motions, to an energy balance, and finally the modelling of non-harmonic oscillations. This requires an open software and hardware system which can be used by the students throughout the problem-solving process and teachers willing to go into more open problems.

15.1 A Lack of Qualitative Understanding

At the end of this summer semester I asked my students of the program for teaching physics at the "Gymnasium" in the sixth and eighth semester, the following question:

> Given a gravitational pendulum with large displacements. Draw a qualitative diagram of:
>
> angle of rotation $\varphi(t)$
>
> angular velocity $\omega(t)$
>
> angular acceleration $\alpha(t)$

when the pendulum is released at the position shown in Figure 1. The students had five minutes time to answer this question in a written form.

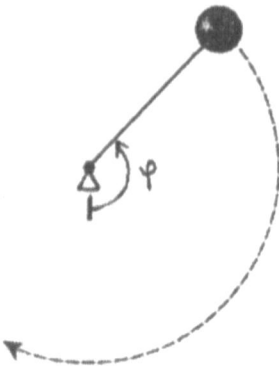

Figure 1   The gravitational pendulum, released with large displacement

How well do you think the students solved this problem? Less than 10% of the students drew the graph of the angular acceleration $\alpha(t)$ with an indentation at the maximum/minimum. The method they commonly employed was to first draw $\varphi(t)$ and then $\omega(t)$, and finally, perhaps, $\alpha(t)$. All graphs had an approximate sinusoidal shape.

Here we are dealing with a basic statement about the fundamental equation of mechanics $F = m * a$, especially for the case of rotation $M = I * a$. The answers given by the students show that they didn't have a deep enough understanding of the facts. Nevertheless, all of them are probably able to state these equations, to interpret them to some degree and to solve corresponding exercises.

Such a lack of understanding is not unique to this case. Problems like this are well known from international publications. The issue is a qualitative solving of problems in the area of physics. Ten years ago, J. H. Larkin et al. [1] had already pointed out the differences in problem solving strategies of novices and experts. They emphasized the importance of qualitative argumentation. The very unsatisfactory way in which the above question was answered shows that necessary physics concepts aren't developed well enough. Furthermore, important connections between these concepts and experiences in applying them are missing.

15.2. What Can We Do to Develop Physics Concepts?

It is crucial for the students to apply and test their own physics ideas and concepts again and again in concrete situations. Besides that, they should be able to question and, if needed, to correct existing concepts.

Specific consequences follow in two areas:

- We should change the problems posed for beginning students. The problems should require the students to penetrate the contents of facts and statements qualitatively.

- During the experiment in the lab sessions, we should call on the students to give concrete explanations for their observations.

In this paper only this second proposition will be dealt with at length. It is obvious that we have to change the present goals of experiments in student labs drastically. Previously in lab we have focussed on understanding and applying the underlying theory and relevant methods of measurement to carry out the experiment. The student generally had to take many measure values and evaluate them. Then he/she had to draw graphs and present results—sometimes in form of equations. These traditional objectives are, without doubt, important. However, they don't seem to enhance the understanding of physics in a sufficient manner.

15.3 Comprehension Experiments in Addition to Measuring Experiments

Up to now it was impossible to implement the additional goal that the students think critically about the course of experiments and the results. The main reason for this was that with the usual technique many single measurements have to be taken and converted in order to be able to draw a graph. This takes a long time. Consequently, until now, to measure and to evaluate were the most important activities during the experiment. The central result of the experiment was to test a formula based on these measurements. When we focus on this we waste a large part of students' learning potential.

In a computer based experiment it is possible to record data directly, instantaneously convert them into relevant quantities and transpose them into the desired graphic representation. When the experiment doesn't proceed too fast it is even possible to simultaneously watch the results of the processing. In this way the student is able to compare his predictions about the course of the experiment to reality as documented in the special graphic representations. This comparison leads to an interrelation of both, which is the crucial element of the proposed lab sessions. The student can make a prediction for the expected outcome based on his/her ideas and concepts. The student is challenged when the expected outcome is different from the result of the experiment. In this case the students have to check their own thinking, their conclusions and possibly also their concepts. It will often be the case that experimental results, particularly when unexpected, will challenge the students to use their ideas in order to reach clarification. Additional experiments with varied conditions can be integrated at once into this process.

In summary, this argument suggests we integrate what we call "comprehension experiments" into the beginners' lab section as well as the usual measuring experiments. Only with such experiments will the students' lab aquire a new quality of learning.

15.4. An Example of Comprehension Experiments: The Gravitational Pendulum

This idea will be illustrated with the example of the gravitational pendulum.

In an experiment emphasizing measurement the students would be told, for example, to measure the time of oscillation T as a function of the maximal angle of rotation ρ_{max} or to measure T as a function of the moment of inertia I.

For a comprehension experiment, however, we propose the following typical steps:

1. Prediction of the course of motion of the gravitational pendulum $\varphi(t), \omega(t), \alpha(t)$

This is the question described above.

2. Qualitative reasoned justification

Here the student should become aware of the ideas and facts on which he/she based the prediction.

3. Putting the real experiment into practice: Recording the course of motion

Next, $\varphi(t), \omega(t)$ and $\alpha(t)$ are to be recorded with an MBL system. As motion sensor we use a traversing wheel with a precision slit disk and differential photo gate, which gives 10,000 impulses per revolution of the pendulum (D. Heuer [2]). Some of the experimental parameters could be varied in this phase (Figure 2).

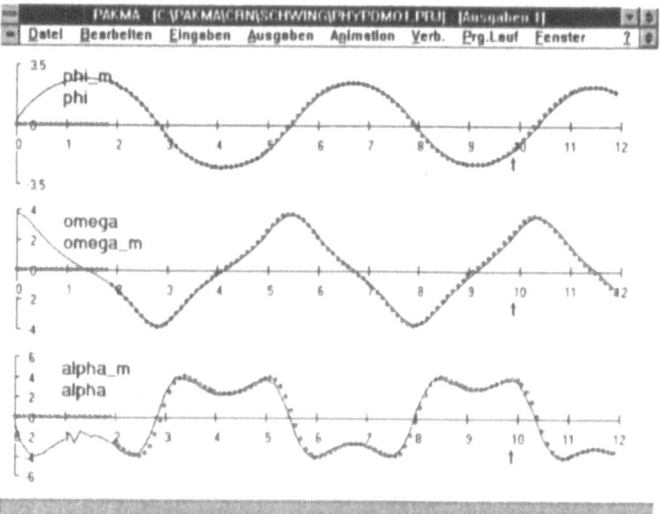

Figure 2 Angular displacement (phi), angular velocity (omega) and angular acceleration (alpha) as a function of time. By pressing a key (here with t ≈ 10s) the actual values of phi, omega, alpha are taken for a model calculation which is made parallel to the experiment (s. marks). This model-calculation corresponds to the real motion to such a degree that the graphs are almost identical.

4. Comparison to the prediction

In steps 1 and 2 the personal ideas of the student are set down and in step 3 the processed "reality" is documented. Now both descriptions have to be compared. When differences occur, especially the student's own reasoning has to be checked and existing ideas questioned. At this point discussion with the lab partner can lead to clarification. In some cases assistance must come from the supervisor of the experiment. Mistakes which are found and misconceptions should be explicitly named and the use of suitable concepts should be explained.

5. Possible analysis of other interrelations and quantities:

A number of other experimental quantities such as omega (as function of phi) or the kinetic energy, as function of time t (Figure 3) or of angle of rotation φ can be examined (Figure 4). If you want to get a deeper analysis of the experiment, you may let the students analyze more interrelations. In this case it is recommended to proceed again as in steps 1 to 4.

This whole procedure, from the prediction to the comparison to the actual results, is characterized by qualitative connections. For a more detailed analysis a quantitative procedure is definitely of interest. Analytic solution methods, however, have roots which are totally different from those of qualitative reasoning. Simple numeric procedures, on the other hand, as are introduced during late sec-

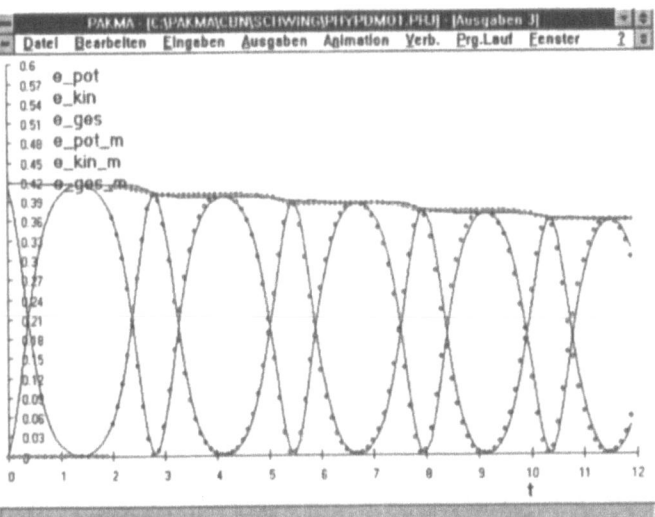

Figure 3 Kinetic energy, potential energy and total energy of the pendulum which performs the motion in Figure 2, shown as a function of time. Model-calculations on real experimental results are somewhat different in this case.

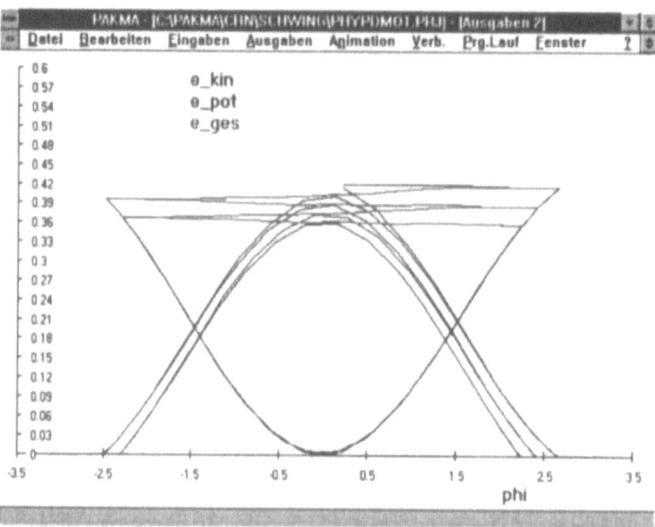

Figure 4   The quantities of Figure 3, here shown as a function of the angular displacement. The movement into the starting position (maximum displacement) is shown as well.

ondary physics education for the creation of models on the computer, provide a means to make qualitative statements more precise. Therefore it is desirable to follow up on the qualitative interpretation with creating simple quantitative models. Typical steps for such an extention of the experiment are described in 6–9 that follow.

6. Simple computation of a model: From qualitative to quantitative

What counts here is not a sophisticated method of iteration but a preliminary understandable sequence of computations. These should be filled into the blank lines of a given core program. In this program only the problem itself is to be formulated. When the variation of physical quantities is included during the runs of the program it shows how well the model describes reality qualitatively. In order to make a detailed comparison, model and experiment should appear on the screen at the same time.

7. Comparison model—experiment

Such comparisons are accomplished very easily and quickly, when model and experiment can be presented to the student simultaneously in real time.

8. Variation of the modelling

At this point the student might refine the simple model by, for example, including terms due to friction. Since models are in general idealisations of physical reality the students can, if interested, make additional more or less realistic assumptions for the modelling and test these.

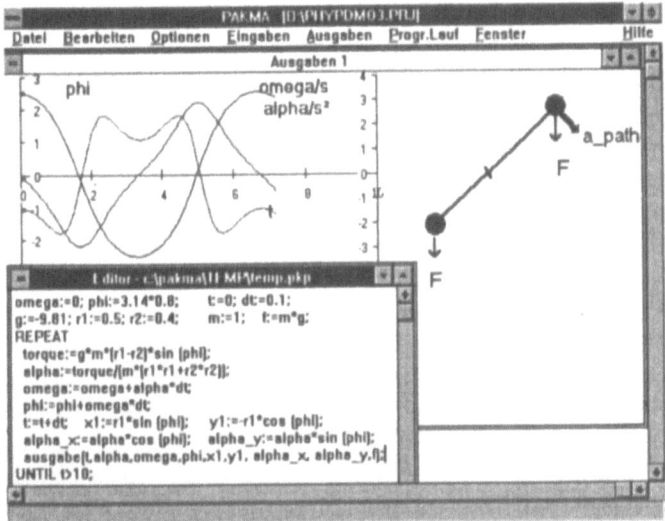

Figure 5 Example of a typical PACMA-Screen. The editor-window shows instructions for modelling the physical pendulum, the output-window results of the modelling: graphs and animation.

9. Comparison experiment—model with a given program (Exact method of iteration)

A simple method of iteration done by the students themselves, as proposed during step 6, will only be acceptable as a first approximation. At the end of the modelling phase the course of the real experiment and the model should be compared again in real time, this time with a ready-to-use program which uses a more exact method of iteration.

To work on a corresponding problem in another physics context is a suitable extension when this experiment is finished. This suggests one final step:

10. Prediction, experiment and modelling of a similar situation as a transfer problem

For example a one-dimensional anharmonic oscillation with other special force-functions (see D. Heuer [3]).

15.5 Suitable Software Is Required

For realization we need, amongst other things, suitable software. This is shown by the example of the gravitational pendulum. First of all, the software system we intend to use should be open, allowing the experiment to determine how the data will be recorded, processed and displayed.

---

[1] PACMA hardware and software is obtainable from Microsystems, Dr. H. Parsche, Neufahrner Str. 21, D-85748 Garching, Germany

Secondly, the system should be easy to manage. To achieve both of these goals, PACMA: Physics Active Measuring, Modelling, Analyzing, Animation[1] was developed at the education department of the Physics Institute, University in Würzburg. PACMA is a measuring, modelling and animation system. Because of this, the modelling can be computed simultanously to processing the real data. Therefore it is possible to visualize and compare the results of both experiment and model-calculations on screen at any time during the experimental proceeding. For that the user draws up a core program by means of an editor. This core program contains only instructions concerning the physics involved in the problem the user works on. All other allocations are done by means of menus and requesters.

In order to enable students having no computational knowledge to work with PACMA, drawing up a core program can be done easily even when having nearly no idea of syntax and structure of any computer language. For that a parser converts the text of the core program so that Turbo PASCAL is able to compile it. Figure 5 gives an example of a typical PACMA-screen as it might be used for experiments with the gravitational pendulum. The editor-window shows part of the instructions modelling the pendulum. In the output-window there are results of the modelling: graphs and animation. More examples of animations are given in D. Heuer [4].

PACMA is available as a PC version for Windows 3.1 and as a version for the Amiga. It is intended that the PC-version of PACMA for Windows 3.1 will be available as part of the CUPLE [5] project too.

15.6 A Challenge to Change Concepts

When we want to introduce such a conception for the introductory lab, we have to be aware that computers and software by themselves won't do the job. The willingness to handle situations and questions which are more open than they used to be is necessary as well. This is not only true for the students but also for us teachers. Maybe this will be seen as the biggest obstacle for what is proposed here. I have a different standpoint: I think this is a challenge. Among other things, learning means to change concepts and thought structures. Consequently the goal is not to exchange old knowledge for new knowledge but to change the existing knowledge with the help of the new insight that was gained. Prerequisite for this is an active interaction. D. Nachtigall [6] states: "Instead of demanding an excessive amount of stored facts, the connections between which are not realized by the students, we have to emphasize critical thinking and understanding" (translated by D.H.). It is too easy to assume that this is primarily the business of secondary schools and teachers. As D. Nachtigall says: "Everything that's been said so far demands, in my eyes, a reformation of teachers' education, first of all at the university" [6].

I think we have to face this challenge.

References

1. Larkin, Jill H., McDermott, John, Simon, Dorothea, and Simon, Herbert A.: "Expert and novice perfomance in solving physics problems," Science 208, 1135 ff. (1980)
2. Heuer, Dieter: Bewegungen "haargenau" messen mit Sonarmeter oder Laufrad, PdN-Ph. 41, Heft 4, S. 4 ff. (1992)
3. Heuer, Dieter: Anharmonische Schwingungen—qualitatives Argumentieren, angeregt durch Experimentieren und Modellieren, PDN-PH 42, Heft 6 (7/93)
4. Heuer, Dieter: Dynamic physical representation of real experiments. Chapter 13 of this book
5. Wilson, Jack M., and Redish, Edward F.: The comprehensive unified physics learning environment: Part I. Background and system operation. Computers in physics Vol. 6, No. 2, P. 202, Mar/Apr 92.
    Wilson, J.M., Redish, E.F.: Trends in computer software for use in physics education. Chapter 18 of this book.
6. Nachtigall, D.: Reform des Lehrens und Lernens im Fach Physik, Mitt HV 40 Heft 4/ 92, S. 269 ff.

16. Teaching Mechanics Through Interactive Video and a Microcomputer-Based Laboratory

A.L. Ellermeijer, B. Landheer, and P.P.M. Molenaar

University of Amsterdam

Abstract. The topic of mechanics has been taught in a secondary school in Amsterdam by a method which starts from students' real-life experiences and uses Interactive Video and a Microcomputer-Based Laboratory (IV/MBL). Student responses were found to be appreciative and their results in tests showed a considerably better grasp of basic mechanics concepts than for comparable students taught by more traditional methods.

16.1 Introduction

Over the past few years, the Physics Education Department of the University of Amsterdam has cooperated with the University of Nebraska-Lincoln (Professor R. Fuller) and Kansas State University (Professor D. Zollman) in a project to bring Interactive Video and a Microcomputer-Based Laboratory into the classroom. Videodiscs developed at the Universities of Nebraska and Kansas, together with hardware, software and instructional materials developed at the University of Amsterdam, have been used to teach the topic of mechanics at Fons Vitae Secondary School in Amsterdam.

In this paper, we report on the first results of this project. Section 16.2 discusses the present views in physics teaching in the Netherlands that provide a background for the project. In Section 16.3, we indicate reasons why Interactive Video and a Microcomputer-Based Laboratory might help in solving existing problems in the teaching of mechanics. The details of the project and its implementation in the classroom are given in Section 16.4; and the results in terms of student appreciation and performance and reactions by the teacher are described in Section 16.5. In Section 16.6, we state our conclusions and recommendations and our plans for the future.

16.2 Current Trends in Physics Teaching in the Netherlands

In recent years, a considerable amount of research has been carried out on students' preconceptions of basic mechanics concepts such as speed, acceleration, force, momentum, energy, etc., not only in the Netherlands but in other countries also [1,2,3]. Universally, certain misconceptions have been found to be very common: if an object is at rest, its acceleration must also be zero; a projectile has zero acceleration at the top of its path; an object moving at uniform speed requires a force to keep moving.

Mechanics teaching should of course correct these misconceptions, but traditional teaching has not been found to be particularly effective in doing so [4]. Many students who have completed mechanics courses have been found to have an insufficient grasp of the basic mechanics concepts mentioned above; they rely on memorisation and reproduction to score in tests; and they are quite unable to apply these concepts in unfamiliar situations.

Students also have great difficulty in reading and interpreting graphs; for example, they are inclined to interpret any graph shown to them as a distance vs. time graph. Since graphs are used extensively in mechanics, this provides a serious handicap to their understanding of the subject.

On top of all that, or perhaps as a result, many students have a strong dislike for the entire subject of physics: they consider it to be too abstract, too mathematical, too hard, dull, and quite irrelevant to their everyday lives. In fact, they do not see any relationship between the theoretical concepts they have heard about in mechanics class and their experiences in real life. Until they learn to describe real-life situations in terms of the correct mechanics concepts, their preschool misconceptions will keep coming back.

In recent years, a new approach has been developed for the teaching of physics in the Netherlands: Attempts are being made to teach the basic concepts in relation to situations familiar to students. These might be situations created in the classroom, through practical work or demonstrations (school contexts); but preferably they should be situations that students already knew before they ever came to the physics class (real-life contexts).

The new secondary school curriculum in the Netherlands (introduced in 1990) emphasises these contexts and encourages teachers to use them as starting points for their teaching. Within the subject of mechanics, this means that traffic has become an important topic, and texts have been written on this basis for different levels of teaching [5,6,7]. The motion of a bicycle or car, cars overtaking, the effects of rolling resistance and air resistance, the need for bumpers and safety belts are now being used to introduce concepts such as speed, acceleration, force, momentum and energy. Initially, the focus of attention is on students having a real grasp of the concepts themselves and being able to apply them qualitatively; only later do numerical problems come in, requiring the use and manipulation of formulae.

16.3 The Need for Interactive Video and a Microcomputer-Based Laboratory (IV/MBL)

We consider practical work to be the cornerstone of good physics teaching. By doing experiments for themselves, students get first-hand knowledge of, and experience with, physics concepts; it gives meaning to these concepts. Successful experiments show physical phenomena to be consistent and reproducible. This motivates students; it builds up both their understanding and their confidence.

In the topic of mechanics, however, the number of experiments that can be carried out successfully and within a reasonable time span is limited. This is especially so for real-life situations. Recording the position of a runner as a function of time is possible, but time-consuming, and therefore tedious. It is very difficult to take adequate readings on the motion of a bicycle or a boat moving in water; rolling resistance and air or water resistance cause deviations from expected behaviour that reproduce badly and fluctuate unpredictably. Recording the motion of a ball, or some other object in free fall, can be done by stroboscopic photographing techniques, but this requires the use of a darkroom.

Air tracks and air tables operate with very little friction, which greatly improves the reproducibility of experiments; this makes them very useful for clarifying basic mechanics concepts. However, these are pieces of equipment that have been developed especially for teaching purposes; they have no place in the real-life context of school children. There is a danger that students will fail to see the relationship between these experiments and real-life situations. Also, to record the position of a glider on an air track by means of light gates is time consuming and gives only a few measured points, so that analysis of the measurements remains difficult and does not always produce clear results.

Interactive Video and Microcomputer-Based Laboratory experiments provide an additional teaching tool that avoids most or all of the above-mentioned problems. In Interactive Video, measurements are taken from video films shown on the screen of a computer; the film can be stopped to allow the student to take readings; and the readings can be recorded, graphed and analysed by the computer. By using films of real-life situations (e.g., sports events) students realise that physics does, after all, have some bearing on real life. They can also take measurements on films of dangerous situations that cannot be reproduced in the classroom in any other way, e.g., car collisions.

In computer experiments on motion, sensors can be used to record the position of the moving object twenty or more times per second, and these data can again be graphed and analysed by the computer. The often tedious work of taking a large number of readings, tabulating them and plotting them in a graph is therefore done by the computer, so that the student can concentrate on the mechanics concepts involved in the experiment.

The computer can also be used to predict the motion of an object starting from given values of the parameters involved, and moving in accordance with a given model, i.e., a set of equations of motion. Students can then be asked to investigate the effect of changing one or more of the parameters. They can also use the model to compare its theoretical predictions with their own experimental results.

IV/MBL experiments provide an open learning environment for students to develop their understanding of physics concepts. Once students have become familiar with the use of the computer, they decide for themselves what they are going to measure, how they should calibrate, and what readings they should take. They can stop at any time to reflect on what they are doing and why they are doing it; they draw and formulate their own conclusions. This means they are following their own learning route.

16.4 The Project Materials and Their Implementation

At Fons Vitae Secondary school in Amsterdam, the topic of mechanics is taught in form 4 (age group: 15–16 years). There are two streams: the VWO stream prepares students for going to universities, and the HAVO stream for going to other colleges. The project materials were tested in one class of each type. We consider these streams to be roughly comparable to, respectively, the "Honours" and "Regular" students in the USA; and this is how we will refer to them in what follows.

Both classes use *Motion* by J. Dekker (parts 1 and 2, ref. 7) as their textbook. As mentioned above, this textbook uses the context approach; traffic is a central theme. Part 1 presents concepts in a mainly qualitative way; formulae and quantitative work are done in part 2. The 4th forms of 1990–91 also used this textbook, but the use of IV/MBL was introduced for the first time for the 1991–92 year groups.

The teacher taking both classes (one of the authors, P.P.M.M.) kept track of the time spent on various classroom activities. Results are shown in Table 1.

Activity		Time Spent (%)
Theory	Teacher-centered discussion	27
	Students work by themselves	17
	Discussions on topics other than motion	6
Experiments without the use of a computer	Demonstrations	5
	Student experiments	4
IV/MBL Experiments	Interactive Video experiments Measuring by means of the computer Modelling on the computer	24
Evaluation	Tests	17

Table 1   Time spent on different classroom activities

The total amount of time allocated to the subject of mechanics was 90 periods (of 50 minutes each at a rate of 3 periods per week). These were taught in two sections: the first section, dealing with part 1 of *Motion,* was taught in September through December. The second section (part 2), in March through June.

To introduce the students to the use of the computer, instructional materials have been developed for each IV/MBL experiment. For the first few experiments, these instructions are rather detailed; but as students become increasingly familiar with the computer, the instructions become briefer, leaving much of the initiative to the student. Also, the first few experiments are of a qualitative kind, and later experiments become more quantitative, in accordance with the textbook used.

In carrying out the computer experiments, the students work in pairs. The maximum number of students in a class was 23. We had three IV sets and four sets of computer-linked experiments available; the remaining students did modelling experiments on computer. Each experiment was designed to take up one lesson period, with some additional homework time for analysis.

Generally, experiments were arranged in groups of three, one of each type (IV, MBL, modelling). All three experiments within one group would deal essentially with the same topic. Students would take turns in doing each of the three types of experiment, in three consecutive periods. A list of experiments offered is given in Table 2. For the IV experiments, we used videodiscs developed in the USA [8]

---

1. Interactive Video Experiments

a. **Runners**: a film of two runners on a football field. Illustrates uniform motion and uniformly accelerated motion.

b. **Dive**: a film of a swimmer doing a somersault dive. To determine the acceleration due to gravity and show the difference between the motion of the (approximate) centre of gravity and other parts of the body.

c. **Baseball**: a film of a batsman hitting the ball. To determine the impulse given to the ball, and to estimate the force acting on the ball during impact.

d. **Sprinter**: the start of a sprinter. To determine non-uniform acceleration.

e. **Bicycle**: a bicycle slowing down under the influence of rolling resistance (and air resistance). To determine the magnitudes of these resistive forces.

f. **Shot-put**: a film of an athlete putting a shot. To illustrate two-dimensional motion and determine the force exerted by the athlete and the resistive force.

g. **Trampoline**: a girl jumping on a trampoline. To show the relationships between position, speed and acceleration, and determine the forces acting.

---

2. Computer-linked Experiments (MBL)

a. **Air Track 1**: motion of an arbitrary object, and uniform motion of a glider on an air track. To record the motion, we use a "Kinegraph" (developed at the University of Amsterdam): an ultrasonic transmitter + receiver + associated software that measure distance by finding the time needed for the ultrasound to travel that distance; the transmitter is triggered by a pulse of infrared radiation.

b. **Boat**: motion of a toy boat through a container with water. To show that the boat reaches a terminal speed, and to investigate how this terminal speed depends on the net force acting.

c. **Air Track 2**: accelerated motion on an air track, using Kinegraph. To investigate how the acceleration depends on the net force acting.

d. **Air Track 3**: collision on an air track, using Kinegraph. To investigate the forces acting during the collision.

e. **Arbitrary Motion**: recorded using Kinegraph. To investigate the relationships between position, speed and acceleration.

f. **Cart**: motion of a cart when given a push. To determine the pushing force and various retarding forces.

Table 2   List of computer experiments (continued on next page)

3. Modelling Experiments

   a. **Runner**: To predict the (uniform) motion of a runner, and to produce graphs of position and speed vs. time.

   b. **Parachute**: To predict the downward motion of a parachutist, and investigate how the terminal speed depends on the mass, and on the parachute's diameter.

   c. **Bicycle**: To predict uniform and accelerated motion of bicycles, and to investigate the effects of rolling resistance and air resistance.

   d. **Baby in Car**: To predict the forces acting on a baby held by a passenger in a car that collides with a heavy truck.

   e. **Arbitrary Motion**: To predict the motion of an object under a variety of conditions, and to compare this with actual motions observed.

   f. **Cart**: To predict the motion of a cart under the influence of a pushing force and several resistive forces.

   g. **Soccer**: To predict the motion of a soccer ball when kicked by the goalkeeper, and to investigate how the distance reached depends on the angle of kick, the initial speed, wind and air resistance.

Table 2   List of computer experiments (continued from previous page)

16.5 Results

16.5.1 Student Response

The appreciation of the course by the students was tested by a questionnaire and through interviews. A questionnaire was handed out at the end of every computer experiment; students were asked to indicate their responses to that experiment by circling a number between 1 and 5. This means that, for an indifferent student, or one who has no opinion, the response would be 3.

The questions, and the average student response to each question (averaged over both classes), are shown in Table 3. (Higher values indicate positive responses.)

Question	Average student response
Did you like the experiment?	3.4
Did you have enough time?	3.2
Were the instructions sufficiently clear?	3.4
Did this help you improve your understanding of physics?	2.5
Did you learn something that will be useful outside the area of physics?	2.1

Table 3   Student response

These figures indicate that students generally liked doing experiments on the computer, and that they were satisfied with the organisation of the computer classes and the instructional materials provided; but they did not seem to think they were learning a great deal, and they did not seem to be aware that they were learning things that were applicable to everyday life. This is remarkable in the light of their test results (see Sections 16.5.3 and 16.6).

The questionnaire also contained an open question, inviting student comments. The comments given, and the interviews conducted with individual students, confirm the general impression stated above. Students appreciated the topics presented and enjoyed taking measurements on sports events. Initially, Interactive Video and computer-linked experiments were more popular than the modelling experiments; but appreciation for the latter increased during the course. The students also liked the opportunity to become familiar with the computer in general and considered this a useful side-effect.

16.5.2 Comments by the Teacher

Initially, the introduction of the computer presented some difficulties. Students had the impression that computers already know everything; so it should not be necessary to give detailed instructions to the computer. Calibration, in particular, was felt to be a tedious task that, surely, the computer could do by itself.

The taking of readings on the monitor screen (in the IV experiments) was not always done accurately, and the written instructional material was not read carefully enough. Students would have much preferred receiving all instructions through the monitor screen, instead of having to go from the screen to the text and back again all the time.

Girls were generally more conscientious in reading texts and taking measurements than boys. On the other hand, they were often shy to handle the computer, while the boys were generally confident in that and enjoyed it. Boys would often use a trial-and-error approach, pressing various keys and moving the mouse around without bothering to read their instructions first.

After a few rounds of computer experiments, however, handling the computer ceased to be a problem. The reading of instructions also gradually improved when students realised it actually saves time to know what you are supposed to be doing before you start doing it. For the brighter students, the rather detailed nature of some instructions did become irritating; once students have become familiar with the computer, the details of the procedure should be left to their own initiative as much as possible.

Initially, the students had problems with the modelling experiments; they found them rather abstract and had difficulty in visualising the processes that the graphs on the screen of their monitor were supposed to describe. Having experimental apparatus alongside the computer helped students overcome this difficulty.

Setting up the experiments that were linked to a computer would often take students quite a lot of time. The teacher, also, found that it took him quite a lot of time to prepare the computer room for use by these classes; and that he was very busy during the classes helping students with a great variety of problems. The instructions handed out to students were found to be very important: they should be clear and concise at the same time. This means that the writing of these instructions can be very time consuming.

16.5.3 Student Test Results

A set of seven tests has been designed to be used with the *Motion* textbooks [7]; these tests had also been taken by the Honours students in the previous year who used that same textbook, but without IV/MBL. The average results of both year groups are presented in Table 4.

Test	Results in '90/'91 without IV/MBL	Results in '91/'92 with IV/MBL	Difference
1	72	82	+10
2	72	78	+6
3	71	73	+2
4	54	58	+4
5	74	79	+5
6	74	82	+8
7	60	68	+8
average	68	74	+6
Results for mathematics, average	68	55	-13

Table 4   *Motion* test results (%)

The test results show that the class that used IV/MBL did a little better than the previous one in every test, in spite of the fact that they used the same textbook *(Motion)* and were taught by the same teacher. The amount of time allocated to *Motion* was 90 periods and this was in fact used by the '90/'91 class; but the '91/'92 class used only 81 periods. One could wonder if the '91/'92 class might, by chance, be generally better than the '90/'91 class; but the teacher felt that rather the contrary was the case, and this is supported by the class results for mathematics (also in Table 4) which were considerably worse on average for the '91/'92 group.

To check the validity of these results, we also set a Dutch translation of the "Force concept inventory test" [9] to our students, and to students at three other schools as well. Results (test scores, in %) are presented in Table 5; the USA results have been taken from ref. 9. The table distinguishes between Regular and Honours students (HAVO and VWO students, respectively, in the Netherlands) and between students taught by a "traditional method" or a context method *(Motion,* in the Netherlands only), with or without IV/MBL.

These test results indicate that students who have been exposed to IV/MBL, on average, do significantly better than those who have not. Using a context approach (as the *Motion* textbooks do) does make a considerable difference by itself, but further improvements can still be achieved by using IV/MBL.

Country	Level	Method	Number of students	Average score (%)
USA (Arizona)	Regular	Traditional	612	48
Netherlands	Regular	Traditional	16	45
USA (Wells)	Regular	with MBL	18	64
Netherlands	Regular	with IV/MBL	19	56
USA (Arizona)	Honours	Traditional	118	56
Netherlands	Honours	Traditional	63	53
USA (Wells & Swackhamer)	Honours	with MBL	93	70
Netherlands	Honours	*Motion* without IV/MBL	49	65
Netherlands	Honours	*Motion* with IV/MBL	20	74

Table 5 Force concept inventory test results (%)

16.6 Conclusions and Recommendations

IV/MBL helps to improve student understanding of basic mechanics concepts. Results establishing the validity of this statement in the USA [10,11] have now been confirmed for students in the Netherlands. This improvement is achieved without any overall loss in time; in fact, the total time spent was less when IV/MBL was used. In view of the rather small samples of students involved, the precise extent of the improvement remains uncertain; further research will be necessary to clarify this. The context approach, implied in the use of IV/MBL, is to be preferred to traditional methods of physics teaching.

Remarkably, the students themselves do not seem to be aware that they are learning more. We believe that this is related to the traditional image of physics as a subject with a heavy emphasis on the handling of formulae and the doing of sums. Students seem to feel that, unless they have done these things, they haven't really done any physics; and they do not seem able to assess the improvement in their own understanding of basic concepts.

In the context + IV/MBL method, the emphasis on formulae and calculations is considerably reduced. This aspect certainly does get attention, but in situations that students can visualise and relate to. Apparently, this makes the students less aware that they are, in fact, applying these formulae!

We should, however, remain aware of the limitations of IV/MBL. In our view, actual experiments are to be preferred if they can be done (i) with reproducible results, (ii) in a reasonable amount of time, and (iii) without danger to people involved. IV is to be preferred for real-life situations that would be very difficult or impossible to measure directly. Computer measurements are to be preferred when great numbers of measurements are needed that would be tedious to perform, or when the total time for the event to be recorded is too short to allow measurement by conventional means. Modelling experiments, too, can be done in these cases; and they can be done also in situations that would be dangerous in actual practice. Initially, students can be expected to have problems with the higher degree of abstraction of modelling experiments; but these can be overcome by linking them directly to an experiment that students can actually carry out.

At present, we are working on a further investigation into the development of mechanics concepts. The existing materials are being used in the new 4th forms, with the aim of obtaining more reliable results. At the same time, we are working to extend this approach to the topic of Vibrations and Waves, which is dealt with in the 5th form. On the technical side, the present programs are being incorporated into one single package, IP-Coach, which will work under DOS. This will make it easier to record IV measurements and subsequently process them. We are also hoping to master techniques (DVI) to allow students to take measurements on videotapes of themselves or their friends; this would further increase student motivation.

References

1. Genderen, D. van (1989): *Mechanica-onderwijs in beweging (Mechanics teaching in motion)*. WCC Utrecht (in Dutch, with summary in English).
2. Halloun, I.A., and Hestenes, D.P. (1985): Common sense concepts about motion. *American Journal of Physics* 53, 1056–1065.
3. Viennot, L. (1979): Spontaneous reasoning in elementary dynamics. *European Journal of Science Education* 1, 205–221.
4. Halloun, I.A., and Hestenes, D.P. (1985): The initial knowledge state of college physics students. *American Journal of Physics* 53, 1043–1055.
5. Dekker, J.A., and Ellermeijer, A.L. (1992): Relative motion for 15–17 year old students. *Proceedings of the GIREP '91 conference on Physics education,* 362–368. Nicholas Copernicus University Press, Torun´, Poland.
6. Genderen, D. van, Gravenberch, F.L., and Kortland, J. (1983): *Verkeer (Traffic),* thematic lesson material for class 4 HAVO; NIB Zeist.
7. Dekker, J.A. (1990): *Bewegingen 1 en 2 (Motion, parts 1 and 2)*. Physics Education Dept., University of Amsterdam .
8. R. Fuller, and D. Zollman (1992): *Physics of sports, Studies in motion, Energy transformations* Videodiscs.
9. Hestenes, D., Wells, M., and Swackhamer, G. (1992): Force concept inventory. *The Physics Teacher* 30, 141–157.
10. Thornton, R.K., and Sokoloff, D.R. (1990): Learning motion concepts using real-time microcomputer-based laboratory tools. *American Journal of Physics* 58, 858–867.
11. Laws, P.W. (1991): Calculus-based Physics without lectures. *Physics Today,* December, 1991.

17. Wanting to Know: Interactive Video Providing the Context for Microcomputer-Based Laboratories

Robert G. Fuller

University of Nebraska-Lincoln

Human beings seem to be naturally curious about the world around them. They seem to want to understand the world and to feel at home in the universe. Our most effective physics courses are those that engage this intrinsic interest of human beings in the world around them. Interactive video can be used to bring this world into our physics classrooms and provide a context for microcomputer-based laboratories.

The essential attributes of intrinsic motivation have been studied by Thomas Malone. He suggested that intrinsic motivation depends upon fantasy, curiosity, and challenge [1].

Some of us were motivated, in the beginning, to become physicists because we loved real-world, story problems. We liked the stories built around physics problems. These stories, or fantasies, can make learning environments more interesting and more fun. A good fantasy helps us apply old knowledge to new situations, and by provoking vivid images a good fantasy can help us remember. We are fortunate in physics. We have a wide variety of visual images from which to select that can be interesting. Interactive video can enable us to offer physics stories where different students can choose different fantasies, or story problems, that may include text, sound, animation, and graphics, as well as full motion video.

Some of us were intrigued by physics because the goal of understanding nature is challenging. It is a challenge which provides a goal whose attainment is not certain. It is a goal which became personally meaningful for us to achieve. Physics used the skills that we were being taught. Understanding nature is a good goal because it allows us to develop a sense of power; once we had accommodated some new knowledge, then we could do more. Interactive video, I believe, provides us with with some wonderful new approaches to this aspect of intrinsically motivating learning. Interactive video enables us, as physics teachers, to provide our students with experiences of variable difficulty and randomness, simulating nature. An appropriate challenge is captivating because it engages our self-esteem. Our students should have higher self-esteem at the end of our physics courses than at the beginning. Proper interactive video experiences can help us empower people and enhance their self-esteem.

Thirdly, some of us were captured by physics because of curiosity. A learning task needs to provide an optimal level of informational complexity for us, as learners, to be attracted to it. If a task is too simple we are not interested. It should be surprising and novel, but *not* completely incomprehensible. We are made curious by both sensory stimuli and cognitive stimuli. Interactive video with images and sound allows us to provide both of these. Interactive video needs to present just enough information to make our existing knowledge seem to be incomplete, inconsistent, or unparsimonious. Then our natural human curiosity helps to motivate us to learn more.

Too often the laboratory experiences we provide in our physics courses are those that only physicists can love. Interactive video lets us bring into our physics classrooms the kind of real-world events that can motivate our students to study and enjoy physics. We can allow them to consider everyday events from sports and car accidents to space travel and sunsets on the moon, using video images.

Using video images we can once again unlock the wanting-to-know spirit that remains in our students. We can set before them the wonder and variety of nature and challenge them to more mature patterns of reasoning.

For us to know physics as we presently do, we had to gradually change our patterns of reasoning and advance to another level of understanding. This is a lifelong process of change. It occurs when what we think we know about nature is not substantiated by our experiences, that somehow nature does not quite make sense. "Knowing" is rooted in our innate desire to want to understand ourselves and our environment. Hence, a primary task of interactive video in knowing physics is to facilitate these ongoing changes in our mental processes as related to concepts in physics. What we need to do with interactive video is not to try to make physics superficially easy, but to reveal its appropriate level of complexity. Thus the interactive video task is to provide a credible reality and a challenge to our existing mental processes, in short, to provoke us into an appropriate level of cognitive conflict and motivate us to continue the process of knowing.

The gradual change in our patterns of reasoning was an interest of Robert Karplus and he suggested that reasoning can be assisted by the use of a classroom strategy he called the Learning Cycle [2]. The Karplus Learning Cycle consists of three phases: exploration, invention, and application.

Exploration features open ended questions and action on concrete materials. During *Exploration* the students learn through their own more or less spontaneous reactions to a new situation. In this phase, they explore new materials or ideas with minimal guidance or expectation of specific achievements. Their patterns of reasoning may be inadequate to cope with the new data, and they may be encouraged to change their patterns of reasoning.

Invention offers unifying concepts based upon the exploration activity. During the *Invention* phase, a new concept is defined or a new principle invented to expand the students' knowledge, skills, or reasoning. This step should always follow *Exploration* and relate to the *Exploration* activities. It will thereby assist your students to change their reasoning patterns. You want to encourage individual students to "invent" part or all of a new idea for themselves, before you present it to the class.

Application uses the invented concepts in additional content areas. During the last phase of the learning cycle, *Application*, students find new uses for the concepts or skills they have invented earlier. The *Application* phase provides additional time and experiences for self-regulation to take place. It also gives you the opportunity to introduce the new concept repeatedly to help students whose conceptual reorganization proceeds more slowly than average, or who did not adequately relate your original explanation to their experiences. Individual conferences with these students to identify their difficulties are especially helpful.

In conclusion, then, interactive video can be used within the structure of the Learning Cycle to provide the motivational setting for microcomputer-based laboratories (MBL) that can be used to foster the the development of more advanced reasoning by our students. Let me close with a few examples of how this might be done. The examples are all video scenes from *Physics: Cinema Classics* [3], described in Table 1.

Exploration	Invention	Application
Trains and Automobiles [Chap, 27, Side C] These scenes show trains and automobiles whizzing past an observer and the sound track clearly illustrates the Doppler effect.	Doppler Effect—a discussion or hands-on activity can be used to introduce this concept.	Lab Experiments The analysis of sounds emitted from a moving object can be analyzed in the lab using MBL techniques. Analysis of the sounds from videodiscs can be done using digitizing audio software .
Light-Swimming Pool [Chap., 66, Side C] The view from an underwater camera is shown as the camera person walks away, underwater, from the edge of the swimming pool.	Ray Diagrams and Refraction—discussion and hands-on activity can be used to introduce refraction and total internal reflection.	Lab experiments using optical benches and MBL techniques can be used to extend the students work with refraction and the index of refraction of materials.
Billiard Balls [Chap. 63, Side E] Very slow motion scenes of the collision between two billiard balls is shown. The collision is seen to violate simple ideas of momentum conservation.	Conservation of Energy and Momentum—discussion and hands-on activity can be used to introduce the concepts of energy and momentum conservation. The equations of motion can be solved for the standard cases.	Motion sensors can be used with MBL techniques to analyze collisions. The slow motion scenes of the billiard ball collisions can be analyzed using digital video techniques.

Table 1   Example video scenes from *Physics: Cinema Classics*

Thus, we can see how the use of the video of real-world events can be used to provide an intrinsically motivating environment in which to ask students to conduct physics laboratory experiments using MBL techniques. These two technologies can be brought together in such a way as to encourage the development of more advanced reasoning patterns by our students, using the Karplus Learning Cycle as a classroom teaching strategy.

References

1. Malone, T., Toward a theory of intrinsically motivating instruction, *Cognitive Science* 4, 333 (1981).
2. Karplus, R., Science teaching and the development of reasoning, *J. Res. Sci Teach.* 14 (2), 169 (1977).
3. *Physics: cinema classics,* a set of 3 double sided videodiscs, AAPT, 1992, now available from Ztek Company, Louisville, KY.

Part IV

Hardware and
Software Systems

18. The Development of a Communication Protocol for School Science Laboratory Equipment

Angela E. McFarlane[1] and David S.C. Thompson[2]

[1] Homerton College
[2] Educational Consultant, N. Ireland

18.1 Introduction

This paper contains the conclusions of a series of seminars involving leading UK manufacturers of school science laboratory equipment and science curriculum developers (referred to below as the Educational Data Monitoring and Control Group, EDM&CG, a list of whose members is shown in Appendix 1). These seminars were initiated at Homerton College, Cambridge, England in 1988 and have continued with a frequency of two or three per year ever since. Since 1990 the seminars have been sponsored by the National Council for Educational Technology (NCET), a non-departmental government body. The objective of the seminars is to review the progress taking place in the development and use of microelectronic based data capture equipment for classroom science activities.

This paper focusses particularly on the issues related to the establishment of a communication protocol between science laboratory equipment and a data capture or display device, usually a microcomputer.

18.1.1 The Need for a Protocol

Since the introduction of microcomputers to UK schools, small numbers of teachers and pupils have been using them to capture and display data during scientific investigations. The value of this work has been recognised by its inclusion in the National Curriculum for Science, which is now a statutory requirement for pupils in maintained schools in England and Wales. All pupils must have experience of monitoring experiments using microelectronic based equipment as part of their science course, starting at 7 years of age. In addition, the recommendations for technology make a similar stipulation for older children. The result is that all primary, middle and secondary schools will have to possess microelectronic resources to monitor and control experiments and make adequate provision for children to have hands-on experience of these resources in secondary schools.

As demand for this type of resource has increased, equipment has evolved from simple buffer boxes and sensors, often built by the end user, with crude software to more sophisticated interfaces, often with intelligence. Each year, it seems, at least one more device, complete with a range of sensors, is released by manufacturers. Initially the interfacing hardware available to schools was extremely heterogeneous and largely incompatible with any alternative sensors, microcomputer or software. However, over the last three years some convergence has occurred in the design of new resources leading to increased interconnectivity.

Initially, no two data capture systems showed any compatibility in the way they presented their captured data to the host microcomputer. Most systems employed:

i)   differing communication ports, connectors, signal levels and form;

ii) differing use of the port lines to manage the communication;

iii) differing formats in which the data are presented to the host.

Although some progress has been made, details of which are given later, no two systems yet employ identical protocols. As a result, data capture software has to be individually tailored (or versioned) for each system: host microcomputer, interface. sensors, experimental application. The development of sophisticated educational software to support this hardware cannot be a commercial venture in itself due to the high costs involved and the nature of the market. Software maintenance and enhancement is often difficult (if not impossible) and expensive. As a result, the majority of software currently on offer does not exploit the newer, intuitive user interfaces found on 16-bit and 32-bit microcomputers.

In addition, the incompatibility of hardware has made upgrading of equipment extremely costly. Often it has been difficult for schools to improve existing systems gradually; rather it has been necessary to replace whole systems in order to improve the specification. In some cases products from the range of equipment from one manufacturer are not compatible, and even now most manufacturers feel obliged to create entire systems when new products are developed. This necessarily makes products expensive for schools, as investment in research and development by manufacturers has to be recouped from sales.

18.1.2 A Way Forward

In order to alleviate some of these difficulties, it would be desirable to establish a communication protocol for school science equipment which is based on the new digital technology. This would create an independent standard to which developers could work when creating future hardware, software and curriculum-support materials. The resulting ability to interconnect different systems would offer a platform for manufacturers to utilise software and curriculum-support materials produced independently, while allowing them to employ their expertise in the production of cost-effective, reliable, robust, state-of-the-art hardware backed by customer service and after-sales support. Manufacturers would be able to formulate a long term strategy for research and development, manufacturing and marketing in the environment of a more stable and measurable market place which will necessarily be expanded in the UK by the requirements of the National Curriculum. This will lead to a coherent development of available microelectronic data-capture equipment so that schools will be able to upgrade parts of their existing systems as and when appropriate. This development, based around an accepted protocol, can only improve the quality of resources available to schools.

In this light it has been suggested that a protocol for communication between school science laboratory equipment be considered as a matter of urgency.

18.1.3 Protocols, the Current Situation

There exist a number of protocols which specify how (sub-)systems may be interconnected. To date, no one manufacturer has developed a communication protocol which has resulted in the emergence of a *de facto* standard for data transport or communication. Other protocols such as IEEE-488, CCITT X.25, Intel bitBUS, etc., seem to have been discounted mainly because they are too expensive to implement for schools. These protocols were written to support sophisticated and complex system interconnection of industrial instruments, and are not optimised to take advantage of the more limited environments offered by school-based data-capture systems.

In the system required by schools, outlined in Figure 1, there are three levels of compatibility:

1) The physical level

2) The data transport level

3) The communication level

At a physical level, connection between a host microcomputer and an interface used to be limited to a connector type and electrical specification, offered by a port available on a BBC model B microcomputer, e.g., user port, parallel port, RS 422

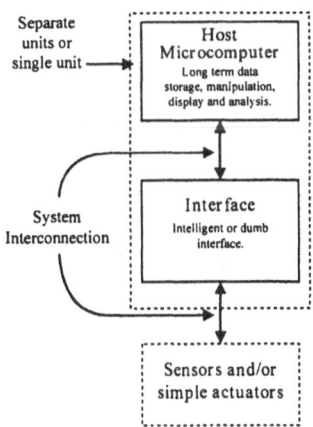

Figure 1 A model of a school data-capture system

port and 1MHz bus. This situation arose due to the popularity and success of the BBC Model B when UK school microcomputers were all 8 bit.

Initially it seemed that connection to the newer 16-bit and 32-bit systems was approached by considering how the connection to an 8-bit system may be appropriately modified, rather than by examining the facilities offered by the newer systems. This meant that moving to a 16- or 32-bit system required an internal card providing these ports, a serious disadvantage for schools. It meant added expense and restricting use of peripherals to the computer(s) with the card. Over the last two years new product development has converged on the use of the RS 232 serial port, a variant of which is found on the vast majority of computers as a standard port. This has resulted in improved compatibility between peripherals and computers. An interface or data logger purchased for one type of computer can be moved to a different computer, but a different lead (due to different pin assignments) and, of course, different software are required.

In existing systems, connection of the sensor/control line to an interface employs a proprietary connector, pin arrangement and electrical specification. At the transport and communication levels cost-effective and simple protocols have usually been developed independently by each manufacturer. As a result, proprietary sensors are specific to one; however, all major systems do support the use of 0-1V sensors via 4mm jack plugs. This means an extensive range of sensors produced by leading manufacturers can be used with all popular data loggers currently available for UK schools.

One standard which is emerging is the use of the Software Independent Data (SID) file format to store captured data. One intelligent data logger will even present data as an SID file to the host computer. The details of this standard are provided in Appendix 2. This file format provides a node through which data can be exchanged between data-capture and data-handling software applications. It is now used by 21 products in the UK and has extended the range of software applications which can be used to manipulate captured data.

18.2 Objectives of the Protocol

18.2.1 The Advantages of a Standard

According to ISO/IEC Guide 2, general terms and their definitions concerning standardisation and related activities, a standard is a

> Document, established by consensus and approved by a recognised body, that provides for common and repeated use, rules, guidelines or characteristics for activities or their results, aimed at the achievement of the optimum degree of order in a given context.

Standards are conventional rules established with the cooperation and consensus of parties concerned, aimed at specification, unification and simplification in all fields with a view to achieving improved efficiency. There are thus three main stages to establishing any standard:

i) Specification—to define characteristics and performance for achieving fitness for purpose.

ii) Unification—to enable interchangeability.

iii) Simplification—to achieve variety reduction, to produce more cheaply.

It is envisaged that a communication protocol would perform all these functions in the field of data capture in schools.

To succeed, any standard must be validated by consensus and represent all relevant interests. In trying to achieve this, the original initiative included all the major UK school science laboratory equipment manufacturers, and frequent consultation with interested parties in manufacturing and/or curriculum development has occurred throughout.

18.2.2 The Advantages of the Protocol

The protocol will achieve:

i) **Machine Independence**—The general purpose microcomputer system supporting the user interface may be one of a wide range of 16-bit and 32-bit microprocessor systems.

ii) **Manufacturer Independence**—The protocol standard should allow differing manufacturers to design and develop products which are fully interconnectable with the system, while allowing each company to concentrate on their own area of excellence.

iii) **Upwardly Transparent System Management**—The software must be able to interrogate its environment and present the available facilities to the user in an intuitive manner. The hardware must not prevent system enhancement at a later date.

iv) **A Framework for Progression**—It should be possible to adopt the protocol standard in a modular manner wherever possible, to facilitate product evolution.

18.3 A Technical Overview of a Data Capture System Specification for Schools

18.3.1 Terms of Reference of the Protocol Study

In drawing up terms of reference shown below the following assumptions have been made:

• New software designed for school data capture systems will probably employ an intuitive, graphical user interface. Elements of such a user interface include windows, pointers, icons and alternatives to keyboard input, such as a mouse, roller ball, touch-screen, touchpad, membrane keyboard, etc. At present such intuitive graphical interfaces seem to be moving to a common 'feel' independent of the microcomputer system on which they are implemented. Such an interface is only achievable on 16-bit or 32-bit microprocessor-based systems with at least 1 MByte of RAM. When considering the requirements of an educational data-capture system, the greatest single cost is the production and maintenance of sophisticated software. It is with this software element of the requirement that the lowest product margin is usually associated.

- The specification of physical, data transport and communication protocols for a school data-capture system will rationalise software production and maintenance costs. It will be possible to provide a modular structure to the software and a user interface which is independent of the microcomputer being used. (As a result the need for teacher in-service training will be reduced due to the familiarity of the data-capture software environment.) The application software and hardware may be made available as a series of compatible modules. Schools may purchase system modules as the need arises or the funding becomes available.

- The compatibility of software and hardware from differing manufacturers which meet the protocol specification will reduce equipment duplication and redundancy. The protocol will provide a homogeneous UK market place, develop teacher and pupil awareness of, and confidence in, data-capture equipment and increase the sustainable size of the market. The development of a homogeneous market place will result in a product range suitable for the world market (providing the earlier premises are accepted), increasing further the sustainable size of the total market.

There follows below an account of the three layers of the protocol which will have to be considered by the protocol study. These are:

i)   The Functional Model

ii) Two Physical Models

iii) The Transport Model

18.3.2 A Functional Model of a School Data-Capture System

The Host Microcomputer System. This is considered to be a general-purpose microcomputer system of 16-bit or 32-bit architecture. The system will provide high level graphics (either monochrome or colour) with an intuitive user interface, e.g., Digital Research GEM, Microsoft Windows, Acorn Archimedes Desktop, Apple Macintosh Desktop.

At the application level the host will:

i)   Interrogate the Intelligent Data Capture System (IDCS) and determine the signal information available, i.e., the number and type of sensors and the state of the control lines.

ii) Provide the user with an intuitive environment in which to specify the data-capture algorithm. This could include the concept of the virtual instrument.

iii) Either command the IDCS interactively while a real-time data experiment is conducted, or download a program to be executed by the IDCS either locally or remotely.

iv) Interrupt and control an IDCS.

v) Provide a platform for long-term data storage, display, analysis, transmission, etc.

The Intelligent Data-Capture System (IDCS). It is most likely that the IDCS will be an 8-bit or 16-bit microprocessor-based system, if independent from the host microcomputer. It may be one of a number of satellite data-capture sub-systems operating under the control of the host microcomputer system. The connection of multiple IDCS units to the host will allow two modes of operation:

i) distributed capture positions as part of one investigation, e.g., monitoring temperature in several rooms

ii) overseeing several individual investigation sites simultaneously, e.g., one host microcomputer to support a class of experiments.

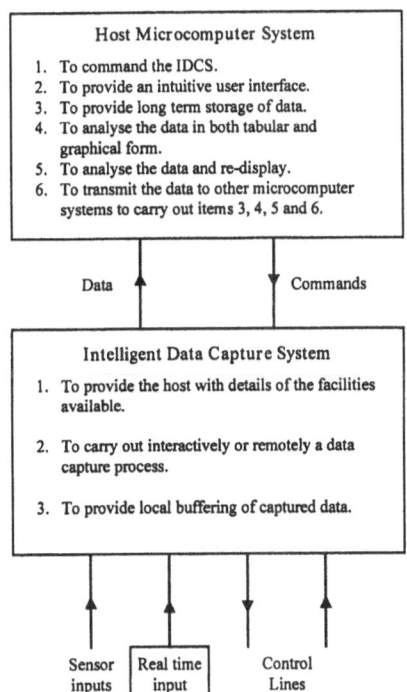

Figure 2 A functional model of a school data-capture system

The IDCS will interrogate the sensor inputs and the state of the control lines. It will make the necessary information concerning sensor type and control-line status available to the host. The IDCS should handle the signal conditioning of the sensor signal input to it intelligently, so as to maximise the utilisation of the available ports. For binary input sensors (threshold detectors) a digital signal path should be provided. For analogue (level) sensors the signal should be so conditioned as to maximise the dynamic range of the analogue to digital converter employed and provide suitable low-pass filtering to reduce the effect of noise.

It is possible that the host microcomputer will act as its own IDCS. This would enable the more exotic data-capture sub-systems to connect directly to the microcomputer bus, increasing the data bandwidth both in word size and word throughput (e.g., capturing voice prints) or allow ports available on the host to be used for simple introductory data-capture experiments (e.g., a simple temperature sensor, a pressure switch, etc.). The host would present this integrated IDCS to the host application as if it were an external IDCS, i.e., the host application will see a homogeneous environment.

18.3.3 Physical Models of a School Data-Capture System

School Data-Capture System Model. For the model shown in Figure 3 it is possible that A and B will form one sub-system, or that B and C will form one system or, in the extreme case, A, B and C will form one complete system.

Host Microcomputer to IDCS Connection. The host microcomputer system will provide the application layer of the school data-capture system model. The applica-

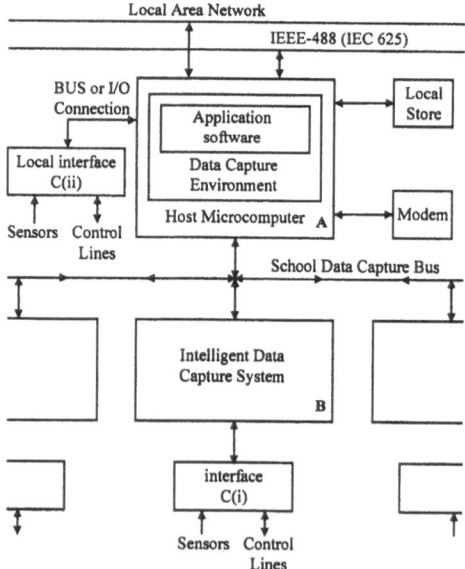

Figure 3 A physical model of a school data-capture system

tion software will run on this application layer, i.e., the application will run in a data-capture environment in the host microcomputer. This application layer must be common to all host microcomputer systems.

The host microcomputer will communicate with data-capture sub-systems in two ways:

i)   directly as part of its systems architecture, either via I/O ports or directly on the microcomputer system bus

ii) from an intelligent data capture system which acts as a slave sub-system connected via a communication bus (other than the I/O ports and microprocessor bus).

The application layer must provide a transparent interface to the application software such that all sensors and control lines appear as part of one homogeneous environment.

In effect, for locally connected sub-systems the host microcomputer architecture behaves in a dual role: to support the application layer and to act as an intelligent data capture system for the locally connected interface.

Where instruments are to be connected to the data-capture system these may be connected to the data-capture protocol converter, to the school data-capture bus or directly to the host microcomputer via an I/O port. Again, these instruments will appear to the application layer in a transparent manner independent of other methods of connection.

The external bus required for communication between the host and the IDCS will be of a master/slave type with the host as the master. The bus will have to be a short-range multidrop type which may be extended to include up to 15 separate IDCS. The host system will be the sixteenth, internal, IDCS. This number will support the maximum likely requirement of a whole class.

At a higher level it is possible that the host will be connected to a local long-term data store such as tape, floppy disc or hard disc storage. More remote data storage may be available via a local area network (LAN) or modem.

For more advanced systems the host may be equipped with an IEEE-488 or the proposed "Field Bus Standard." This would enable the school data-capture system to behave as an instrument on an IEEE-488 system or as a sub-system gateway into a larger field bus system. It is not suggested that this would be an early consideration, but care must be taken whenever possible to allow such a development to take place in the future. There are clearly possibilities of extending the use of the protocol for schools into the industrial sector via this route.

The Intelligent Data Capture System (IDCS) to Sensor/Control Line Connection. The IDCS will control a local interface to the sensor inputs and the control lines. The function of the interface is to allow a standard physical connection to be employed between the IDCS, sensors and control lines. The physical connection will be defined by both connector type(s) and electrical specification. The interface will also provide the local signal conditioning (when this is available) as required by each sensor input, input protection and electrical isolation of a sensor where required, e.g., for sensor inputs attached to a pupil.

The connection between the interface and the IDCS will not be defined other than to specify that it must provide the system with the communication ability required for a fully implemented system.

The control lines are simple digital I/O lines employed to start and stop control subsystems. The input (acknowledge lines) are to allow the control system to indicate that it has responded to the last command. High level control functions are not included in the specification of the data capture system.

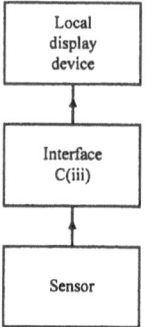

Figure 4 A physical model of a direct measuring instrument

Direct Measurement Model. A model which will allow sensors to be used as direct measuring instruments beside or in place of their classical counterparts is shown in Figure 4. For this directly measuring instrument model the interface may form part of the local display device.

18.3.4 The Transport Model of the School Data-Capture System

Communication Channels. With reference to the physical model it can be seen that two communication channels exist;

i)   The bus employed between the host microcomputer and the IDCS.

ii) The connection between the sensors/control lines and the IDCS via the interface.

The Host–IDCS Communication (The School Data Capture Bus). If compared to the International Standards Organisation (ISO) Open System Interconnect (OSI) seven layer model shown in Figure 5, the school system uses only three layers, the others being null layers.

7. Application	7. Application
6. Presentation	6. Null
5. Session	5. Null
4. Transport	4. Null
3. Network	3. Null
2. Data Link	2. Data Link
1. Physical	1. Physical
ISO OSI 7 layer model	Schools Data Capture Bus model

Figure 5 Layered protocol models

1) The Physical Layer—defines the electrical and functional interchange which establishes, maintains and disconnects the physical link between two ends. Interface standards include RS 232-C, RS 422, RS 485.

2) The Data Link Layer—contains the functions which transfer data reliably over a single communication link and includes control functions between the end nodes. Data-link protocols include: IBM bisync (character oriented), DEC DDCMP (byte-count oriented) and ISO HDLC and IBM SDLC (bit-oriented) protocols.

7) The Application Layer—provides the application software with a machine-independent environment in which to communicate with the system, in this case the data-capture system. Examples of such environments are: Microsoft Windows, Acorn Archimedes Desktop, Apple Macintosh Desktop and Digital Research GEM.

The definition of the physical and data link layers of the school data-capture bus must in no way define or limit the data which is to be transported by this communication bus. The physical and data-link layers provide a vehicle for reliable, transparent communication of data of any type.

In this way the school data-capture bus will not limit possible developments or future enhancements of the data-capture system.

The IDCS–Sensor Communication. This will require definition of a physical connection system or possibly a list of allowed connection systems capable of communicating the same information.

According to the physical connection system chosen, the communication of the sensor type information may be either serial or parallel. It is assumed that the sensor signal information will be in the form of an analogue potential difference.

The Communication Protocol. The communication protocol should define the format and interpretation of the commands and data carried by the transport system.

The areas covered by the communication protocol should be:

1. Sensor type (Sensor to IDCS)

2. IDCS command language (Host to IDCS)

3. Format of data received from the IDCS (IDCS to host)

1) Sensor type—This will require the drawing up of an agreed list of school-oriented sensor systems with associated interpretation or calibration charts. The list of sensor types must include at least one default-type direct voltage input. A unique sensor classification will be assigned to each sensor type in the list.

2) IDCS Command Language—This will require the specification of the commands to be carried out by the typical IDCS in a set of control language primitives.

3) Data Format—To ensure that the data capture system is transparent to the application software it is essential that data is transferred in such a manner as to allow it to be completely interpreted irrespective of the means of capture—real-time data, pseudo-real-time (buffered fast capture) or remote and previously captured data.

18.3.5 Work in Progress

As mentioned above, the nature of the protocol is such that it can be developed and implemented in a modular fashion. The establishment of the SID file format was the first of these, and the convergence on the use of serial communication via the RS 232 port, the support of 0-1 V sensors, and self-identifying sensors have all contributed to improved interconnectivity. Certainly the current state of development provides a more fertile ground for improving compatibility than was available four years ago.

The area currently being addressed is that of communication between software in the host computer and the data-capture system. An outline specification has been developed and a project based at Homerton College has developed Windows software which communicates with a number of devices via dynamic link libraries. In developing this software a common interface between application software and data-capture system drivers has been prototyped.

18.4 In Conclusion

The case for a communication protocol for school science data-capture systems is clear: improved product compatibility will lead to reduced development and maintenance costs for manufacturers and increased quality of resource provision for schools. This means enhanced experience of practical science for children, which is, after all, the whole point of all this technogobbledegook!!

Some progress has been made in the areas of connection to the computer and interchange of captured data. At the moment there are still several tensions which hold back progress; the production of a protocol is expensive, and ultimately potential profits from such a specialised product area are perceived as small. Manufacturers with sophisticated data loggers already on sale are not in a position to make radical changes to existing products to accommodate new protocols. However, the maintenance of existing software, as new sensors and new features are added to current products, is proving a real headache. On balance, the route of producing utilities such as dynamic-link libraries (DLLs), separate from the costly main application, to handle sensor identification, calibration, communication with the data logger, etc., is promising and can easily be updated and replaced. These utilities provide the solution. Developers of sophisticated software need only include one routine to 'talk to' the DLLs to be able to access the captured data. This must lead to lower development and maintenance costs all around, which is good for everyone.

References

Atkinson, J.K.: Addressable transducer protocols, Proc. *CAPTEURS '84*, Paris, (Paris: CIAME) pp. 307–16, 1984

Atkinson, J.K.: Communication protocols in instrumentation, *J. Phys. E: Sci. Instrum.* 20, 1987

Atkinson, J.K.: Transducers—principles and practice, *NEMEC*, 1989

Biodata Ltd: *Microlink data acquisition catalogue*, 1989

Black, U.: *Computer networks—protocols, standards, and interfaces*, Prentice-Hall International, 1987

Cater, C.: Field bus. In: *Measurement + control*, Vol. 21, 1988

Freer, J.: *Computer communications and networks*. Pitman, 1988

Halsall: *Introduction to data communications and computer networks*, Addison-Wesley, 1986

Hopper, Temple & Williamson: Local area network design, Addison-Wesley, 1986

IEC 625 parts 1 & 2, *An interface system for programmable measuring instruments (byte serial, bit parallel)*

INTEL, *8-Bit embedded controller handbook*, 1989

INTEL, *BITBUS interconnect serial control bus specification*, Serial No. 280645-001

INTEL, *Distributed control modules data book*, Serial No. 230973-004

INTEL, *Embedded control applications handbook*, 1989,

INTEL, *Microprocessor and peripheral handbook*, v.l & 2, 1989

Maine, A.C.: *Interface standards for computers*, an IEEE monograph, 1986

Measurement Ltd: *Interfacing catalogue*, 1989

National Instruments: *IEEE-488 control, data acquisition and analysis for your computer*, Spring/Summer edition, 1989

Penfold, R.A.: *Midi projects*, Bernard Babani Publishing Ltd, 1986

Philips, (Mullard), I^2C-bus compatible ICs, *Philips Components technical handbook*, book 4, parts 12a &12b, Integrated circuits, 1989

Purser, M.: *Data communications for programmers*, Addison-Wesley, 1986

Rapid System Inc.: *Rapid systems PC instrument catalog*. Datalab Sales and Service, 1989

Texas Instruments: *Linear and interface circuits—product applications*, Vol. 1, 2 & 3, 1986

Appendix 1

Membership of the Educational Data Monitoring and Control Group

Mrs. Mary Webb	Wheathampstead Education Centre
Mr. Laurence Rogers	University of Leicester
Mr. Reg Jones	E & L Instruments Ltd
Mr. Stephen Allen	Educational Electronics
Mr. David Palmer	DCP Microdevelopments Ltd
Mr. David Bell	Acorn Computers Ltd
Mr. Richard Phillips	Shell Centre for Mathematics Education
Mr. Tony Kiddle	Economatics (Education) Ltd
Dr. John Crellin	Philip Harris Manufacturing
Mr. David Duff	Unilab Ltd
Mr. Terry Richer	Irwin-Desman Ltd
Mr. Colin Watkins	Research Machines Ltd
Mr. Fred Daly	Technical Director, NCET
Dr. Angela McFarlane	Homerton College
Mr. David Squires	King's College, London
Mr. Stephen Cousins	SCC Research
Mr. David Thompson	Educational Consultant
Mr. Keith Hemsley	NCET
Mr. Richard Orton	CLEAPSS
Mr. A.I. Kicks	TICST
Mr. Peter Claxton	LEGO Dacta
Mr. Dominic Savage	British Educational Equipment Association
Mr. Peter Watson	WPA
Mr. David Conway	Unilab Ltd
Mr. David Evans	NCET
Mr. Behrooz Chini	The Concept Keyboard Company
Mr. Merlyn Kline	Minerva Software
Mr. Nick Swift	Economatics (Education) Ltd
Mr. John Wardle	NCET
Dr. Colin Rouse	Felingwm Systems & Software
Mr. David Headey	Computer Education Unit, Oxford

Appendix 2
The Software-Independent Data Format

Background

The use of data logging equipment to capture data is now well established in school science, with a prominent place in the UK National Curriculum. There is a growing realisation that captured data can be further investigated with the aid of the powerful data-handling software now widely available in schools. Traditionally, data-capture software developed for schools did not support data export, thus the data could only be explored within the original data-capture program. When developers, within the forum provided by the Educational Data Monitoring and Control Group (full membership is shown in Appendix 1), began to discuss resources to facilitate the transfer of data between software packages, the issue of file format for data was closely examined. CSV (comma separated values) is used widely in data-handling software but was perceived to present two obvious difficulties:

i)   It is an extremely loose specification with many fundamental variations. (In the last year the File Formats Working Group of Acorn Computers Ltd, Cambridge, England, has attempted to tighten the definition of CSV for Acorn RISC OS computers, but on other platforms the definition remains unclear).

ii)   CSV files only contain data, with little supporting information about the origin or type of data. The most that can be offered is field names. Thus it is difficult for software to handle the data intelligently, and data files stored on disk contain no easily read information to indicate what the data refer to, a serious issue when managing data storage, especially in schools.

In order to solve both these difficulties, across all computer hardware platforms, the Software-Independent Data (SID) Format was specified by the Educational Data Monitoring and Control Group (EDM&CG) and published in the first National Council for Educational Technology Technical Bulletin—*resources for data monitoring and control for science education*—in 1991. The SID file format offers a common clearing house for data moving between different software applications. It has been adopted by 21 products at the time of writing, some of which are entirely new, some of which have been revised to include SID support. The range includes both data-capture and data-handling packages. These packages may create and/or read SID files. Some use SID as their internal file format, others use simple utilities which change the proprietary file format to or from SID. In this way any developer creating a software application to capture or handle data can ensure that data files can be imported and exported between the new and existing packages.

The Status of SID. Prior to publication in 1991 a description of the agreed format was circulated for comment to group members and other interested parties. That consensus was published as a proposed specification by NCET in a discussion paper,

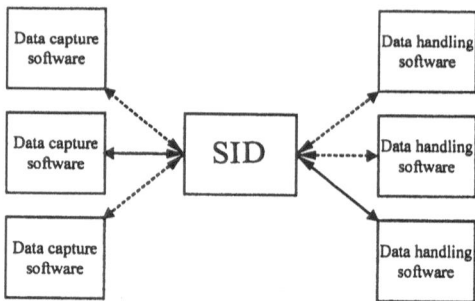

Figure 6   The role of the SID format in data import and export

entitled *Software-Independent Data Format—a file format for school science which facilitates data transfer between data capture and data handling software*. A small number of minor, but significant, changes were made subsequently in response to comment from a wider audience. The final specification was published in the 1991 NCET Technical Bulletin—*resources for data monitoring and controller science education. Since that point the SID specification remains unchanged and should be regarded as stable*. That full specification follows here.

The Software-Independent Data Format

Overview. The file format is known as the Software-Independent Data format, SID. The MS-DOS file extension is .SID.

The file type is registered with Acorn Computers Limited, Cambridge, UK, as &C7D, name SID. The RISC OS icon for a SID file is a square with the letters horizontally across it.

An SID file is an ASCII (text) file which consists of a header followed by a data file. The data format is based on values separated by commas, with the addition of a header to allow for transfer of additional information.

The header is compulsory for an SID file. It consists of a list of 'commands', each command starting with two percent symbols to identify it. The minimum requirement is two commands, the identifier which must be the first command, and the data size which will be the second command. All commands used must be listed, in lines, preferably in the order given below, before the data block. Each command occupies one line terminated by a carriage return and line feed. Software reading the file should ignore any command it does not recognise and have sensible defaults such that all commands, other than the identifier and data size, can be considered optional. It is essential that software does not crash if an SID file header contains more information than can be used by the importing package. Everything following the header is data.

SID is totally case independent, i.e., users should not assume case dependence at any point.

When writing, a carriage return (CR) followed by a line feed (LF) must be used to terminate all lines. The CR and LF therefore acts as command line separator in the header and record separator in the data block.

Parameters will be separated with a comma. It is also possible that some data fields may be blank, thus will be expressed as two commas without an item of data between them.

NB: The minimum specification for an SID file is a two-command header followed by data as values separated by commas. To be SID compatible, any software reading an SID file should accept such a file without hanging, and treat any other commands as optional, using them if present but having sensible defaults where they are absent.

Use of ASCII Characters. An SID file contains only ASCII characters from the decimal set 10 (LF), 13 (CR), and 32 to 126 inclusive. NB: This does not include ASCII 9 (TAB) or ASCII 127 (delete or backspace). It should be noted that the following ASCII characters have special significance:

- A comma, ASCII 44, can only be used to separate fields.

- White space, ASCII 32, is transparent and used only for clarification. See also 'Expressing numbers'.

- %% as the first two non-white space characters of any line denote a command

Note that the symbols < and > are not part of the command, they are shown in this document to clarify layout.

Other non-alphanumeric characters, e.g., single or double quotes, are textual characters with no special meaning and can only be used in strings.

Writing Data. There are only two data types, number and string. It is assumed that data is numeric unless otherwise stated, hence the provision of the special field unit string in the SID specification. Strings must be identified using the special field unit string. Thus data files including data as strings must include in the header the line:

    %%fieldunit, fieldnumber, string CR LF

This allows data filters to screen out string data in packages which cannot handle them.

Data values are separated by commas. Some data fields may be blank, thus will be expressed as two commas without an item of data between them.

Each record occupies a new line and lines are terminated using a carriage return (CR) followed by a line feed (LF).

It is normal practice for a field in a record, in any data file format, to be of one type (or a union of types). An SID file can also, in line with normal practice, only contain one record structure; e.g., the data block

    6,car,7,12 CR LF
    7,van,5,8 CR LF   Valid Data
    5,moped,8,12 CR LF

where field 2 is declared as a string in the file header, is within the SID format, as it has one consistent record structure. But the data block

```
car,van,moped CR LF
6,7,12 CR LF
7,5,8 CR LF                    Invalid Data
5,8, 12 CR LF
```

is not within the SID format as the record structure is inconsistent.

Expressing Numbers. Numbers are expressed as decimal values. The characters +, -, ., 0, 1, 2, 3, 4, 5, 6, 7, 8, 9 are accepted. The use of the + is optional, numbers are assumed positive unless otherwise stated.

Exponentials are not supported as they cannot be expressed using the above terms.

Non-integer numbers must be expressed using a decimal point where decimal fractions are present. Integers may make optional use of the decimal point, e.g., 1.0, 1.00, 1. and 1 should all recognised as identical.

Number fields may be blank where no value is recorded.

No limit is declared for the length of a numeric data value in the SID specification. Users are advised that when writing or reading any numeric value or string it is impractical to allow more than 255 characters.

The Structure of an SID File

Part 1: Standard Commands for Inclusion in the Header. Note that the symbols < and > are not part of the command, they are shown in this document to clarify layout.

Commands used should be listed in the order shown.

%%identifier, <file type>

This *must* be the first command, used to identify the file/header. The file type string is SID. *NB: The whole file, including file type, is case independent, i.e., SID, sid and Sid should all be accepted.*

%%datasize, <number of records>,<number of fields>

This is the other compulsory command. *NB: It is recommended that commands be listed in the order shown, making this the second command. The data file may not contain any records but have a field structure; therefore datasize values of 0,0 or 0,n (where n represents any positive integer greater than 0) may occur.*

%%title, <title>

This title will generally appear at the top of any display or window. The maximum length is 30 characters.

%%filedescription, <description>

The description is that of the data contained in this file and is not formatted in any way. Some software may need to truncate this to 200 characters, so this is the maximum length.

%%fieldname, <field number>, <field name>

Where the field number is an integer from 1 upwards, there will be no theoretical limit to the number of fields (or channels in data-logging terms). The field name is usually used to describe the sensor or the parameter measured. It should be limited to a maximum of 15 characters.

%%fieldunits, <field number>, <field units>

Field units are a string; software reading this string will normally display it with the field name. Software writing this string will usually use standard units of measurement. (See *Special field units* below.) However, they will have to be exactly specified by the developer if software is to recognise them and allow conversion to other suitable units. Non-ASCII symbols, such as that normally used for degrees in temperature measurement, will have to be avoided and written in full. This string has a maximum of 15 characters.

%%fielddescription, <field number>, <field description>

The description will be limited to 200 characters.

%%starttime, <HHMMSS>

The start time of the data capture is expressed as a six digit number HH is for hours, MM for minutes and SS for seconds, e.g., 145707 for 2.57 and 7 seconds pm.

%%stoptime, <HHMMSS>

The stop time of the data capture is expressed as a six digit number. HH is for hours, MM for minutes and SS for seconds, e.g. 025807 for 2.58 and 7 seconds am.

%%startdate, <YYMMDD>

The start date of the data capture is expressed as a six-digit number. YY is for the year, MM the month and DD the day, e.g., 921204 for 4th December 1992.

%%stopdate, <YYMMDD>

The stop date of the data capture is expressed as a six-digit number. YY is for the year, MM the month and DD the day, e.g., 031206 for 6th December 2003 (it is assumed no data will predate 1950).

%%interval, <seconds>

The interval between readings may be specified. The presence of an interval command represents a compressed field of time data. An interval of 0 may be used to signify time-independent data. If channels are sampled at different rates then the minimum interval should be listed, a record sampled at a larger interval may therefore not contain data for each field. In such cases an unused field will be blank, i.e., will appear as two commas without an item between them. Intervals of less than one second will be expressed as a decimal, e.g., 10 milliseconds = 0.01.

%%maxmin, <field number>, <maximum value>, <minimum value>

Maxmin refers to the maximum and minimum useful values in a data set. NB: The actual data may fall within this range or outside it. E.g., a thermistor may give readings in response to a temperature range of -20 to 150 degrees Celsius; however, the readings may only be reliable in the range 0–100. In this case the maxmin command might read:

%%maxmin,l,+100.0,0 CR LF

However, the actual data values read might fall in the range 25–100. The maxmin command is useful for filtering out readings outside the reliable range and for setting the limits of a graph axis, for example.

%%comment, <string>

Comments may be added as required.

Special Field Units. Developers are encouraged to use standardised field units, enabling software to recognise appropriate ones. These must be expressed using standard ASCII characters. Where possible these units will be given as normal SI units, but attention is drawn to the fact that the SID format simply specifies a string. Defining standard units clearly lies beyond the scope of this document but developers' attention is drawn to the Association for Science Document, SI Units, Signs, Symbols and Abbreviations for guidance in the use of appropriate units.

A few examples of what might be found in the field unit string are given below:

Parameter	SI unit	Alternative string
Time	s	second(s)
Length	m	metre(s)
Frequency	Hz	hertz

Some special cases of non-SI units have been identified; these are listed below.

Specials: time   Indicates that the field contains data in the form HHMMSS

            date   Indicates that the field contains data in the form YYMMDD

            mark   Indicates that the field contains only data marks of value 0 or 1 (1=Marker) ASCII 48 (a space or 'no mark') or 49 (a mark)

            string   Indicates that the field is data which has no numeric interpretation

Part 2: The Data which Follows the Header.

Record 1 field 1, Record 1 field 2, ..., Record 1 field N CR LF
Record 2 field 1, Record 2 field 2, ..., Record 2 field N CR LF

Numeric values will be stored at the maximum available resolution. Data handling software reading these data may limit the resolution used.

New Optional Commands. NB: The minimum specification for an SID file is a two-command header followed by data as values separated by commas. To be SID compatible, any software reading an SID file should accept such a file without hanging, and treat any other commands as optional, using them if present but having sensible defaults where they are absent.

Proprietary command prefixes. A new development in the use of the SID file format is the use of proprietary command prefixes. This has arisen through the extension of the use of SID for proprietary data files, not merely as a node for the exchange of data files. This extended use is extremely welcome, but to avoid the possibility of two new commands being introduced with the same name but differing meaning, proprietary command prefixes have been established and must be used before any new command.

Proprietary command prefixes should be a minimum of two characters, preceded by %% and ended with an underscore, e.g., the new LogIT command 'sensor' would be written thus:

%%LogIT_sensor,1,14 CR LF

and the new Educational Electronics command 'sensorname' would be shown thus:

%%EE_sensorname,Light CR LF

ALL proprietary command prefixes MUST be registered with the Convenor of the EDM&CG. Those registered so far are;

Prefix: %%DH_
Owner: Data Harvest Group Ltd
Contact: Mr. S.D. Allen at Educational Electronics
Use: Future Data Harvest products

```
Prefix:     %%NCET_
Owner:      National Council for Educational Technology
Contact:    Keith Hemsley, NCET
Use:        Future software

Prefix:     %%EE_
Owner:      Educational Electronics
Contact:    Mr. S.D. Allen at Educational Electronics
Use:        Future Educational Electronics products

Prefix:     %%PHE_
Owner:      Philip Harris Education
Contact:    John Crellin at Philip Harris
Use:        DL plus and future products

Prefix:     %%SCC_
Owner:      SCC Research
Contact:    Steve Cousins at SCC Research
Use:        SCC Products

Prefix:     %%LogIT_
Owner:      LogIT Project
Contact:    Steve Cousins at SCC Research
Use:        LogIT specific features

Prefix:     %%GBX_
Owner:      Minerva Software
Contact:    Merlyn Kline at Minerva
Use:        Graph Box Professional

Prefix:     %%HC_
Owner.      Homerton College IT Unit
Contact:    Angela McFarlane at Homerton
Use:        Software products
```

When a proprietary command prefix is used in an SID header, it is recommended that a comment relating to ownership is also included. Developers should not adopt other users' proprietary commands without close liaison with the owner.

This development of the use of SID does not weaken the SID format. It has always been the case that software reading SID must be able to ignore any commands it cannot use. However, the extensions allow developers to use one file format for data which can be used to convey all the information needed in proprietary software, and from which data and a chosen subset of header information can be extracted reliably by other software.

Comments on the Presentation of Time-Related Data. Much debate has taken place over the presentation and use of data relating to elapsed time between data readings. The SID specification is quite clear in what it allows:

There are two ways of including these data in an SID file, by use of the %%interval command (which effectively represents a compressed field), and/or as a field of individual time data.

Any software reading an SID file must be prepared to deal with either method of presenting these data where present. What follows is simply a consideration of related issues which may be helpful when deciding how to implement the use of SID files.

The debate has focussed on the issue of selecting data to be used for plotting the x coordinates on a graph. This is particularly pertinent to much of the likely use of SID files.

If data are time dependent, time is likely to be the variable used on the x-axis. Many applications creating data arrays of time-dependent data show elapsed time as field 1. However, the data may not be time dependent, time may not be field 1, and/ or the user may wish to plot against an alternative field. If a software application importing SID files to be displayed on a graph is to perform predictably without limiting user options unnecessarily, there are several options open to the programmer, some of which are outlined below:

i)   Where no %%interval command is included in the header, data may be plotted automatically against field 1 (often time) or according to the record number.

ii) Where %%interval is included in the header, this may be used to generate the x coordinates automatically. Programmers must remember that if elapsed time is also included as a field, they will generate a line on the graph of elapsed time v interval time. If the field containing the time data carries a field name including the word time, it could be recognised before plotting.

iii) The selection of data used to plot the graph can be left to the user; the data file could be interrogated by the software to determine the possible options which are then made available for user designation as x or y coordinates.

When writing SID files it has become general practice to use %%interval *or* include time-elapsed data as a field (usually field 1). Programmers may wish to take this into account when deciding on defaults for reading or writing SID files.

Main Points of the SID Specification

1) An SID file must begin with a header consisting of a minimum of two command lines, %% identifier and %% datasize. All other commands are optional but should be used in the order shown in the following section. New commands must have a registered proprietary prefix.

2) When writing an SID file a carriage return (CR), ASCII 13, and line feed (LF), ASCII 10, must be used to terminate every command line and every record in a file.

3) A line in an SID file may be either:

   i)   a command line in the header

   ii) a record in the data section

iii) a blank line containing only CR LF (e.g., where no data exists, or if a reading has been missed in a single field record where elapsed time is described using the interval command)

Lines are made up of two types of *string elements*: textual string elements or numerical string elements.

A textual string element may contain a mixture of ASCII characters ASCII 33 to 43 and ASCII 45 to ASCII 126 inclusive. A textual string element must not contain ASCII 13 (CR), ASCII 10 (LF), or ASCII 44 (comma).

The white space character, ASCII 32, is used to separate string elements for clarity, e.g., to separate words in a comment string.

Only command line strings can have the symbols %% as the first characters in the line.

A numerical string element is used to represent a numerical value in a header command line or as an item of data in a record.

A numerical string element may contain only the ASCII characters ASCII 43, 45, 46 and 48 to 57. NB: This excludes ASCII 32, the white space character. If ASCII 43 or 45 (+ or -) is included it must only appear once at the beginning of the string; ASCII 43 and 45 are mutually exclusive.

The period character, ., ASCII 46, must be included when the numerical value represented by the string has a decimal fraction as part of its value.

Comma, ASCII 44, can be used *only* to separate parameters in a command line or fields in a record. In either case the presence of white space characters before and/or after the comma will have no special significance, e.g.,

        %%identifier,sid CR LF

is the same as

        %identifier , sid CR LF

Example SID Files

White space is used for visual clarity; the positions of carriage return (CR) and line feed (LF) are shown.

Using a Full Header with No Proprietary Commands.

```
%%identifier, SID CR LF
%%datasize, 6, 3 CR LF
%%title, pH and Temperature CR LF
%%filedescription, Jo Bloggs pH and temperature experiment
        1st Oct 90 CR LF
%%fieldname, 1, Time CR LF
%%fieldunits, 1, Seconds CR LF
%%fielddescription, 1, Time from start (readings every 10 seconds)
        CR LF
% %fieldname, 2, pH CR LF
% %fieldunits, 2, CR LF
%%fielddescription, 2, Standard glass pH probe CR LF
%%fieldname, 3, Temperature CR LF
%%fieldunits, 3, degrees C CR LF
%%fielddescription, 3, Chemical resistant temperature sensor CR LF
%%interval, 10 CR LF
%%starttime, 153000 CR LF
%%stoptime, 153120 CR LF
%%startdate, 901001 CR LF
%%stopdate, 901001 CR LF
%%maxmin, 3, +100.0, -10.0 CR LF
0,7,25.6 CR LF
9.9,7,25.6 CR LF
19.8,7.1,25.7 CR LF
29.7,7.6,25.1 CR LF
39.9,7.5,25.0 CR LF
49.9,7.4,24.9 CR LF
```

Using the Minimum Permitted Header

```
%%identifier, sid CR LF
%%datasize,9,3 CR LF
0,7,25.6 CR LF
10,7,25.6 CR LF
20,7.1,25.7 CR LF
30,7.6,25.1 CR LF
40,7.5,25.0 CR LF
50,7.4,24.9 CR LF
60,7.4,25.0 CR LF
70,7.3,25.3 CR LF
80,7.3,25.4 CR LF
```

Using a Partial Header, Showing Data Channels Sampled at Differing Rates

```
%%identifier,SID CR LF
%%datasize, 9,3 CR LF
%%title, pH and Temperature CR LF
%%fieldname, 1, Time CR LF
%%fieldunits, 1, Seconds CR LF
%%fielddescription, 1, Time from start at 10 second intervals CR LF
%%fieldname, 2, pH CR LF
%%fieldunits, 2, CR LF
%%fielddescription, 2, Standard glass pH probe (readings every 20
        seconds) CR LF
%%fieldname, 3, Temperature CR LF
%%fieldunits, 3, degrees C CR LF
%%fielddescription, 3, Chemical resistant temp. sensor (readings
        every 10 seconds) CR LF
%%interval, 10 CR LF
0,7,25.6 CR LF
10,,25.6 CR LF
20,7.1,25.7 CR LF
30,,25.1 CR LF
40,7.5,25.0 CR LF
50,,24.9 CR LF
60,7.4,25.0 CR LF
70,,25.3 CR LF
80,7.3,25.4 CR LF
```

Using Time as HHMMSS

```
%%identifier, sid CR LF
%%datasize, 5,4 CR LF
%%title, !PriSM data CR LF
%%fieldname, 1, Time CR LF
%%fieldunit, 1, time CR LF
%%fieldname, 2, pH CR LF
%%fieldunit, 2, ratio CR LF
%%fieldname, 3, Temperature CR LF
%%fieldunit, 3, degrees Celsius CR LF
%%fieldname, 4, eventmarker CR LF
%%fieldunit, 4, mark CR LF
%%interval,10 CR LF
140743,7,25.6, 0 CR LF
140754, , 25.6, 1 CR LF
140802, 7.1, 25.7, 1 CR LF
140812, ,25.1,0 CR LF
140823, 7.5, 25.0,1 CR LF
```

Using a Text String Field

```
%%identifier, sid CR LF
%%datasize, 5,2 CR LF
%%title, Pupils ages CR LF
%%fieldname, 1, Name CR LF
%%fieldunit, 1, string CR LF
%%fieldname, 2, age CR LF
%%fieldunit, 2, value CR LF
Brian, 6 CR LF
Paul, 8 CR LF
Jane, 7 CR LF
Barbara, 6.5 CR LF
Steven, 5.75 CR LF
```

19. Software: Integration, Collaboration, Standards, and Progress

Jack M. Wilson

Rensselaer Polytechnic Institute

19.1 Introduction

Any discussion of the software standards for microcomputer-based laboratories (MBL) must take into account the proliferation of technological tools for use in physics. Among these are: spreadsheet physics, microcomputer-based laboratories, programming for problem solving, videotapes, videodiscs, CD-ROMs, simulations, symbolic mathematics, and modeling [1]. We must avoid the trap of becoming "the person with a hammer to whom all things look like a nail."

Many of these tools have been incorporated into the physics classroom as physicists began to accept the growing body of evidence that showed how effective these tools could be. While early use of computing tended to concentrate on "programmed learning" or tutorial uses of computers, today much software in use in physics is based on general-purpose software tools. Some popular tools are actually commercial applications which have been adapted for use in physics education, while others were created specifically for educational use. Spreadsheets and computer algebra systems are examples of the former while microcomputer-based laboratory systems and programming tools for problem solving are representative of the latter.

The physics education literature is filled with descriptions of how these tools are used individually in various physics classes. There is far less discussion of how these tools can be used in an integrated fashion or how different tools might be appropriate for particular education contexts. Those of us who work with microcomputer-based laboratories are not immune to this malady.

The CUPLE Consortium was born of the conviction that the development of powerful educational software in physics had reached the point where many talented people were building excellent materials in idiosyncratic ways and that the physics community would benefit from a unified approach to the development and distribution of new materials [2]. From the beginning the philosophy of the CUPLE (Comprehensive Unified Physics Learning Environment) project has been to incorporate existing approaches and materials while integrating these materials to work together with a common user interface, common file structures, and common mechanisms for data exchange. Thus integration, collaboration, and standards are among the most important features of the project.

19.2 Standards

The early inspiration for the CUPLE system was provided by the development of the NeXT computer system, but it quickly became clear that, if we were to hope for widespread use, we would have to build the system for either the Macintosh or the IBM and compatible Windows-based systems. Either of those would provide the common graphical user interface that could allow the integration of such disparate tools. Although the Mac interface appeared to have the edge in 1989 when this decision was made, the group elected to build CUPLE for the Windows environment because the hardware was available from a wide variety of manufacturers at approximately half the cost of a Macintosh. Either the group was prescient or incredibly lucky, as Windows has become the de facto standard for modern personal computing.

The choice of Windows brought with it access to standards that have become widely supported by a variety of manufacturers. The Common User Access (CUA) standard requires that all applications use a similar approach to menuing and to selection and activation of objects. Dynamic Data Exchange (DDE) allows each of our CUPLE programs to communicate with the others as well as with a variety of popular commercial applications. For example: the CUPLE video tool is able to take space and time measurements of video, load the Excel spreadsheet, create a new sheet, insert that data into the appropriate cells, and present and organize the results.

The Multiple Document Interface (MDI) standard is adhered to by all CUPLE applications that create multiple documents that need to be managed by a single window. The Window on Physics (WinPhys) modeling and simulation programs often create several graphs for a particular situation. MDI allows those graphs to be moved, resized, tiled, cascaded, minimized, and maximized in a consistent fashion. Students do not have to learn a new interface for each tool.

In order to control videodisc players, videotape players, video cameras, audio, and digital video sources there must be standards for communication. We have adopted the Windows Multimedia Extensions (MME) for all of our interactions with video and audio devices. Where no standards had yet been developed (for example, control of videotape players or MBL devices), we developed prototype standards modeled on the existing standards. Communication with these devices is funneled through standard Dynamic-Link Libraries (DLL) that can be shared with other programs.

CUPLE needed to have techniques for embedding one application within another and being able to launch that application along with any required drivers and programs. Using the facilities within Windows we created visual representations of these objects that can be embedded into documents and linked to the original objects. Objects that could be embedded include video, audio, mathematical modeling, spreadsheets, MBL data acquisition, graphics, text, and access to databases. Windows is now providing some of that capability with a new standard for Object Linking and Embedding (OLE).

We elected (as have many others) to use comma-delimited text files to store all data created by each application. The first line contains the information about the application that originated the data file, the first item in the second line is the number of data sets, and the remainder of the second line contains additional information about the data set that may be appropriate to the application. This allows data created by the video tool to be loaded into the MBL tools and compared to data loaded from a simulation as well as data collected by the motion detector.

An authoring environment was created to provide a common look to all hypertext materials created for CUPLE. The authoring environment might be referred to as a "physics processor" since it appears to the user to be similar to a word processor. The author simply brings up the CUPLE template under the Toolbook hypermedia system, selects author mode, and begins to type. The author can use the standard Toolbook palettes to add new text fields, graphics, colors, or buttons to the page. The CUPLE extensions to the standard menu allow the author to select Video, WinPhys, Laboratory, MBL (data acquisition), Demos, or String-and-Sticky tape and the appropriate icon and code is added to the book. The author is prompted to enter the name of the specific video clip, program, etc., and the functionality necessary to do this is automatically added to the page. The net result is a unit that looks very much like other units and uses the same icon and button approaches.

The adoption of these (and other) standards enabled us to integrate the disparate tools that make up the full toolbox of the physics teacher.

19.3 Spreadsheets

Spreadsheets are one of the most popular tools for physics education. Lotus 123 is popular on the older MS-DOS systems while EXCEL is the most popular on the newer MS-DOS/Windows systems and the Macs. Other popular spreadsheet programs include Quattro and WingZ, a newer spreadsheet with spectacular 3D graphics and the ability to run the same software across all three major platforms (DOS/ Windows, Macs, and UNIX workstations). There are three books that anyone interested in spreadsheet use in physics should have: *Spreadsheet Physics*, by Misner and Cooney (Addison Wesley 1991); *Dynamic Models in Physics*, by Potter, Peck, and Barkley (N. Simonson 1989); and *Wondering about Physics*, by Dykstra and Fuller (Wiley 1988). The books are listed in decreasing level of complexity and coverage. For simple problems, spreadsheets are fast and easy for the students to learn and use. For more complex problems, they can be as difficult—or even more difficult—to use than the programming tools mentioned later. Spreadsheets are significantly slower than either the programming approaches or the mathematical tools mentioned below for large calculations. The choice of tools for problem solving in physics depends very heavily on what the students already know and what they may be expected to need to know later. For students who are not going on in a technical field, spreadsheets are the clear choice. For those who will go on and will need to do programming as part of their education, problem solving with programming might be more appropriate.

19.4 Symbolic Mathematics

Many mathematics departments are experimenting with the use of symbolic mathematics packages, like Maple, MathCAD, or Mathematica, in the teaching of calculus. At Rensselaer all 1,100 freshman calculus students take weekly calculus classes taught in classrooms equipped with UNIX workstations running the Maple symbolic mathematics program. Although Rensselaer may be unusual in its scale, other universities are moving toward a similar model. These tools have begun to infiltrate the physics curriculum as well, allowing students to perform powerful mathematical manipulations with a minimum of effort put into learning the package. In general, they are fairly easy to use to define functions, to do symbolic mathematics such as differentiation and integration, and to display the results as beautiful two and three dimensional graphs. Learning to use the built-in programming languages can be much more challenging. When more numerical approaches to mathematics are desired, physicists frequently select packages like MatLab which is particularly good for situations requiring linear algebra and is heavily used by engineers.

19.5 Programming for Problem Solving

The M.U.P.P.E.T. utilities [3] developed for use with Turbo Pascal have proven to be a popular way to incorporate problem solving through programming into the undergraduate curriculum. Programs written with these utilities have come from Europe, Australia, and Japan as well as a number of U.S. universities. A later version of these utilities, Window on Physics (WinPhys), allows object-oriented graphics-based programming in Windows 3.x for students with only a basic understanding of the PASCAL language, similar to that gained in many freshman Computer Science courses.

When problem solving through programming is a desired objective for undergraduate courses, utilities like M.U.P.P.E.T. or WinPhys can be a way of providing it without requiring the students to invest too much time in the usual programming interface questions. Otherwise, a spreadsheet or symbolic mathematics package can meet the problem-solving goals. The following books are excellent sources of good problems for undergraduate and graduate physics courses: *Computational Physics* by Steve Koonin (Benjamin Cummings 1986), *An Introduction to Computer Simulation Methods* by Gould and Tobochnik (Addison Wesley 1988), *The M.U.P.P.E.T. Manual* by Redish, Wilson, and Johnston (Physics Academic Software to be published), *Numerical Recipes in PASCAL* by Press et al. (Cambridge 1989), and *Pascal Applications for the Sciences* by R. Crandell (John Wiley 1985).

19.6 Hypermedia Tools

It is becoming clear that graphical user interfaces (GUI's) will be the standard user interface on all new computer systems—Windows 3.x on the MS-DOS, System 7 on the Macintosh, and Motif on UNIX systems. GUIs make the use of some computer tools far easier for students, but they can make the task of software developers much more difficult. Prototyping tools, such as HyperCard and Toolbook have been developed to overcome such difficulties. These hypermedia tools have spawned many new hypermedia and multimedia packages that are useful in physics instruction. One example would be *The Essence of Physics,* a HyperCard-based computer animation collection and study guide for physics that has been in use in selected classes at Harvard [4]. Another would be the CUPLE project, which has set out to create a hypermedia (Toolbook) and multimedia resource that can be used in a variety of traditional and innovative formats [5,6]. Combining text, graphics, animations, microcomputer-based laboratories, problem solving, and on-computer-screen video, CUPLE aspires to be an expandable "publishing" system that will attract and incorporate newly created materials. CUPLE has been offering faculty workshops at each annual American Association of Physics Teachers (AAPT) meeting for the last few years as well as longer summer workshops.

19.7 Video

Video pioneers such as Robert Fuller and Dean Zollman have developed several excellent videodiscs such as *The Puzzle of the Tacoma Narrows Bridge Collapse*, *The Physics of Sports, Automobile Collisions*, and many others. The recent release of the AAPT-produced *Cinema Classics* set makes available on videodisc many of the classic films of physics including the best of the single-purpose films. A collection of *Physics Lecture Demonstrations* was produced by a team that includes Jearl Walker of Cleveland State, Richard Berg of Maryland, and John Davis of the University of Washington. With sponsorship of the Annenberg/CPB project, the National Science Foundation, and the IBM Corporation, California Institute of Technology produced a collection of videotapes, and then videodiscs, from the television program *The Mechanical Universe*. The availability of such high-quality video in easily available high-quality formats has encouraged much more widespread use of video than in the past.

Video is an example of how CUPLE attempts to embrace the results of other projects without duplicating those projects. CUPLE contains databases of the information needed to link video to the other materials and in many cases already has the links in place. In order to activate the links, one must acquire the original video from the distributor. Now that compressed digital video has become the standard for computer-based video, we will be challenged to come up with creative ways to maintain copyright control while enabling the use of compressed digital video from computer hard disks.

19.8 Computer-Based Video

The advent of hypermedia and multimedia tools has brought together the use of video and computing materials in an integrated package. The CUPLE video tool allows the student to play video from a videodisc, videotape, video camera, or compressed digital video directly onto the computer screen. By overlaying computer graphics on the video, students are able to make measurements of the spatial and time dependencies of various phenomena. At Rensselaer, the CUPLE video tool is used to follow the motion of a golf ball dropped from the top of a ladder and recorded during class. The data acquired by the video tool is then exported to a spreadsheet and posted on the campus network where it becomes one of the week's homework assignments for the 1,000 students per year who take Physics 1. The solution is often sent back to the professor by E-mail. A week later the experiment is repeated with a ping pong ball and the analysis has to take into account the air resistance term. At Dickinson, students use a similar video tool to analyze the fall of a bungee jumper.

The EDUCOM award-winning *Workshop Physics* program is a related example of an innovative curriculum that combines microcomputer-based laboratories, hypermedia-based materials, and an approach to the classroom based upon the laboratory rather than the lecture [8]. *Workshop Physics* has used some of the CUPLE tools including the video tools and has provided the inspiration for the MBL portion of CUPLE.

19.9 Simulations

The number of simulations in use in physics instruction is so large that one cannot do justice to them in a short article. Many of the Computers in Physics Award winners were simulation programs including: *Lenses*, *Orbits*, *Chaos Demonstrations*, *Spacetime Physics*, *Thermo*, and others [9]. Physics Academic Software, a joint effort of the American Institute of Physics, the American Association of Physics Teachers, and the American Physical Society, has many well-reviewed simulations available. The best simulation materials provide great flexibility for the user to investigate changes in the underlying models and to look at the effects of different mathematical approaches to solving the models [10]. *Interactive Physics* and *Physics Explorer* are Macintosh-based applications that allow users to construct their own "worlds" or simulations with a point-and-click interface [11]. They can be helpful for simple simulations but may not be suitable for more complex tasks.

The Computer in Upper-level Physics group (CUPS) is bringing together a large number of university physicists to create simulation and tools programs for the advanced undergraduate physics courses [12]. CUPS is using the M.U.P.P.E.T. tool kit with substantial extensions, but they are not designing to a graphical user interface such as Windows. As the CUPS programs become available in coming years, physicists will find them to be excellent examples of fine physics at the junior and senior level.

Any of these simulation programs can be embedded into any of the CUPLE modules and can be called by all other programs.

19.10 Microcomputer-Based Laboratories

Beginning with the AAPT microcomputer-based laboratory (MBL) workshops and materials developed by Tinker, Ingoldsby, et al., and refined by Layman, DeJong, et al., in the early 1980s, use of the microcomputer as a laboratory instrument has been one of the most popular uses of computers. Today the direct descendants of these materials have found their way into commercial packages that provide an inexpensive way to use the computer for data acquisition, analysis, and visualization. Careful evaluation of student learning [13] has convinced the community that MBL is an indispensable part of a physics laboratory.

The two most popular low-cost interfaces that plug into the popular computers are probably Vernier Software's Universal Laboratory Interface (ULI) and IBM's Personal Science Laboratory (PSL). Each of these comes with the appropriate software for acquiring and graphing data in real time. Ron Thornton's Tools for Scientific Thinking Project and Priscilla Laws' Workshop Physics project were the major contributors to the ULI software and curriculum materials. The low-cost interfaces are not fast enough or powerful enough for many physicists, who opt instead for data-acquisition boards like those from National Instruments that provide the performance they need at somewhat higher cost.

Each of these interfaces is supported by the CUPLE environment. The ULI (and soon the PSL) was fully integrated by the author into the object oriented WinPhys modeling system. These object-oriented tools allow faculty or students to create their own PASCAL programs that run within Windows, follow the standards described earlier, and provide sophisticated data-visualization and analysis tools. Knowledge of PASCAL is desirable for working with these materials, but no knowledge of the Windows API is required. Within the WinPhys ULI objects, the author has attempted to create standards for communicating with MBL devices that are modeled upon the MME described earlier. The ULI support is a graphic illustration of how standards (combined with object-oriented modules) can allow students and faculty to create sophisticated programs that meet industry standards without becoming experts in programming.

It would be desirable for those who are involved with MBL software and hardware to begin to meet together to discuss standards for interchange of data and communication with hardware. An earlier presentation by Angela McFarlane has illustrated how such a process could lead to increased cooperation among developers in the U.K. We are fortunate to be able to build upon the work of industry committees that have already developed standards for interaction with other kinds of devices. Adoption of an international standard would be extraordinarily difficult, but our own experience has shown that true integration, collaboration, and cooperation are possible only when such standards are agreed to. We are fortunate to be able to build upon the work of industry committees that have already developed standards for interaction with other kinds of devices.

References

1. Wilson, J.M.: Computer software has begun to change physics education, *Computers in Physics* 5(6), 580 (Nov./Dec. 1991).
2. The CUPLE Project: A hyper- and multi-media approach to restructuring physics education. A chapter in: *Sociomedia: multi/media, hypermedia, and the social construction of knowledge*, E. Barrett (ed.), (MIT Press, Boston, MA, 1992).
3. MacDonald, W.M. et al.: The M.U.P.P.E.T. Manifesto, *Computers in Physics* 1(1), 23 (July/Aug. 1988).
4. Mazur, Eric: A hypermedia approach to teaching physics, *AAPT Announcer* 21(2), (May 1991).
5. Wilson, J.M., and Redish, E.F.: The comprehensive unified physics learning environment: Part I: Background and system operation, *Computers in Physics* 6(2), (Mar./Apr. 1992).
6. Wilson, J.M., and Redish, E.F.: The comprehensive unified physics learning environment: Part II: Materials, *Computers in Physics* 6(3), (May/June 1992).
7. Laws,P.: Workshop physics: Replacing lectures with real experience. In: *Proc. of the Conf. Computers in Physics Instruction.*, E. Redish and J. Risley (eds.) (Addison Wesley, Reading, MA 1989).
8. Laws, Priscilla: Workshop physics: Learning introductory physics by doing it, *Change*, 20 (July/Aug. 1991).
9. Donnelly, D.: CIP's first annual software contest: The winners, *Computers in Physics* 4(5), 540 (Sept./Oct. 1990).
10. Wilson, J.M., and Redish, E.F.: Using computers in teaching physics, *Physics Today* 42(1), 34 (Jan. 1989).
11. Tam, P.: Courseware review: Review of "Interactive Physics," *The Physics Teacher* 29(6), 383 (Sept. 1991).
12. Ehrlich, R. et al.: Project to develop computer software and texts for nine upper-level physics courses, *AAPT Announcer* 20(4), 37 (Dec. 1990).
13. Thornton, R.: Learning physical concepts with real-time laboratory measurement tools, *Am. Journal of Physics* 58(9), 858 (1990).

20. IP-Coach—A Useful Tool for Universities in Developing Countries

V.J. Dorenbos and G.H. Dulfer

Vrije Universiteit Amsterdam

20.1 Introduction

At the end of the 1980s, in view of its potential benefits, we decided to pilot the science interfacing package IP-Coach in some of our educational development cooperation projects. IP-Coach runs under MS-DOS and transforms an IBM-compatible microcomputer—equipped with a special interface board (UIB-board)—into a versatile measuring instrument.

The first try-out was in a workshop for staff members of Universitas Kristen Satya Wacana in Salatiga in Indonesia. As a follow-up, a few of its undergraduate physics students used the package for data acquisition in physics experiments. Data analysis was done with a commercial spreadsheet, which the students had previously used for other purposes. Staff and students were enthusiastic and readily acquired the necessary knowledge and skills for most basic operations.

Later on, and interface board (UIA) and software (IP-Coach) were introduced during a Regional Workshop on 'Computers in Education' in Botswana in December 1991. Background information was provided on the use in education and the participants were offered the possibility of acquiring hands-on experience. The package provoked a number of positive reactions and enthusiasm among the users.

Hence, we decided to prepare a comprehensive description and evaluation of IP-Coach for interested staff at partner universities. Our attention was not directed towards application of the package in secondary or high-school education, but in universities and teacher-training colleges. Although the use of computers is spreading rapidly in countries like Indonesia and in southern Africa, we do not expect a rapid nationwide introduction of computers in secondary education in most of these countries, where many students in the schools do not even have physics textbooks at their disposal. If introduced, students and staff will most probably start with computer awareness courses rather than with microcomputer-based laboratories (MBLs).

The prospects of fruitful application of IP-Coach in tertiary education in developing countries and the lack of an English language review stimulated us to aggregate our findings in a review article entitled *IP-Coach, an Open Science Interfacing Package for IBM-Compatible Computers.*

20.2 Short Description of IP-Coach

The basic IP-Coach package consists of a hardware and a software part. The essential hardware part is the so-called UIB-board, an interface board which should be built into the computer. It can be complemented with a Measuring Console which facilitates the connection between sensors and the interface board. IP-Coach consists of a *Shell* with the following application programs:

a.   Programs for data acquisition and real-time display: *Demometer, Multiscope, Step Measurement* and *Signal Analysis*;

b.   Programs for data analysis and data processing: *Processing* and *Calcsheet;*

c.   The program *Modelling* in which the results of actual measurements can be compared with the outcome of a user-programmed model;

d.   The program *Control* for controlling digital inputs and outputs;

e.   Some utility programs, the most important of which are *Experiment Editor* and *Detection.*

Additional hardware equipment which is available include among other things a Signal Amplifier; a System Board designed for logic, measurement, control and automation practicals; a Filter Box; and a Control Box, a console with four digital and four analogue inputs and eight digital outputs for automation processes.

20.3 Positive Findings

IP-Coach has a number of positive aspects. Basically, the package has been designed as an *open learning environment*, which offers the user a large degree of freedom to select application modules and to determine their mode of operation, while allowing the teacher to adapt the environment to the specific needs of the learner. For the latter, the application program *Experiment Editor* is available which can be used to design pre-planned experiments, whereby the instructor can hide menu options to prevent students from getting distracted or spending an unreasonable amount of time on a side-track.

It is an *integrated package* of which the most important feature is its consistent user interface, which is the same for all application programs. Furthermore, it is easy to switch between application programs, e.g., by using specially defined shortcut keys such as <F10> to exit immediately from any application program to the main menu of the Shell. Finally, the exchange of data between application programs is easy and in most cases runs smoothly. For instance, for analyzing a measurement done with *Multiscope* one can easily switch to *Processing*, whereby data are automatically carried along.

The package also offers possibilities to do *'off-line' data analysis and modelling*. This means that experimental data from a classroom demonstration can be distributed to the students and analyzed by them on computers which do not have to be equipped with the interface board.

The *use of macros* facilitates the application of the package very much.

Finally the built-in possibility of a *pre-trigger time* is very convenient. As an example, we present the measurement of an induced emf by a magnet falling through a coil. With *Multiscope* this experiment can be performed very easily by using a trigger condition at (for instance) 50% of the positive slope of the signal and using a pre-trigger time of 50% of the total measurement period. As soon as the trigger condition is met, the program draws the measured data, including those prior to the trigger moment. This allows one to measure the desired curve without any external trigger signal of the moment the magnet starts falling (see Figure 1). Setting up and executing this experiment takes less than 10 minutes.

20.4 Some Limitations

Most limitations and inconsistencies we found in version 3.1 of the package have been improved in the current version 4. The only limitation left is that the possibilities for data analysis and data processing of the package might be too limited for certain experiments at tertiary level. In such cases, we recommend exporting data to an advanced business spreadsheet.

20.5 Recent Use of IP-Coach in Universities in Asia

At Universitas Kristen Satya Wacana in Salatiga in Indonesia, IP-Coach 4 is currently used by fourth-year physics students, who are doing their own (small) research projects. Examples are a 'falling ball' experiment (time measurement), a spectroscopic experiment (measurement of light intensity) and a calorimetric experiment (temperature measurement). Also the data analysis is performed with IP-Coach.

In the Chemistry Department of the Faculty of Science and Mathematics it is used in the second year for the physical chemistry practicals for the investigation of the properties of an oscillating reaction (a simultaneous measurement of electrode potential, light transmission, pH and temperature).

In 1995, the 'Physics Development Project' at the University of San Carlos in the Philippines has started. An important aim of the project is the improvement of the BSc courses offered by its Department of Physics. This will include among other things a complete revision of the laboratory courses and the introduction of computer-controlled experiments with IP-Coach.

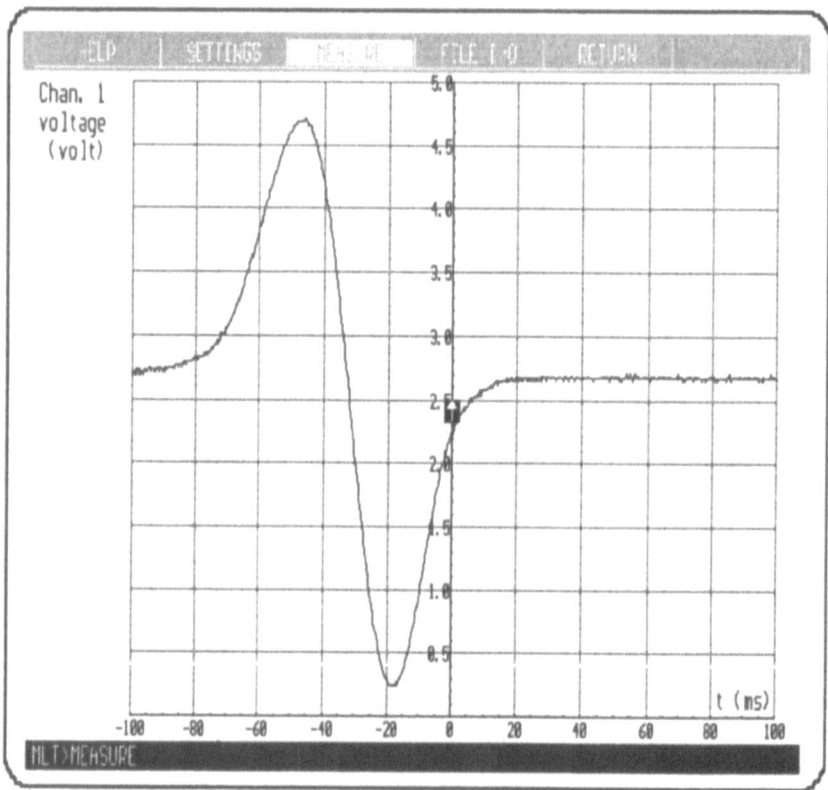

Figure 1: The induced emf produced by a magnet falling through a coil measured with *Multiscope*.

Notice the pre-trigger time of 100 ms.

20.6 Concluding Remarks

Our own findings and those of Szydlowski provide evidence of the enormous potential of UIA-Board and IP-Coach for experimental physics at undergraduate level.

Some years ago, Robert Tinker dreamed of

> a shoebox filled with a few dozen transducers that could measure a very wide range of physical phenomena. One or more of these could be plugged into a universal interface and be recognized by a single, powerful software package.

With IP-Coach his dream has come true. The relatively low cost of the UIA interface board and IP-Coach, and the modest hardware requirements make the package an excellent choice for universities in developing countries.

The work of Laws, Thornton and Sokoloff in mechanics using a motion detector provides challenging opportunities for the use of MBLs in physics education. It would be interesting to investigate whether a sequence of classroom activities in which students' predictions, demonstration experiments and classroom discussions are closely linked would also yield an important gain in students' understanding. This would be a feasible option in low-budget situations (i.e., in the absence of a complete micro-computer-based lab for groups of students) by requiring only one computer and related equipment.

Although a standard Graphical User Interface and a unified learning environment like CUPLE in principle show greater promise than further developments on the DOS platform, it is important not to forget the low-budget segment of the market where the purchase of large numbers of powerful microcomputers is not (yet) possible and/or the need to use (and not discard) available hardware is essential.

On several occasions during this workshop the need to convince university staff of the benefits and importance of MBLs, e.g., in mechanics education, was mentioned. The use of MBLs in student investigations demands the availability of staff members with a reasonable degree of research experience and flexibility. In many situations in developing countries, traditional labs do not require such capabilities of the staff; experiments are standard and do not change over a great many years. Student investigations, however, do not have a predictable course and outcome, and hence put higher demands on the staff in charge. Therefore, these staff members need opportunities to gain experience and to reinforce their background in the field of microcomputer-based investigations.

References

The Centre for Development Cooperation Services of the *Vrije Universiteit* Amsterdam functions as the linking pin between the university (mainly its faculties) and partner universities in Asia and southern Africa. One of the main fields of cooperation is science education, in particular the preparation of students for entry into science-based studies (pre-entry science courses) and the in-service training of secondary science teachers.

Dorenbos, V.J. & Dulfer, G.H. (1992). *IP-Coach, an open science interfacing package for IBM-compatible micro computers*. Workshop paper.

Ellermeijer, A.L., *The training server: STOLE, Scientific and Technical Open Learning Environment. A proposal for DELTA*. University of Amsterdam. Reprint for the workshop.

Hartsuyker, A., van Bart, C. & van Zandbergen, P. (1992). *Courseware development and educational research in the Dutch approach to microcomputer-based laboratory (MBL) experiments in chemistry education*. Workshop paper.

Laws, P., Sokoloff, D. & Thornton, R. (1992). *New mechanics. An MBL curriculum to teach Newton's laws*. Workshop paper.

Szydlowsky, H. (1992). *Microcomputer-based laboratories at Mickiewicz University.* Chapter 23, this volume.

Thornton, R. & Sokoloff, D. (1990). Learning motion concepts using real-time microcomputer-based laboratory tools. *American Journal of Physics 58*, 858–867.

Wilson, J.M. (1992). *Trends in computer software for use in physics education*. Workshop paper.

21. The CALIOPE: A Computer-Assisted Laboratory Instrument Oriented to Physics Education

A.M. Gonçalves and A.A. Melo

Faculdade de Ciências de Lisboa

Abstract. We are developing a learning environment for the physics didactic laboratory at different educational levels. The core is based on a computer-assisted workbench. Using a modular and expandable design, we have built some probes, and are developing others, for different phenomena: light, sound, magnetic field, nuclear radiation, etc. These probes are connectable to an MS-DOS computer. The end-user interface is based on Object Oriented Programming (OOP). This methodology allows a quick development of general purpose instruments (timers, data loggers, etc.) or the elaboration of specific applications. The acquired data could always be analysed through a standard program such as a spreadsheet. This approach is being used in teaching mechanics and electromagnetism to Physics university students at an introductory level, and we are extending the environment to the basic and secondary school.[1]

21.1 Introduction

Recent inquiries into the level of physics knowledge and laboratory skills of our first year university physics students revealed the persistence of well-known pre-conceptions and a defective laboratory knowledge (Pereira and Gonçalves, 1988 and 1991). This occurred to an extent one would not have expected as a result of the high-school curricula. Three main causes may be identified:

- unsuitable didactic equipment considered obsolete by many of those involved in the teaching–learning process;
- teachers' insufficient laboratory knowledge;
- absence of motivation towards laboratory work.

The technological evolution in measurement processes and the increase of programmable components in consumer electronics are clearly affecting the way one views laboratory physics teaching (Thornton, 1988; Mackenzie, 1988; Stuessy et al.,

---

[1] CALIOPE is the former MOSAIC project at Lisbon University, financed by Instituto de Inovação Educacional under contract n° 19/91

1989; Stein et al., 1990). Computers play a useful role in the teaching process: in real-time data acquisition, in graphical presentation of acquired and processed data, and in non-laboratory environments, allowing for a good combination of experimental data and simulation methods. Many traditional experimental demonstrations for which high schools are usually equipped may be improved, or even deeply changed, when computer assisted (Collings and Greenslade, 1989, Cordes, 1990).

Our own experience in the university teaching sphere has convinced us that a computer-assisted laboratory teaching strategy serves two complementary educational objectives better than traditional methods:

- perception of physical phenomena as the source of scientific concepts and laws;

- a practical introduction to the use of technology which is fundamental to the understanding of home, office and factory instruments today.

The first objective is aimed at upgrading the student's scientific culture, with long-term results in his aptitude to learn more and better, and is justified by epistemological arguments about physics as a mainly experimental science, and by the constructivist theory about the learning process. The second objective is directed toward development of the student's knowledge of how to integrate more easily into the modern work-force, and derives from socio-cultural assumptions in a world dominated by information technologies. Using a computer in school-related activities also provides motivation for a number of students in the learning process.

In this context, we are developing and validating a learning environment—The CALIOPE: A Computer Assisted Laboratory Instrument Oriented to Physics Education—devoted to different teaching levels: basic, secondary or introductory university. The basic considerations about this system are given in the next section.

The hardware component of this project consists of several basic modules directly connected to virtually any digital and/or analog computer interface available commercially. These basic modules allow the connection of different sensors and/or transducers, covering a wide range of physical quantities to be measured. The architecture of this low-cost data-acquisition system is described in Section 21.3.

To get the maximum flexibility from the hardware, we developed a universal user interface, simulating actual instruments on the computer screen. The acquired data is recorded in different formats, allowing the use of specific or general-purpose programs for data analysis. The global philosophy of this user interface is presented in Section 21.4.

Specific activities, devoted to the high-school teaching environment, are being developed. These activities, taking advantage of this computer-assisted measuring equipment, will be validated in the actual classroom. Some of them are described in Section 21.5.

21.2 The Aims of CALIOPE

From a pedagogical point of view, the laboratory must allow the student to deal with the nature of physical phenomena used to detect other events and with the physical interpretation of the recorded data. The use of a computer does not obscure these main objectives. The student must retain the connection between the primary physical phenomena used in the detection process and the numerical data recorded by the computer. The connection is more effective if the student can see a real image of the measured data on the screen. Moreover, the computer tends to perform the whole process very quickly, drastically reducing the time taken by data collection. The student will therefore have more time to focus on data analysis.

Generally, the use of a computer in the laboratory is crucial when we need:

- to record events as a function of time, as we find quite often in simple mechanics studies;
- to represent the acquired data graphically and immediately, which happens in almost all situations;
- to deal with a great volume of data;
- to control the data acquisition as a function of certain trigger events.

We use the computer only to assist in the experimental work, not as a substitute for the student's own decision making.

The global project was shaped to provide a uniform hardware/software platform for different laboratory situations, allowing the use of a very limited number of measuring instruments, all of them with a similar end-user interface. The basic software itself deals only with the process of recording data directly from the different sensors used. The analysis of these primary data must be done explicitly by the student afterward.

So, we are trying to reduce to a minimum the need for specific programs devoted to special data acquisition. We want to stress the data-acquisition process as it is, even if we use a computer to assist this action.

We are currently integrating other components into the system, including programs for special data analysis and numerical simulation.

21.3 Hardware Architecture

We have taken into account the fact that MS-DOS computers are dominant in our secondary schools. The existing machines were bought mainly for informatics teaching, and they are largely XT or 286-ATs, old-fashioned models not able to run modern operating systems like Windows. In the near future, new machines will be acquired for informatics teaching, and we expect to get school councils to apply the old ones to less demanding laboratory work. However, these expectations impose some restrictions on the CALIOPE's hardware–software target platform. We need to develop applications that will run under DOS and that will support interfaces compatible with an 8-bit bus.

To reduce development work, we decided to use commercially available digital interfaces and/or analog-to-digital interfaces. To perform the external connections to the probes easily, we built connection boxes with power supplies, using three levels of hardware complexity.

An external connection box is dedicated exclusively to the connection of photogates and ultra-sonic sensors to a digital interface, as needed in the mechanics lab. Another external box, connected to an analog-to-digital interface card, provides enough space to install up to eight transducer circuits. A more sophisticated box is used as a host for a dedicated board built around a flash ADC, to emulate a digital storage oscilloscope (Melo, Gonçalves and Gonçalves, 1990).

The block diagrams of these arrangements are represented in Figures 1, 2 and 3.

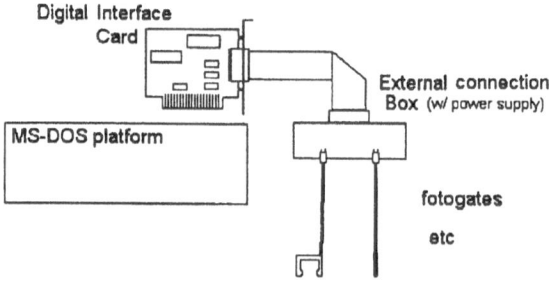

Figure 1   Hardware block diagram of a computer assisted Multi-Event Timer. The external box allows the connection of position sensors based on single and triple photogates. The connection of a sonic sensor is also planned.

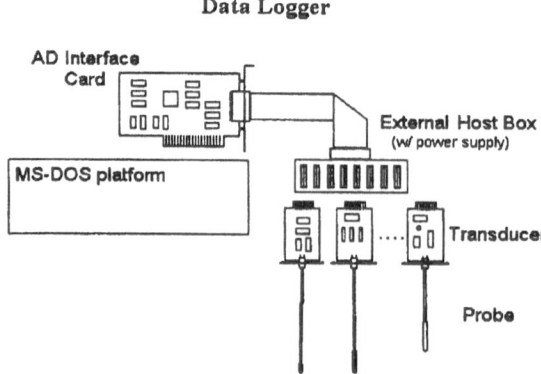

Figure 2   Block diagram of a Datalogger. The external box holds up to eight transducer cards, each one connected to a specific probe such as a temperature sensor, Hall probe, pick-up coil, etc.

Storage Oscilloscope

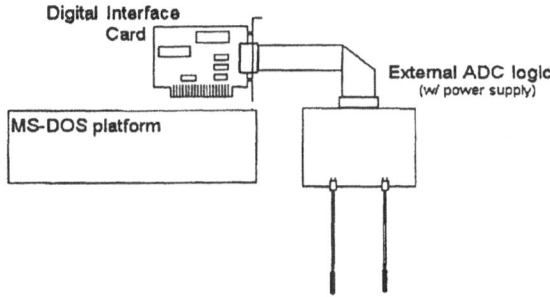

Figure 3 Storage oscilloscope block diagram. The external box has two analog input channels and one flash ADC card.

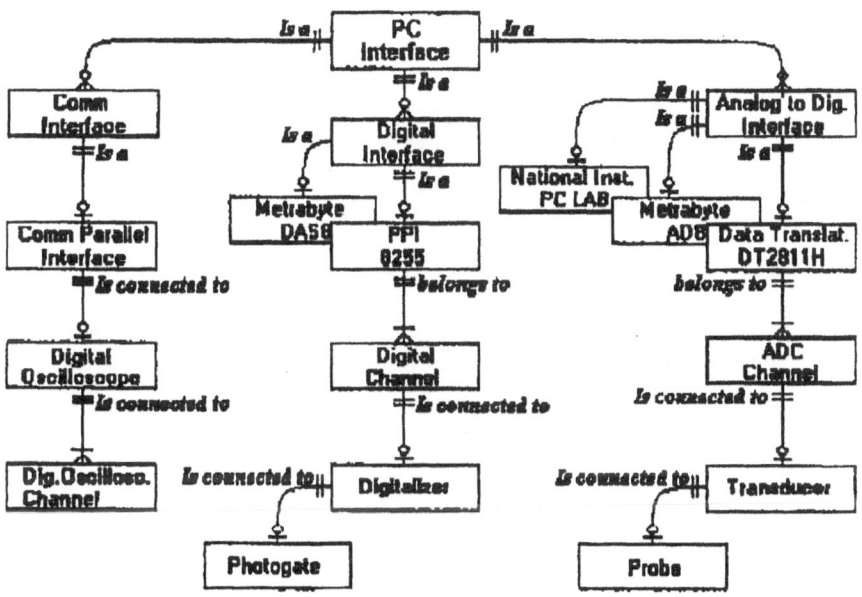

Figure 4 Hardware data model used in the definition of the C data structures

Using the C language, we have described all the hardware modules, through data
structures and associated functions. In Figure 4 we show the corresponding data model
as produced by a CASE program of system analysis.

Figure 5 End-user interface data model

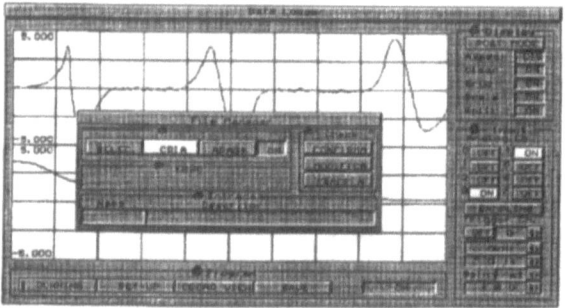

Figure 6 Datalogger end-user interface with file manager running in a different window. This is the main instrument used at the university in the basic electromagnetic lab.

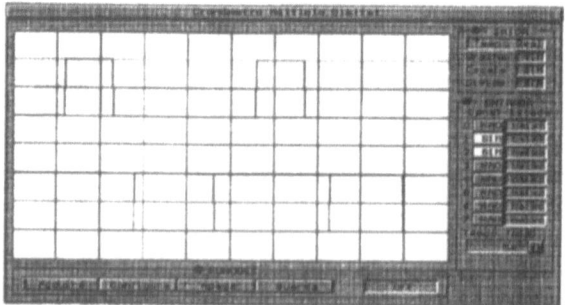

Figure 7 Multi-Event Timer end-user interface. This is the main instrument used at the university in the introductory mechanics lab.

21.4 Graphical User Interface

The end-user interface simulates an instrument panel on screen. (See Figures 6 and 7.) In Figure 5 we show one of our software instruments: a Datalogger. Currently, we are completely re-coding the software in C++, an Object Oriented Programming (OOP) language, to profit from its natural capabilities to describe objects such as those involved in this project. To describe a real-life object, like a software button in a simulated instrument panel, or a temperature probe (connected to a transducer, which is connected to an analog channel, belonging to an analog card), it is more natural to use specific instances of appropriate classes. These classes contain not only the usual data structures as we know them from procedural languages, but also the actions associated with the object. This philosophy enables us to define a set of classes (programming objects) which form not just a tool-box, but a true workbench.

This new approach enables us to develop new instruments quickly, as well as to develop applications oriented for data acquisition and analysis in specific conditions, such as those arising in physics teaching at secondary schools. The authors intend next to evaluate the impact of this computer-assisted lab workbench in general physics teaching.

21.5 Laboratory Examples

Photogates are probably among the oldest position sensors available in the didactic mechanics laboratory (Melo, 1986). This does not mean that this sensor is fully explored. For instance, all the commercially available models employ an infra-red emitter–detector matched pair. The use of invisible radiation obscures the immediate perception of the physical process employed. To make it more understandable for beginners we use visible red light instead. Also the lack of space–time resolution was recognized before (Mosca and Ertel, 1989). The signal usually connected to the digital computer interface is the signal produced by the detector. This signal is proportional to the radiation received, and consequently, an analog signal. This makes this signal inappropriate as a time marker, and it should be replaced by a digitized signal obtained at the analog half value (Melo, Gonçalves and Martins, 1987). Also the full analog value must not be saturated, which means the half intensity value is reached at a half entrance shadow. These improvements in the circuitry design are crucial for the accuracy of the measurement. We are studying a simple use of these photogates to evaluate directly the velocity and acceleration of a body with a decreasing linear dimension of the object. These measures will allow the introduction of these notions as a limit, but as derived from experimental data rather than as a mathematical abstraction (Melo, Gonçalves and Pereira, 1992).

The use of photogates connected with our Multi-Event Timer allows the recording of data for a long time in a form very easy for any student to use. As an example, we show in Figure 8 the results for a simple pendulum, reporting the isocronism of the period, and the loss in velocity at the same time.

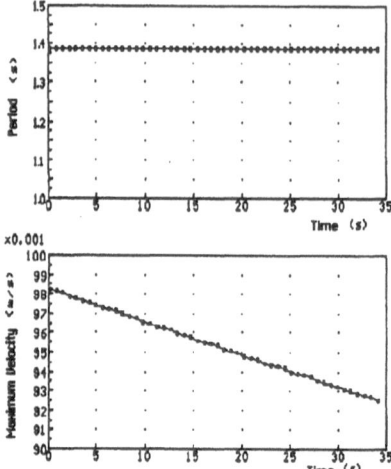

Figure 8 These graphs represent the period and the maximum velocity of a simple pendulum over time.

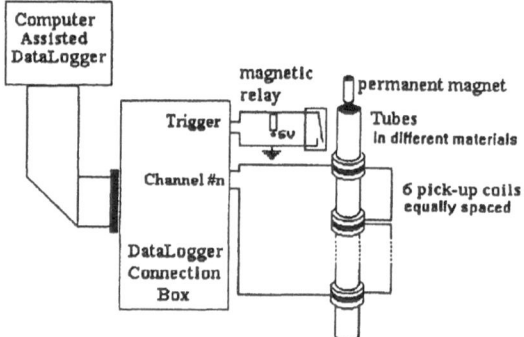

Figure 9 Schema of the apparatus used to study Faraday's and Lenz's law

These sensors can be used to improve the performance of any existing experiment. One of the most obvious is the air track where we can separately record the
velocities involved in multiple two-body collisions, observing the linear momentum
conservation and the kinetic energy dissipation. However, our efforts are going towards the design of apparatus specially conceived to be fitted with photogates.

Our Datalogger can be used in a simple experiment, as in the study of a body
cooling, where the sensor is directly connected to the computer interface card (Marques
and Gonçalves, 1990). However, we can use it in a more complex way, like in the
electromagnetic lab to study Faraday's law. Following some previous ideas (Nicklin,
1986) we assemble around a tube several pick-up coils serially connected (Figure 9).
We can use tubes from different materials. A permanent magnet falls freely inside the

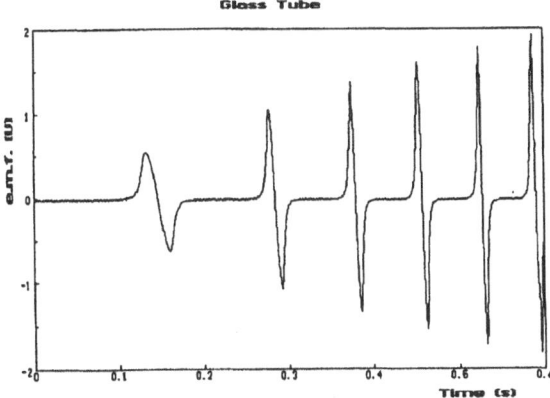

Figure 10 Data recorded by the Datalogger using a glass tube in the apparatus represented in Figure 9

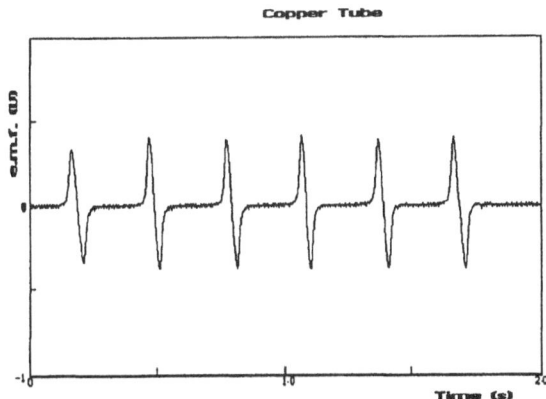

Figure 11 Data recorded by the Datalogger using a copper tube in the apparatus represented in Figure 9

tube. The pick-up coils detect when the magnet passes through. When the tube is made from an insulator material, the recorded signal is as that represented in Figure 10. This kind of curve is not useful for studying a falling body as suggested by Fox et al. (1988). On the other hand, from the knowledge of falling body mechanics laws, the curve is useful to infer the electromagnetic Faraday's laws.

An even more exciting set of results is obtained when the tube is made from copper (see Figure 11). Now the interpretation of the curve implies that the magnet falls with a constant velocity, as predicted in the Lenz law. More experimentation can be done covering a wide range of experiments dealing with the basic electromagnetic laws.

In practical use at university level the student is advised to perform data analysis with the help of a standard spreadsheet, using an automatic procedure to get the final results. It is our aim to oblige the student to perform all the required steps in the analysis process explicitly. For instance, based on data like those in Figures 10 and 11, the student must infer the magnet equations of motion and relate the magnet velocity at the pick-up coil positions with the maximum of the induced voltage. He must also compute the magnetic flux variation for the several pick-up coils in the apparatus of Figure 9 and relate this variation with the velocity and the position of the magnet. During this process he needs to perform several regression analyses and numerical integration (Melo and Gonçalves, 1990).

The practical laboratory situations in the secondary school must be less demanding in data analysis than those described. However, we suggest a similar procedure in all the cases where any kind of analysis is required.

Simulation programs are developed to provide more insight into situations not available in the laboratory such as the electrostatic and magnetostatic fields. In these cases it is possible to draw equipotentials and fields lines, and trajectories of charged particles in these fields.

21.6 Conclusions

The short time devoted by teachers and students to experimental work is a well-known characteristic of today's physics teaching. We must fight against this tendency. Our contribution is CALIOPE, a Computer Assisted Laboratory Instrument Oriented to Physics Education. The use of computer-assisted instrumentation in the lab tends to improve many of the classical experiments. Other experiments, not normally done because of the imprecision of classic measuring methods or the time required for collection and preparation of data, now become possible. Performing data analysis in the computer immediately after data collection leads to a significant increase in the efficacy of student experimental work. The use of common hardware and end user software interfaces shortens the teachers' and students' training time.

We believe that the CALIOPE workstation accounts for only a fraction of the total cost of a modern school physics laboratory because of its ability to be connected to a wide range of equipment. We do not need to multiply the measuring instruments.

References

Collings, Peter J., and Greenslade, Thomas B., Jr.: Using the computer as a laboratory instrument, *Phys. Teach.* 27(2), pp. 76–84 (1989)

Cordes, Albert E.: Using computers in the physics laboratory, *The J. Comp. Math. & Sci. Teach.* 9(3), pp. 53–63 (1990)

Fox, J., Gaggini, N., and Eddy, J.: Simple free fall apparatus, *Phys. Teach.*, 26(2), pp. 108–109 (1988)

Gonçalves, António M.: Estudo experimental das leis de Faraday e de Lenz. In: *Trabalhos Práticos de Física Experimental I*, Dep. Física, Faculdade de Ciências, Universidade de Lisboa 1990

Mackenzie, J. Scott: Issues and methods in the microcomputer-based lab, *The J. Comp. Math. & Sci. Teach.*, 8(3), pp. 12–18 (1988)

Marques, José, and Gonçalves, António M.: *Medição de temperatura assistida por computador*, Dep. Física, Faculdade de Ciências, Universidade de Lisboa 1990

Melo, António A.: *Trabalhos Práticos de Mecânica*, Dep. Física, Faculdade de Ciências, Universidade de Lisboa 1986

Melo, António A., Gonçalves, António M., and Martins, Miguel M.: Laboratório de Mecânica Assistido por Computador: uma experiência ao alcance de todos, *Gazeta de Física*, 10(1), pp. 10–18 (1987)

Melo, António A., Gonçalves, António M., and Gonçalves, Manuel A.: *Osciloscópio com memória assistido por micro-computador*, Comunication to FISICA 90, the 7th National Conference on Physics, Lisbon, Sociedade Portuguesa de Física 1990

Melo, António A., Gonçalves, António M., and Pereira, M. Helena: *Conceitos de velocidade e aceleração: abordagem experimental por meio da razão de diferenças finitas mensuráveis*, Communication to FISICA 92, the 8th National Conference on Physics, Vila Real, Sociedade Portuguesa de Física 1992

Mosca, Eugene P., and Ertel, John P.: Photogates: an instrument evaluation, *Am. J. Phys.*, 57(9), pp. 840–844 (1989)

Nicklin, R.C.: Faraday's law—quantitative experiments, *Am. J. Phys.*, 54(5); pp. 422–428 (1986)

Pereira, M. Helena, and Gonçalves, António M.: *Pre-concepções em Mecânica: dimensão do Problema*, Report, Dep. Fisica, Faculdade de Ciências. Universidade de Lisboa 1990

Pereira, M. Helena, and Gonçalves, António M.: *Atitudes de alunos perante a Faculdade, a Ciência, o Laboratório e os Computadores*, Report, Dep. Fisica, Faculdade de Ciências. Universidade de Lisboa 1991

Stein, Joanne Striley, Nachmias, R., and Friedler, Y.: An experimental comparison of two science laboratory environments: traditional and microcomputer-based, *J. Educ. Comp. Res.*, 6(2), 183–202 (1990)

Stuessy, Carol L., and Rowland, Paul M.: Advantages of micro-based labs: electronic data acquisition, computerised graphing, or both?, *J. Comp. Math. & Sci. Teach.*, 1989 (Spring), pp. 18–21 (1989)

Thornton, Ronald K. (1988): Tools for scientific thinking: learning physical concepts with real-time laboratory measurement tools, *Proceedings of the conference on computers in physics instruction*, Raleigh, NC, edited by Redish, E.F. and Risley, J., Addison-Wesley 1989

22. Bremer Interface System: Didactic Guidelines for a Universal, Open, and User-friendly MBL System

Horst P. Schecker

University of Bremen

Abstract. The "Bremer Interface System" (BIS) is an integrated MBL-environment for high-school physics. The system consists of a versatile hardware adapter, graphics-oriented data-logging software and a science spreadsheet. Special features of the system are *universality*, i.e., applicability over a wide variety of contexts; *openness*, i.e., user-control for context-free configuration; and *user-friendliness*, realized by a direct manipulation, graphical interface. This paper explains the didactic guidelines and shows some applications.[1]

22.1 Categories of MBL Materials

Microcomputer-Based Laboratories (MBLs) have become a main application of computers in physics education. A great number of adapters and programs are available. Some examples are described in this book.

The materials can be grouped into three categories:

- low-cost, easy-to-build adapters, driven by simple BASIC programs (e.g., [1]),

- lab packages, including sensor, interface and software for a specific group of experiments (e.g., [2]),

- multi-purpose data-logging apparatus with versatile data-evaluation software (e.g., [3]).

Materials of the first category—besides being cheap—are very flexible. They can be adapted to special demands by re-designing the hardware or changing the program. However, this requires a certain expertise in electronics and computer languages, which most students and many teachers do not possess, and which does not contribute much to develop *physical* understanding.

Materials from category two tip the scale to the other side. The hardware can be used as a black box. You simply connect the components and start the program. You

---

[1] This paper is based on the research and development project 'Computers in Physics Education' funded by the Minister for Education of the Federal Republic of Germany and the Senator for Education, Sciences and Arts of the State of Bremen.

can choose from a set of predefined data-logging jobs while dialog windows lead you through the necessary configuration steps. Various graphing options are at hand and the data are automatically processed—e.g., by calculating velocity from incoming position–time data. It takes little time to learn how to use these materials. Problems may arise if you want to use other sensors than those included in the lab package or calculate quantities that are not listed in the given selection.

Modern multi-purpose systems try to join together the advantages of both categories: high flexibility and convenient usage. The hardware interface has versatile inputs for analog and digital signals, e.g., a wide range of voltages, which allow connections to different probes. The software must support the user in adapting the system to his specific needs by giving control over the configuration of data-logging jobs. Data processing has to be open for user-defined operations on the data. These demands tend to make the system rather complex and confusing. Convenient usage depends on proper software ergonomics.

While multi-purpose hardware has been on the market for several years, there was a considerable lack of appropriate software. Most suppliers of educational lab apparatus focused on hardware and did not invest sufficient time (and money) in software development. Some of the reasons were as follows: a) it is easier to write short special-purpose programs than a comprehensive user-friendly software environment; b) the operating systems lacked appropriate features; c) the suppliers believed that teachers prefer closed lab packages to open hard- and software toolboxes. This situation has recently begun to change. Several examples for advanced MBL software are included in this volume.

22.2 Didactic Guidelines for the Development of BIS

The "Bremer Interface System" (BIS) aims at fulfilling the demands of the third category. The decision to develop this type of MBL system at the University of Bremen was taken under four guidelines which form the basis for our work on computers in physics education (cf. [4]): universality, openness, transparency, and user-friendliness.

22.2.1 Universality

The number of computer materials employed in upper secondary education should be restricted to a limited set of tools that can be repeatedly applied to physical problems. A real computer tool is something you get used to work with in different contexts and domains, as with certain algebraic or analytical methods. The opposite pole would be to use many different programs and adapters, each with only limited scopes. As universal tools we propose modelling systems (like *Stella* [5]), simulation labs (like *Interactive Physics* [6]), and multi-purpose MBLs (like BIS).

22.2.2 Openness

A computer-based approach to the solution of a physical problem should derive from a concrete teaching situation. Hard- and software tools must be as open as possible for students to transfer their ideas onto the computer. Instead of choosing from a list of possible tasks they should have the chance to define a task on their own and adapt the computer to their needs.

Openness is a chance for the user and at the same time a challenge—he *can* formulate a modelling or a measurement task on his own, but he is also *forced* to do so. He cannot lean on the computer to ask for the necessary inputs. The user him/herself has to reflect on the physical phenomenon to decide which quantities have to be considered, which parameters seem to be appropriate, and how the data are to be processed and displayed.

22.2.3 Transparency

Transparency goes hand in hand with *openness*. Because it is the user who defines the measurement task, he is aware of the parameters and knows what is being done with the data. If he wants to produce a velocity–time graph from position–time data, he has to process the data accordingly. This is different from just clicking on a button named "velocity graph."

Transparency does not hold in the same way for MBL hardware. An all-purpose adapter has to make use of advanced electronics. Any modern signal amplifier, oscilloscope, or even demometer is more or less a black box. This is not unique for micro-based data-logging. If necessary, one can illustrate the principal method of converting analog to digital signals by simple demo circuits. But this is learning *about* MBL, not learning *with* MBL.

22.2.4 User-friendliness

In the beginning of computer usage in physics education a universal, open, and transparent use could only be realized by working with computer languages. But even today only few students and by far not all the teachers possess the necessary expertise (cf. [4], Vol. 4). The need for computer-specific knowledge to employ the machines on physical problems was one of the reasons why so little use was made of the materials.

For a long time universality and openness on one side and user-friendliness on the other seemed to be contradictory demands. Research about man–machine interaction and carefully designed user interfaces have changed this situation. Flexibility of computer applications is no longer bound to self-programming. Advanced operating systems like Macintosh OS and Microsoft Windows 3.x enable the computer novice to handle a complex interface system while concentrating on the physical questions to be investigated.

User-friendliness is not just gadgetry. It should form a central feature of educational soft- and hardware. Software design has to visualize universality and openness of an interface system in such a way that students and teachers can make creative use of them.

Some concrete features of BIS under the guideline *user-friendliness* are as follows:

- The user interface is compatible with the simulation and modelling programs used in class. Like all Macintosh programs it consists of menus, windows, dialogues, and alerts. This family resemblance makes it easy for the student to learn the handling of BIS.

- The user interface has a real-world analogy. The screen of the transient-recorder module, e.g., was designed in analogy to a real scope, with software buttons in the place of turning knobs for adjusting the voltage range.

- The hardware adapter is completely controlled by relays. All settings are defined within the user interface and then transferred to the adapter. Apart from the on/off switch the adapter has neither buttons nor knobs. So the computer screen becomes the front-end of the MBL apparatus.

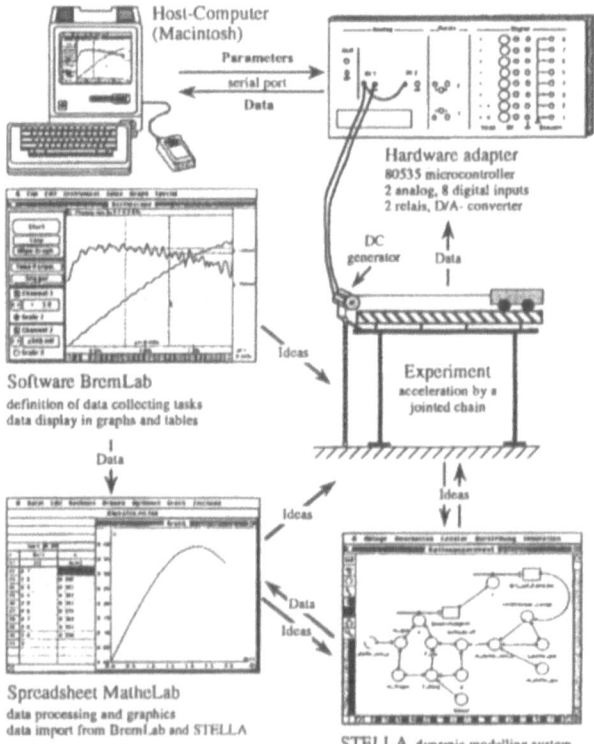

Figure 1 Bremer Interface System (BIS). Overview of the components.

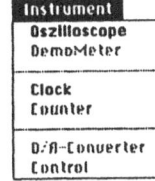

Figure 2   Instrument menu

- The user interface works on a direct-manipulation basis. Parameter definition and graph display are joined in one window. The user is kept informed about all the parameters during the measurement. He stays in control of the process and can stop it at any point. There is no need to learn any commands. All possible actions are "just a mouse-click" away. The keyboard can be disconnected and put aside.

22.3 The System Components

22.3.1 Hardware Adapter

The hardware adapter combines all input and output units together with the power supply in a single case. Its main component is a Siemens SAB50535 micro controller board. This board stores the BASIC and Assembler routines for configuring the input/output periphery and handling the incoming measurement data.

The in-case periphery consists of:

- 2 analog signal amplifiers (-3 mV to +300 V)
- 8 digital input units
- 8 digital outputs
- 2 relays: 8A/220V~
- D/A converter: output +1 V or +10 V

All the inputs are safe against connecting wrong signals. Standard sensors like photogates can be connected directly. Low-cost detectors for light, temperature, and pressure will soon be available.

The micro controller receives a set of measurement parameters from the host computer (e.g., channel numbers, voltage ranges, time span) and configures its periphery accordingly by relays. Data logging is carried out independently. The data are transferred to the host via the serial port.

22.3.2 User Interface BremLab

BremLab is a graphical software environment for defining measurement tasks and displaying the results in graphs and tables. The graphs can be scanned, zoomed and calibrated. Data and parameters can be stored on disk. Graphs and tables can be printed out and transferred to other programs, e.g., to MatheLab (see below). BremLab can be used as a stand-alone application for evaluating stored data sets.

Work with BremLab is based on an *instrument-metaphor.* The software modules are not designed for *experiments* like "charging a capacitor," but for types of measurements: transient voltages (Oscilloscope. DemoMeter), events (Counter) or time intervals/time points (Clock). The user her/himself has to decide which module is suited to the quantities he or she wants to measure, which parameters (e.g., trigger conditions and voltage ranges) have to be set and which values are appropriate. BremLab provides convenient options for the students' own decisions. The set of parameters can be stored in an experiment-specific file. Thus computer-based experimenting does not free students from reflecting on the physics by too many automatic settings. User-friendliness should not lead to an over-simplification of computer based experimenting.

Figures 3 and 4 show the oscilloscope and clock modules. BremLab is available for Apple Macintosh computers. A Windows 3.x version for MS-DOS computers is being developed.

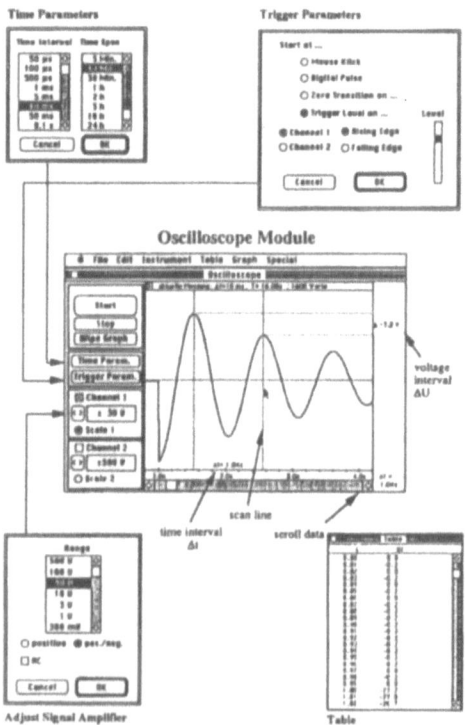

Figure 3 BremLab Oscilloscope module

22.3.3 Science Spreadsheet MatheLab

Tables and charts·of commercial spreadsheets like *Excel* are suited to business purposes. Each table cell has to be defined, and graphing puts more weight on 3-D cake-
graphs than on a flexible calibration of the axes in a scatter diagram. In contrast
MatheLab was designed for mathematics and science instruction, where you mainly
work with columns of data and scattergrams. In MatheLab a column is defined by
writing a formula into its head. You do not need to copy the formula down the cells.
The columns can import data from BremLab and the dynamic modelling tool *Stella*.
All standard mathematical functions are available. The user can define additional
ones himself.

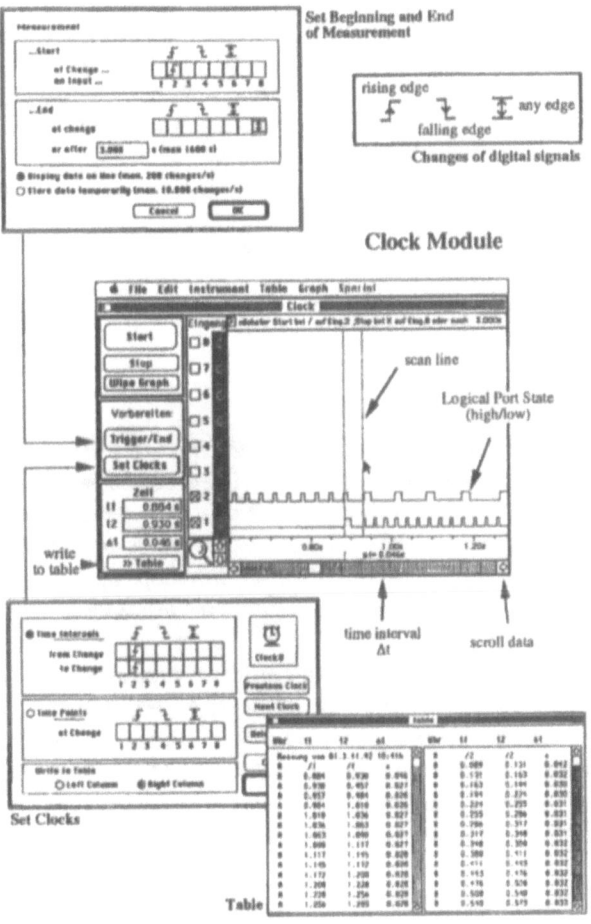

Figure 4 BremLab Clock module

Defining and scaling graphs is very convenient. Direct manipulation options include scanning, linear regression, and graphical integration. Functions can be plotted into the diagrams.

While designing the Bremer Interface System there was an extensive discussion about the question of integrating elaborated data-processing options into BremLab directly. There were two *practical reasons* for adding MatheLab as a separate spreadsheet:

- BremLab would become rather complicated and might confuse the user. Within an advanced operating system, like Macintosh System 7, BremLab and MatheLab can stay separate tools while the user easily switches from one to the other.

- MatheLab can be used as a separate tool in mathematics and physics apart from processing experimental data, e.g., for plotting functions and working out equation-based dynamic models (in contrast to icon-oriented modelling with *Stella*).

Our main consideration however was a *didactic* argument: Collecting data and processing data are different activities. They should be clearly distinguished in an experiment. The student should be aware that the computer does not measure energies or temperatures but that the signals arriving at the adapter are events and voltages over time. The student should be aware of the steps taken after sampling the raw data, e.g., how to calculate the kinetic energy of a pendulum from angle-time data derived from potentiometer readings. Again the guideline *transparency* plays an important role.[2]

22.4 Sample Applications of BIS

22.4.1 Induction

An introductory phenomenon for induction is the rise of a voltage when a coil rotates between the poles of a magnet. Such an experiment can be carried out with a single wire loop (7 cm x 1.5 cm) between the poles of a U-shaped magnet (see Figure 5). Although the voltage only is in the order of 1 mV, it can be directly measured with the BIS adapter and monitored with BremLab's oscilloscope module—even if you turn the table slowly with your hand. Figure 6 shows the result for an angular velocity of about 1 revolution per second.

The results of seven experiments with different angular velocities were scanned from the BremLab diagrams and the data pairs typed into a MatheLab table (Figure 7). Graphing U_ss against Omega clearly shows a linear relationship. The last column calculates the flux density B ($U\_ss = 2 \cdot B \cdot A \cdot \omega$) which is assumed to be homogeneous in the area where the small coil rotates. The average value for B results in about $3 \cdot 10^{-2}$T (compare magnetic field of the earth $3 \cdot 10^{-5}$T).

---

[2] MatheLab was developed by Heinz Weißgerber, Wissenschaftliches Institut für Schulpraxix Bremen, in cooperation with Hans Andraschko and Werner Lorbeer, Zentralstelle für Programmierten Unterricht, Augsburg. Menus and dialogs were partly translated into English for this paper.

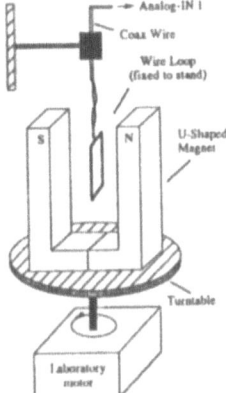

Figure 5 Induction experiment

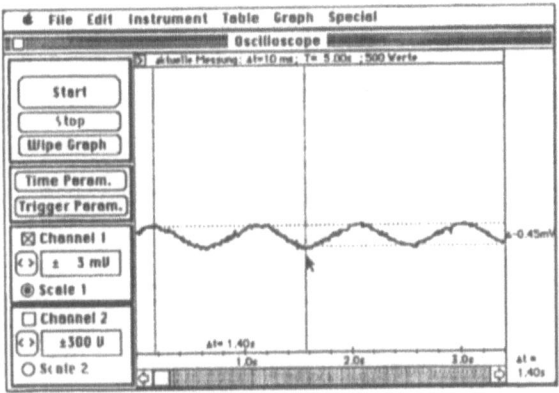

Figure 6 Collecting data with BremLab, Clock module

22.4.2 Gravitational Acceleration

The gravitational constant g can be gained from a simple experiment where you drop a "comb" through a photogate. The "comb" is a transparent plastic ruler onto which stripes of black cardboard were glued with fixed widths and distances. The photogate can consist of a bulb and a photo transistor directly connected to a digital input (see Figure 8).

If you drop the ruler directly above the photogate it takes only a fraction of a second to pass. High→low changes at the digital input have a frequency of more than 200 events per second. The clock module of BremLab has to be configured accord-

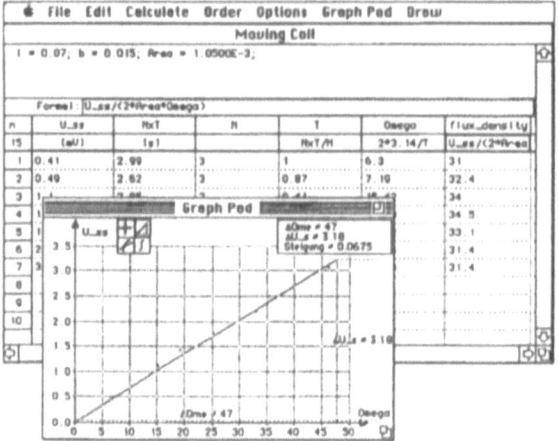

Figure 7   Processing data with MatheLab

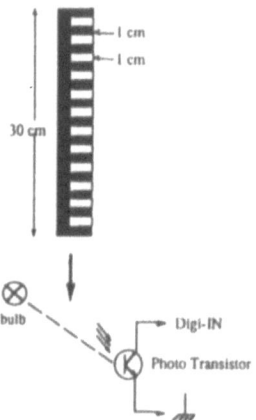

Figure 8   Gravitational acceleration experiment

ingly (see Figure 9). The photo transistor is connected to channel 1. The measurement is triggered by any change on this line; it ends one second later.

More important than the triggered/end parameters is the definition of clocks. Up to seven clocks can be defined by any combination of changes on the eight input channels. In our example two clocks monitor the experiment. Clock A reads differences from rising and falling edges on channel 1 (i.e., the darkening of the photo transistor). Clock B measures the periods passing from falling to rising edges. The "comb" having 13 teeth thus results in 27 values. The data are written to the BremLab

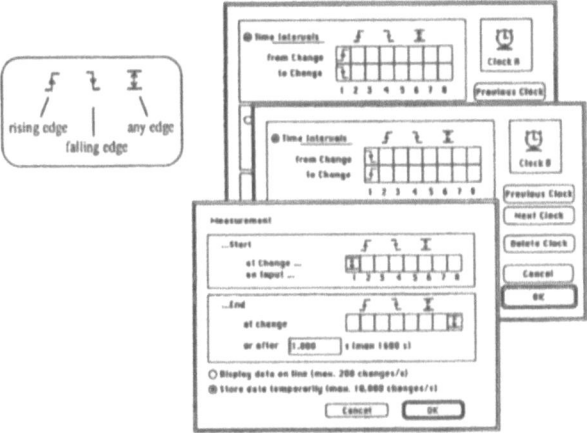

Figure 9 Setting clocks and defining trigger/end parameters

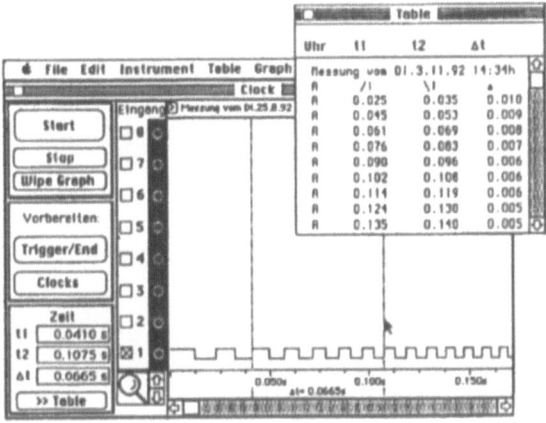

Figure 10 Collecting data with BremLab, Clock module.

table. The port state diagram graphically displays the changes. The diagram can be scanned to read additional time intervals.

For further evaluation the BremLab table is exported to MatheLab (see Figure 11). Interval velocities v are calculated from $\Delta s/\Delta t$ ($\Delta s=1$ cm tooth-width) and assigned to the middle of the time interval. The graph pad plots v against t which shows a linear relationship between the quantities. Linear regression can either be done in a direct manipulation way by drawing a gradient triangle with the mouse—MatheLab displays the parameters of its gradient (9.94) in the upper right corner—or by applying

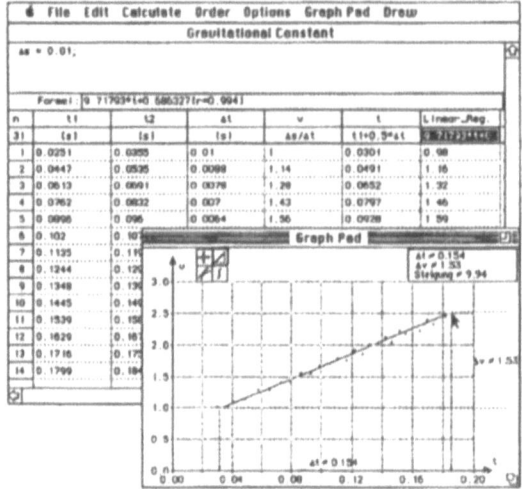

Figure 11 Processing the data with MatheLab.

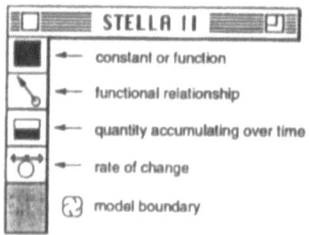

Figure 12 Structure elements of *Stella* models

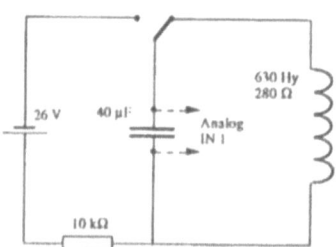

Figure 13 Oscillatory circuit

the operation "linreg (x,y)" in the table. The latter leads to $\Delta v/\Delta t = 9.72$ m/s^2, different from the "by hand" evaluation but also a quite satisfactory value for g.

22.4.3 Measuring and Modelling an Oscillatory Circuit

The didactic concept developed at the University of Bremen for computers in high school physics argues for a wide use of icon-oriented dynamic modelling systems (cf. [7]). They can be applied to model phenomena from nearly all domains of physics, like mechanics, fields, waves, and nuclear physics ([4], Vol. 2). They fulfill our criterion of *universality.* Dynamic modelling is *open* for individual problems and solutions. The physics is absolutely *transparent,* because the student himself formu-

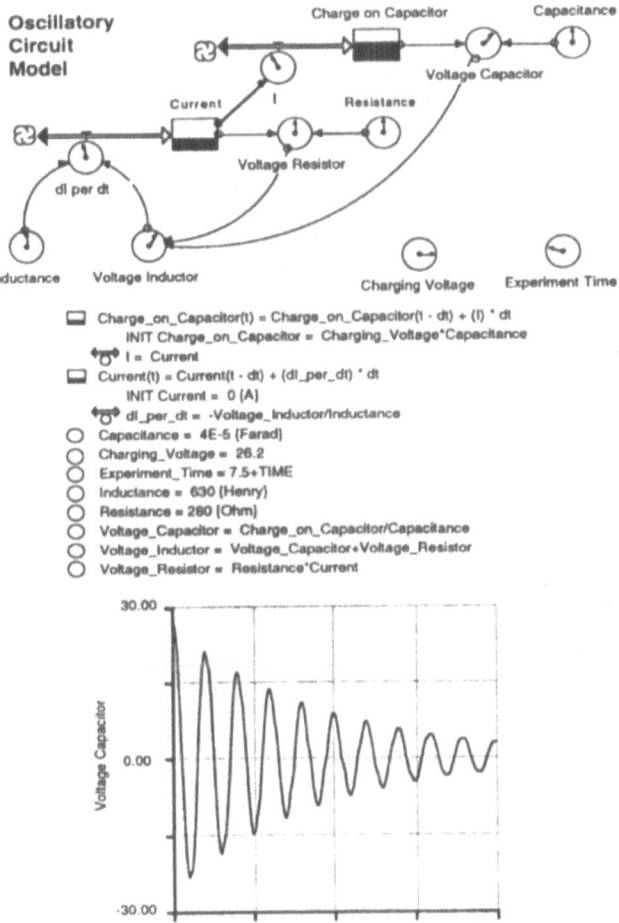

Figure 14 Stella model of the oscillatory circuit. Icon-oriented model, model equations, and graph pad.

lates the conceptual features of a dynamic model graphically on the screen—like in a concept map. Finally, modelling environments like *Stella* or *MODUS* [8] are very *user-friendly*. There is no need to learn a computer language or numerical methods.

Stella models have four types of structural elements: *state* variables, their *rates of change, functions and functional relationships*. These can be placed and connected on the screen with the mouse. The computer translates this iconic representation into a corresponding set of raw equations. The student fills in initial values, constants, and the specific form of a functional relationship. The difference equations are generated automatically. Convenient menus make the scaling of simulation runs with graphs and tables fast and easy.

It is essential that the models are worked out interactively with the modelling system in class—either in student group work or in a classroom discussion. Our research has shown positive effects of this approach for physics teaching (cf. [4] Vol. 4).

Although Stella is not a direct part of BIS it can be easily integrated by exporting simulation data to MatheLab. Measured data and results of simulation runs can be compared in the table and graph pad. This may lead either to changing the model by adapting the parameters, changing the structural features of the model, or revising and repeating the experiment.

The final example shows how measuring and modelling can be linked together. A capacitor is charged and then discharges through an inductor in a closed circuit. Due to its components (C=40 μF, L=630 Hy) the circuit should oscillate at a frequency of 1 Hz.

The voltages monitored across the capacitor are shown in Figure 15. The graph shows a slightly lower frequency than 1 Hz. Damping is quite obvious. Exporting the data to MatheLab can help to decide whether it is exponential.

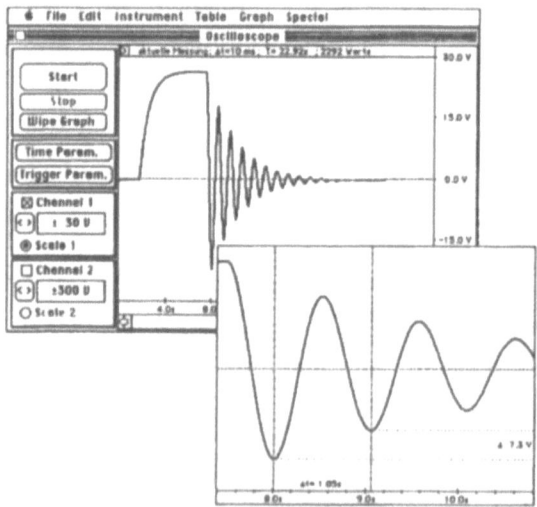

Figure 15 Collecting data with BremLab, Oscilloscope module, zoomed data inserted

In this paper we cannot describe the modelling process of the oscillatory circuit with *Stella* in detail (cf [4] Vol. 2). Figure 14 contains the graphical model, the equations, and the result of a simulation run.

The simulation run shows a much smaller damping than the measured oscillation. The deviations become obvious when both data sets from BremLab and Stella are exported to MatheLab and displayed in one graph pad (Figure 16). As already mentioned, the frequency of the measured oscillation is somewhat lower than the one calculated from the values for capacitance and inductance given on the components and the results from the simulation run based on these values. This deviation can be explained from slight tolerances of the components.

The damping differences between model and experiment are more striking. If the model is assumed to be structurally correct, i.e., its physics is properly modelled, then the value of 280Ω printed on the inductor as its resistance must be wrong for the circuit. A better value for R can be gained from processing the measured data with MatheLab, so that the exponential decrement of the amplitudes can be determined (damped oscillation: $U=U_0 \cdot e^{-k \cdot t} \cdot \sin \omega t$; with $k=R/2L \Rightarrow R=2L \cdot k$). Column 3 in Figure 17 introduces U1* as the absolute value of U1 and at the same time eliminates U1*=0 to avoid difficulties with forming ln(U1*).

The graph in Figure 17 presents ln(U1*) over t. With the aid of the graph pad toolbox a straight line is drawn through the maxima. MatheLab automatically displays its gradient (0.4). The above relationship for the resistance of the circuit results in R≈500Ω. If you go back into the model and change the parameter "Resistance" accordingly—perhaps also slightly adjusting "Capacitance"—model and measurement agree satisfactorily.

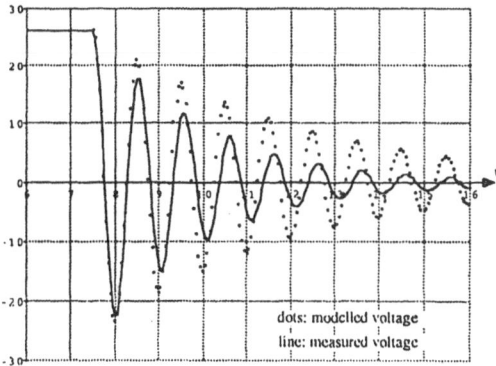

Figure 16 MatheLab graph pad with data from simulation and measurement

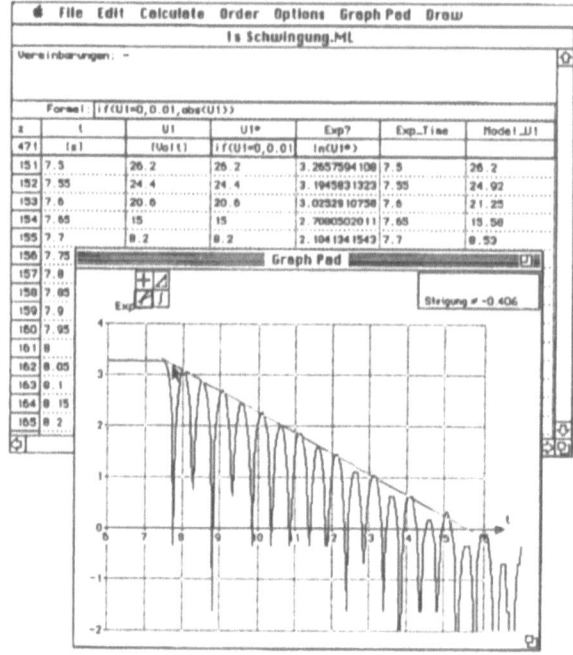

Figure 17 Processing the data with MatheLab

The example shows how measuring and modelling can go hand in hand, provided that the tools are flexible, so that they can be adapted to the specific needs of the user. BIS supports the investigation of oscillatory circuits in more detail than can be shown here. A group of university students worked out that the additional damping does not result from an Ohm's resistance of about 500Ω (instead of 280Ω) but from energy losses due to re-magnetizing the iron core of the coil during a period. Hysteresis loops were recorded by BremLab's oscilloscope mode and evaluated by MatheLab. The *Stella*-model was then altered to include the effect.

22.5 Conclusion

The Bremer-Interface System has been used with students on the university and high-school levels. Several schools in Bremen are already equipped with the system. Schools can usually only afford to buy one interface (the hardware costs about $600; BremLab and MatheLab are free). There are two ways to handle this problem in student experiments. The first is to put computer and interface on a trolley and go from desk to desk during a lab session to make the recordings. A second way is to record data in a demo experiment and spread them on the network for evaluation in the computer lab. The BIS software runs without a hardware adapter connected.

Experience shows that only a short time is required for students to become familiar with the user interface or to connect probes. Choosing physical parameters for data logging was more difficult. Problems in using BIS, for instance, arose when students (or teachers) had no clear idea about an appropriate time scale for a voltage measurement, e.g., the induction caused by a magnet falling through a coil.

A promising way to use computers for enhancing physics education is to combine measuring and modelling. Supported by easy transfer of data from one tool to the other, data from an experiment and predictions of a model can be directly compared. Discrepancies stimulate the reconsideration of theoretical assumptions that go into the model or the modification of experiments. New ideas can quickly be realized and tested. This method has been sucessfully tested at the university level. Trials in schools have begun.

References

1. Wynn, V.T. & Pittard, S.: Some single experiments in medical physics on the heart using a BBC micro. *Physics Education,* 26,127–131 (1991)
2. *MacTimes* Roseville, CA PASCO 1993
3. Ellermeijer, A.L.: *An introduction into IP-Coach.* University of Amsterdam, Centre for Microcomputer Applications 1992
4. Niedderer, H., Bethge, T. & Schecker, H.: *Computers in physics education* (4 Vols.): Vol. 1: Didactic concept, Vol. 2: Materials for modelling and simulation, Vol. 3: Bremer Interface System, Vol. 4: Empirical research. University of Bremen, Institute of Physics Education 1992
5. *Stella II.* Hanover, NH: High Performance Systems 1990
6. *Interactive physics.* San Francisco, CA: Knowledge Revolution 1991
7. Schecker, H.: Learning physics by making models. *Physics Education,* 28, 102–106 (1993)
8. *MODUS* graphical modelling software (PC). Duisburg: Comet Verlag 1992

23. Some Experiments in Physics Education: Using a Force Sensor Connected to a Computer

Örjan Nilsson

Jönköping University

23.1 "The Measuring Tool"—A Data Collecting and Handling System

A data collecting and handling system called "The Measuring Tool" is used in phys-ics education at many Swedish secondary schools ("gymnasium"). It was developed at the Institute of Technology at Jönköping University. The system is used with IBM PC and compatible computers. It consists of a 12-bit AD-converter card that is mounted inside the computer, an interface box and software. Data collection can be performed in two channels. The maximum sampling rate is 70,000 per second, sufficient for sound studies, for instance. The ADC card also contains digital I/O for the connec-tion of lightgates, a GM counter or a stepper motor. The software is menu driven and very easy to handle. The collected data are presented in tables and diagrams. Differ-ent standard mathematical functions can be fit to the points in a diagram by the method of least-squares. There is also a modelling routine so that recorded data may be com-pared with a mathematical model. Numerical integration and derivation can be per-formed as well. For use with "The Measuring Tool" lightgates and sensors for force, magnetic flux, pressure, sound, temperature, etc., have been developed. The force sensor has proved to be especially useful, and a short description of the force sensor and a few examples of its use are given below.

23.2 The Force Sensor

Four strain gauges are mounted on a stainless steel blade. Two gauges are glued to each side of the blade and connected in a bridge configuration to an operational amplifier. There are two models of the force sensor. In a larger model, the steel blade (dimensions 200 mm x 20 mm x 1 mm) is mounted firmly on a wooden base together with the amplifier, which provides an output signal of about 0.2 V for each newton of force applied to the end of the blade. The signal from the amplifier is fed to the AD converter. In a smaller model, intended to be mounted on moving objects, the dimensions of the steel blade are 60 mm x 10 mm x 0.8 mm, and the strain gauges are connected to the amplifier with a flexible cable.

23.3 Some Experiments Using the Force Sensor

23.3.1 The Impulse-Momentum Equation

The force sensor (large type) is placed above an air track together with a lightgate (see Figure 1). As a glider on the track collides with the steel blade, the varying force is recorded by the computer. The speeds before and after the collision are determined using the lightgate to measure the passing time.

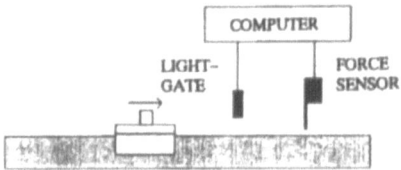

Figure 1 Arrangement of force sensor, air track and lightgate in impulse-momentum experiment

After an experimental run the students get a printout of the lightgate passing times and a table of force as a function of time. After weighing the glider and measuring the width of the card on the glider, which interrupts the light beam, the total change in linear momentum can be determined. The force is plotted as a function of time and the impulse is calculated by estimating the area under the graph. Different masses and initial velocities can be used by different groups of students. The results are put together and the impulse-momentum equation, $\Delta mv = \int F \, dt$, is verified.

The computer can of course be used to calculate both the change in linear momentum and the impulse, as well as to give a graph of the force as a function of time (see Figure 2). But in order to let the students participate actively, the manual treatment of data as outlined above is recommended.

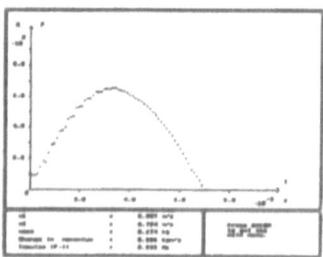

Figure 2 Force as a function of time in impulse-momentum experiment

In the experiment above, the passing times were determined automatically by the computer using a lightgate connected to one of the digital inputs of the AD-converter card. By connecting the lightgate to one of the analogue inputs and the force sensor to the other, a manual and perhaps more accurate determination of the passing times may be performed. In Figure 3 the recordings from the two channels are shown. The measurement started when a narrow "triggering pin," placed in front of the card on the glider, passed the lightgate. It is clearly seen that it takes some time for the lightgate voltage to change level. This observation makes it natural to discuss how the passing times should be determined in order to be as accurate as possible. From the graph of force versus time it is also obvious that the steel blade oscillates after the collision, a fact that might initiate a discussion of what happens with energy during the collision process. It may be pointed out that in this study of the impulse-momentum relation, Newton's third law of motion is utilized. The *measured* force in the collision, i.e., the force on the firmly mounted sensor, is assumed to be the same as the force on the glider (see collision studies below).

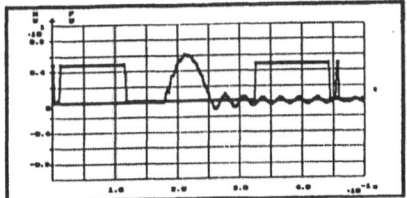

Figure 3   Dual-channel measurements in impulse-momentum experiment

23.3.2 Collision Studies

By mounting force sensors of the smaller type on two vehicles, for instance two gliders on an air track, and performing measurements in two channels, the forces acting on the two vehicles during a collision can be studied. Figure 4 shows a recording where the mass of one of the gliders is about twice that of the other glider. It is obvious that the force versus time is the same for both gliders although the masses

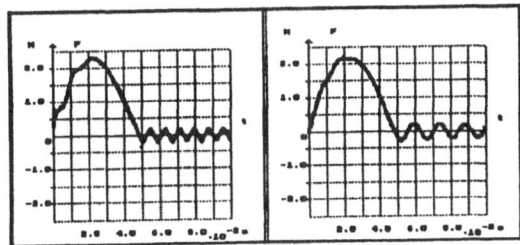

Figure 4   Measuring the effect of glider mass in collision studies

are different. By varying the masses and the initial speeds of the two vehicles, the validity of Newton's third law of motion is easily demonstrated.

23.3.3 Harmonic Oscillations

A body is fixed firmly to the end of the steel blade (Figure 5). The body is set into oscillation and the force as a function of time is recorded (Figure 6).

Figure 5 Arrangement of mass and steel blade in harmonic oscillation experiment

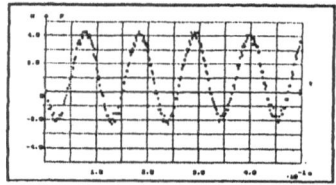

Figure 6 Measuring body oscillations in harmonic oscillation experiment

If the mass of the body is known and the mass of the steel blade is neglected, the vibration amplitude can be estimated. Using the relations of simple harmonic motion and the relation $F = m \cdot a$, the vibration amplitude can be expressed as

$$A = \frac{F_{max}}{m \omega^2}$$

In the recording of Figure 6 the mass m was 100 grams. The force amplitude F_{max} is approximately 3 N, the angular frequency $\omega \approx 10$ s^{-1} and thus the vibration amplitude A becomes approximately 7 mm.

Using a simple model, neglecting the mass of the steel blade and the damping, the theoretical vibration period may be estimated using the formula

$$T = 4\pi \sqrt{\frac{ml^3}{Ewt^3}}$$

where m is the mass of the body, l the length of the free part of the blade, w and t the blade width and thickness respectively, and E the modulus of elasticity. The present experiment was performed with $m = 100$ g, $l = 140$ mm, $w = 20$ mm, $t = 1.0$ mm and $E = 210$ GPa. This gives the theoretical value $T = 0.10$ s, which is in agreement with the experimental value.

23.3.4 Damped Oscillations

Using the same setup as above but with another time between the recordings, the pattern of damped oscillations may be recorded (see Figure 7).

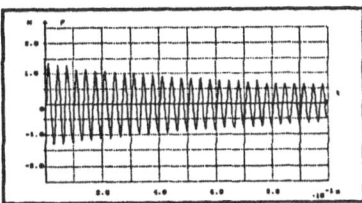

Figure 7   Damped oscillation pattern

The mathematical expression of free oscillation with weak damping may be written as:

$$x = Ae^{-\lambda t} \cos (\omega t + \varphi)$$

If x_1 and x_2 are two successive amplitudes, the logarithmic decrement k can be derived:

$$\frac{x_1}{x_2} = \frac{e^{-\lambda(t+T)}}{e^{-\lambda t}} = e^{-\lambda T} = e^{-k} \quad \text{and thus} \quad k = \ln\frac{x_1}{x_2}$$

In order to get a more accurate value, two amplitudes with a separation of, for instance, ten periods can be measured. In this case

$$k = \frac{1}{10}\ln\frac{x_1}{x_{11}}$$

By mounting cardboard sheets of different sizes on the vibrating blade, the relation between damping and the area of the cardboard sheet may be investigated.

23.3.5 Coefficient of Friction

The force sensor is mounted vertically on a wooden plate that can move on a horizontal surface (see Figure 8). While pulling in a wire connected to the force sensor, the force as a function of time is recorded (see Figure 9).

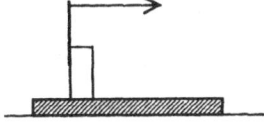

Figure 8   Force sensor placement for coefficient of friction measurement

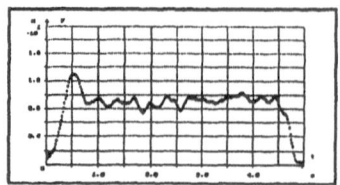

Figure 9   Force as a function of time in friction coefficient experiment

From Figure 9 it is obvious that the force immediately before the body starts moving is higher than the medium force during the movement. The mass was 4.2 kg and thus the friction coefficients just before and during the movement can be determined to 0.3 and 0.2 respectively.

23.3.6 Centripetal Force

One end of a wire (length 1–2 m) is fixed to the force sensor. In the other end of the wire a body is swinging like a pendulum (see Figure 10). In Figure 11 the signal from the force sensor, i.e., the *vertical* component of the force as a function of time, is shown.

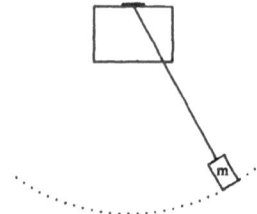

Figure 10 Arrangement of pendulum and force sensor in centripetal force experiment

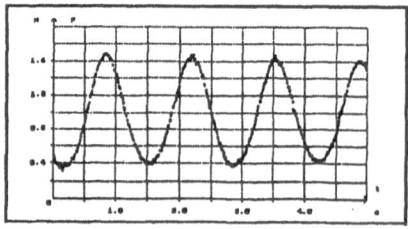

Figure 11 Force as a function of time in centripetal force experiment

The vertical component of the force is

$$F_V = mg\left(3\cos^2\alpha - 2\cos\alpha_{max}\cos\alpha\right)$$

where a is the angle of deflection. From the recording of Figure 11 it may be concluded that the mass was 100 g and the maximum deflection angle 50 degrees.

23.4 Discussion

The use of microcomputers in the laboratory opens up new opportunities for learning and understanding. Data from fast events like collisions are easily recorded, and the fast data recording makes it possible to explore details in the pattern of recorded data that otherwise may not have been seen.

A very important part of physics is the development of theoretical models and the comparison of experimental results with what is predicted by a specific theoretical approach to the problem. In "The Measuring Tool," the experimentally received data may be exported to the Excel or Graph-in-the-Box spreadsheet programs. Using the facilities of these programs for calculation and presentation, development of theoretical models and comparison of theory with experiment is easily done.

Another aspect of the automated data recording is that the use of different sensors shows important technical applications of basic physical concepts. Most young people find the practical use of theoretical knowledge very stimulating. Thus a discussion of the working principles of a specific sensor followed by an exploration of experiments in which the sensor may be used will certainly increase the students' interest.

24. Microcomputer-Based Laboratories at A. Mickiewicz University

Henryk Szydłowski

A. Mickiewicz University

Abstract. A brief history of teaching of computer science and application of microcomputers for educational purposes in general high schools and at universities is given.

One of the on-line experiments (thermal conduction study) prepared for university student laboratories is discussed. The information on Microcomputer-Based Laboratories (MBLs) organized at A. Mickiewicz University within the MAPETT-TEMPUS project is included and ways of using MBL in the education of physicists and physics teachers are proposed. A few interesting experiments designed to be performed in MBLs are presented.

24.1 Introduction

The idea of applying microcomputers in physics teaching directly followed the appearance of the first ZX86 computers in Poland. Already at that time the teaching and calculation programs to be used at schools were written and popularized even through the radio. Unfortunately, unreliability of the keyboards and of the tape recorders used as external memory made teachers reluctant to use computers at school. The appearance of subsequent computer generations such as ZX Spectrum and Junior (Poland) did not improve the situation. The first computer of satisfactory reliability was Amstrad of Schneider and Atari, which was soon eliminated by new generations of IBM and IBM compatible computers.

The application of microcomputers in teaching started with simulation programs. New programs were written, and programs imported from Western European countries were modified. One group includes typical educational programs dealing with atomic phenomena that couldn't be observed in a laboratory, e.g., radioactive decay. Another group comprises programs simulating experiments which cannot be performed at school, e.g., light diffraction on slits. Still another topic that lends itself to computer programs is behavior of gases already known from the university textbooks by F. Reif [1]. In addition, programs testing the knowledge of students, or programs designed for individual student or small group work were written. An example created in our institute is a set of programs on uncertainties of measurement [2].

24.2 Computer Science in Polish Schools

Let me now briefly mention the problems related to teaching of computer science elements at general high schools. At the beginning (1985) computer science was introduced to some high-school syllabi in an artificial way in the sense that students had no chance to use computers, which were unavailable at schools. For this reason the subject was treated only theoretically, was boring for the students and could only discourage them. Later the schools organized computer laboratories equipped with Spectrum or Junior computers and the students were taught the fundamentals of computer programming. The teachers of this subject were usually mathematicians and only those who were computer fans themselves could succeed in getting students really interested in this subject. Unfortunately, most teachers do not use computers and are usually reluctant to introduce new ideas to school. Sometimes teachers of physics make an exception as they realize the need to use computers in their work with students.

The situation at universities is much better as each research worker has already gained some experience in working with computers. However, the actual teaching of computer science is also dependent on the interest of computer fans.

24.3 On-line Experiments: Thermal Conductivity

In university centers, the idea of employing microcomputers in "on-line" experiments, i.e., for automation of measurements, appeared very early. To this effect, interfaces and software, adjusted to measure one or a few physical quantities and to control this experiment, were constructed. Unfortunately, this equipment was very expensive and highly specialized such as, for example, a set for measuring thermal conduction, constructed in our institute [3]. This set, shown in Figure 1, includes a thermally well-isolated metal rod. Through the length of the rod at certain intervals thermocouples were attached, permitting a continuous measurement of temperature. With this set it is possible to measure thermal conductivity λ in the metal rod by using two methods. One of these methods is based on the equation of thermal conductivity

$$UI = \lambda \frac{T_1 - T_n}{h_n} \frac{\Pi}{4} d^2 \qquad (1)$$

where:

UI - denotes the power consumed by the electric heater,

T_1 - denotes temperature at the thermocouple denoted by number 1 located close to the heater,

T_n - denotes temperature at the nth thermocouple, n,

d - denotes the diameter of the rod.

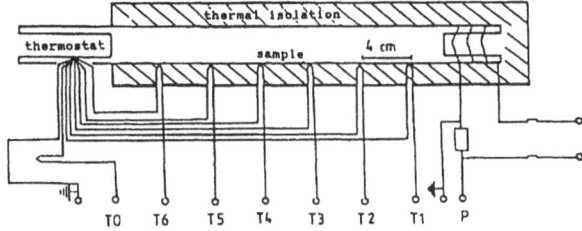

Figure 1   Construction of the thermal conductivity measuring device for metals

The method boils down to the measurement of temperature gradient

$$\frac{T_1 - T_n}{h_n}$$

in a rod with its one end heated to a high temperature and the other cooled by an ultrathermostat to keep it at a constant low temperature (Figure 2).

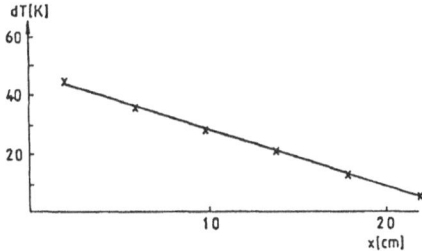

Figure 2   The temperature gradient in a rod

The second method involves observation of temperature waves. The temperature waves were induced by subsequent turning on and off of the heater. Thermal impulses propagate towards the cooler end and are registered by successive thermocouples. The resulting diagram is presented in Figure 3. From the phase shifts at successive thermocouples the thermal conduction coefficient can be calculated according to the formula:

$$\lambda = \frac{(\Delta l)^2 c \rho}{2 \Delta t \cdot \ln(A_1 / A_2)} \qquad (2)$$

where:

c-   denotes the thermal capacity of the rod of density r,

df - denotes the phase shift of temperature waves reinstated by the thermocouple located at distance Δl,

T -   denotes the period of the thermal wave,

A_1 - denotes the amplitude of the waves diminishing with distance from the heater.

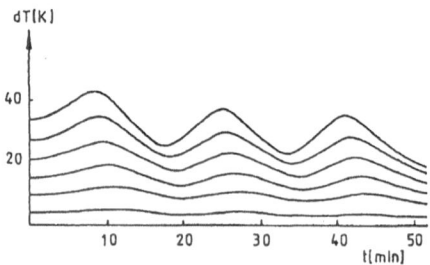

Figure 3 The temperature waves created in the thermal conductivity measuring device

This experiment is not fully automated, as the student decides when to turn the heater on and off, when to conclude the measurements, and in the case of temperature waves, which points on the experimental curve are used to calculate thermal conduction. Special software was written for running and controlling the experiment and for carrying out numerical calculations of measurement results and their uncertainties. This set has two additional devices which use the same interface and are designed for measuring thermal conduction of poor conductors and of powdered materials.

24.4 Microcomputer-Based Laboratories

In the last couple of years, in a few Polish university centers' school teacher training centers, several attempts have been made to construct universal interfaces which would permit using a computer as a universal measuring device. A few prototypes have been constructed. Unfortunately, there is not sufficient software to use them; therefore they are of limited value to the average high-school and university teacher. In comparison with the high rate of technology development in Western Europe and in the USA, our pace of research and discovering things "anew" is very slow, partially due to our modest means. Consequently we always lag behind the West in this respect. One way of bridging the gap for our laboratory and for many other research centers in our country is through the aid we get through the MAPETT-TEMPUS project. This project, initiated in 1990, enabled our institute to take and adopt two independent solutions of the Universities in Amsterdam [4, 5] and Kiel [6], to buy equipment for two student physical microcomputer-based laboratories, and to spread these solutions to other university centers through the institution of resonant universities. Additionally, the Amsterdam program has been introduced to a few Polish high schools, i.e., eight pilot high schools have been supplied with individual sets. Each of the schools is connected with one of the universities.

24.5 MBL in Teaching Students of Physics and Physics Teachers

At present we have to assume that our freshmen do not know the fundamentals of computer science and cannot use computers. A few students show advanced knowledge of computer science but comprise a small percentage of the first-year students. Therefore we have to prepare the educational program starting from the basics, such as learning the keyboard functions. We started teaching computer science in 1987 with one laboratory equipped with IBM XT computers. Most of the time was devoted to teaching the fundamentals, but the students were also taught computer programming in different languages.

Starting of the MAPETT project opened new perspectives. In 1991 two other laboratories were organized. One of them is based on the solution proposed by Amsterdam University and the other on the solution proposed by Kiel University. Instructions to the laboratory experiments were translated into Polish, as was the IP Coach program description.

In the first year of work (second year of MAPETT) the laboratory based on the solution proposed by Kiel university has been used to introduce the beginners to the fundamentals of using computers in physics. This training is guided by Lincke's manuscript [6, 7]. From this bulk collection of experiments we have chosen for the beginning only a few [8]:

- The operation of interfaces for measurement and control.

- Reading out the A/D-converters: measuring voltages, slow storage oscilloscope (1- and 2-channel), fast storage oscilloscope with trigger.

- Switching relays and reading ports: traffic lights, control of water level, models of D/A and A/D converters, charging and discharging a capacitor (2 point control).

- Operating with stepping motors.

- Measurements of time using PIT and light gates: period of the pendulum, falling ladder.

The Kiel solution was also used to perform experiments illustrating behaviors of waves and vibrations on the introductory physics laboratory level. The following experiments proposed by R. Lincke [7] were available:

1. Resonance in an AC-circuit

2. Resonance of the pendulum

3. Oscillations of large amplitude

4. Damped oscillations

5. Interference of the sound waves

6. Velocity of sound waves

7. Oscillations of coupled pendulums

The laboratory based on the Amsterdam university solution has been used to perform experiments in the field of electricity and magnetism, and as a didactic laboratory for future teachers of physics. In this laboratory the following experiments were available for first- and second-year physics students:

1. Electromagnetic induction
2. Electric discharge and capacity of the condenser
3. Electric conductivity; temperature dependence
4. Magnetic hysteresis of the ferrimagnetic materials
5. Electromagnetic waves
6. Characteristics of the diode and transistor

Polish didactic materials for students of all physics experiments were prepared.

24.6 New Experiments in MBL

In the second year of the MAPETT project operation we organized another laboratory for the students working on their M.Sc. theses and for teachers who wish to test the possibilities of computer application in teaching. As a result of their work some new experiments were elaborated and special equipment was constructed.

- Experiments with an air track based on the IP Coach. In these experiments, many solutions for measuring time, velocity and distance were used, especially constructed light gates, a pulley with equidistant holes, an ultrasonic position meter, registration of velocity by tacho-generator, etc. A set of experiments necessary for teaching kinematics, dynamics, conservation of energy, laws of collisions, and conservation of momentum on the high-school level were completed.

- Examination of the free fall. A special instrument with four light gates was constructed (Figure 4). The highest gate is connected with a device which starts the free fall of the sphere and the measurement of the time. The next three light gates register time at the points separated by the distance h_1, h_2 or h_3 from the highest one. The instrument enables us to examine the very simple phenomenon of free fall of spheres made of various materials.

- Measurement of the hysteresis loop. The sample made of a ferro- or ferrimagnetic material is located inside two coaxial coils (Figure 5). The first one is a magnetizing coil powered by alternating current, and the second one is used as a detecting coil. A special high-power generator is used as a power supply for the magnetizing coil. The electric current is proportional to the voltage on the resistor R_M (R1 in the figure) connected in a series with the magnetizing coil. This voltage is connected to channel 1. The signal induced in the measuring coil is integrated by the RC integrating circuit and applied to the channel 2. The use of a multiscope enables us to present both signals on the computer screen. By processing the data, we receive the hysteresis loop shown in Figure 6. The instrument designed and constructed

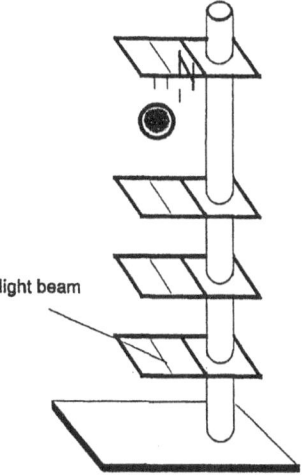

Figure 4 A set for free-fall experiments

allows us to take quantitative measurements of the hysteresis loop, coercion force, magnetic susceptibility magnetization, etc., for ferro- and ferrimagnetic materials.

- Examination of the transformer. A full set of measurements for a transformer consists of voltage and current in both windings. The electric circuit is shown in Figure 7. A four-channel multiscope enables us to complete all necessary measurements. Processing enables us to calculate all parameters: power on input and output coils, phase shift, energy consumption, etc.

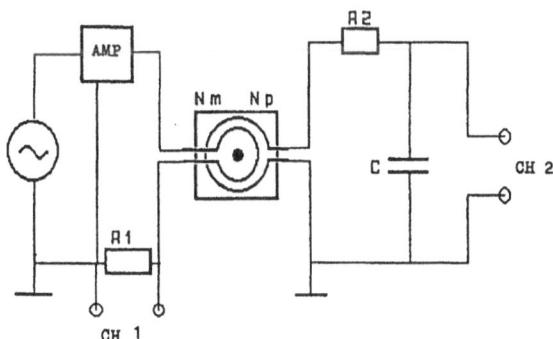

Figure 5 Electric circuit for measurement of the hysteresis loop

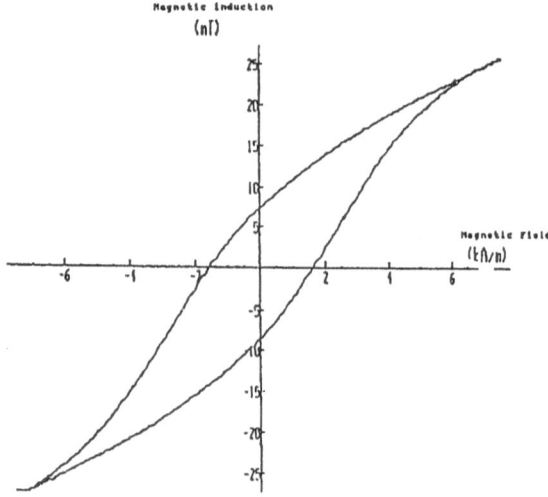

Figure 6 Hysteresis loop obtained in MBL experiment

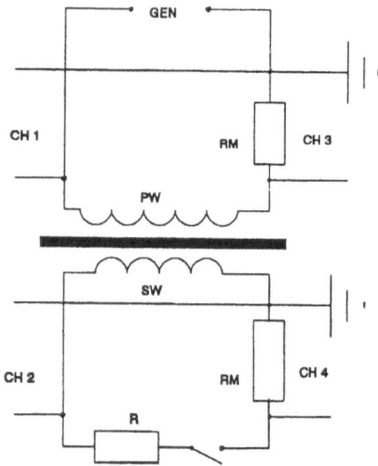

Figure 7 Electric circuit of the transformer

References

1. Reif, F.: *Statistical physics*. New York: McGraw-Hill 1967
2. Szydłowski, H., and Zawieja, B.: *Uczace i obliczeniowe programy komputerowe do statystyki matematycznej* (Teaching and calculation software for mathematical statistics). Zielona Góra: Pedagogical University 1990
3. Szydłowski, H., and Smuszkiewicz, R.: On line measurement of the thermal conductivity coefficient in the physics laboratory for students, *Postepy Fizyki* 42, 335 (1991)
4. *IP Coach 3 Manual*, CMA, Amsterdam 1992
5. Mulder, C., and Nevel, K.: IP Coach and the educational software crisis, *Technology and School, Report of the PATT Conference in Poland*. Zielona Góra: Pedagogical University Press 1991, pp. 93–97
6. Lincke, R.: Computer assisted experiments on physics and control, Manuscript, Kiel 1991.
7. Lincke, R.: Physicalische Experimentale fit PCs Loseblattsammlung. Geislingen an der Steige: Neva GmbH
8. The laboratories were organized by:
   Leszek Wołejko, Dr.—using computers in physics,
   Grazyna Dudziak, M.Sc.—electricity and magnetism,
   Henryk Szydlowski, Dr. habil.—vibrations and waves.

25. Microcomputer-Based Laboratory— The Observation of Light Diffraction and Interference Patterns

Ewa Mioduszewska

University of Gdańsk

Abstract. This article illustrates the use of a lab computer for MBL modelling and observation of light diffraction and interference patterns.

25.1 Introduction

Using microcomputers in the physics laboratory extends students' power to observe and model. The Microcomputer-Based Laboratory is a laboratory "tool" that can help students learn [2]. Students using MBLs are like scientists: they construct their own knowledge about the physical world and build essential physical concepts based on real experiments [3][4].

One important physics lab experiment is the observation of light-diffraction patterns. This experiment is generally based on measuring the relative positions of light intensity maxima and minima. By applying a microcomputer with appropriate hardware and software in this experiment, we can provide a powerful learning experience.

25.2 Experiment

To perform this experiment we used a PC/AT microcomputer with a light sensor, an angle sensor, a Universal Interface Adapter (UIA) board, the measuring console and the software package for the UIA-board, *IP Coach* [1]. This experimental setup and package is a product of CMA, the Centre of Microcomputer Applications at the University of Amsterdam. To produce diffraction patterns we used light from a low power He-Ne laser (Polish school laser type LG-200). To detect spectra we applied a simple device, shown in Figure 2, including a light sensor to measure the intensity and an angle sensor to measure the position. The sensors connect to the computer via the UIA-board which also performs analog-to-digital conversion. Before the actual experiment, we calibrated the sensors.

In our experiment, the calibration of the light sensor is relative. The brightest reading is set to 100 while the background (the minimum intensity) is set to 0.

The position sensor is a mass hanging over an angle sensor.

After the initial calibration of equipment, which can be saved on disk before the experiment, we can observe light diffraction and interference patterns. The experimental apparatus and the program *Multiscop* allows the observation of diffraction patterns obtained for a single slit, a double slit, and a grating with *n* slits. The results for single and double slits are illustrated in Figures 3 and 4.

More advanced students can compare experimental results with the theoretical light intensities generated from modelling software. The CMA software package, *Modelling*, was used to generate Figures 5 and 6.

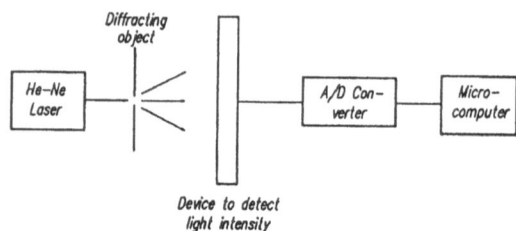

Figure 1   Schematic diagram of the experimental apparatus

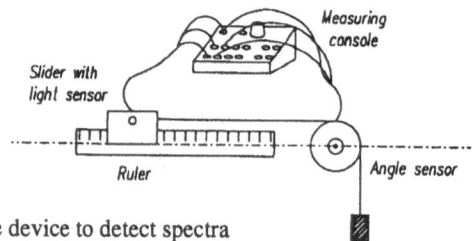

Figure 2   The device to detect spectra

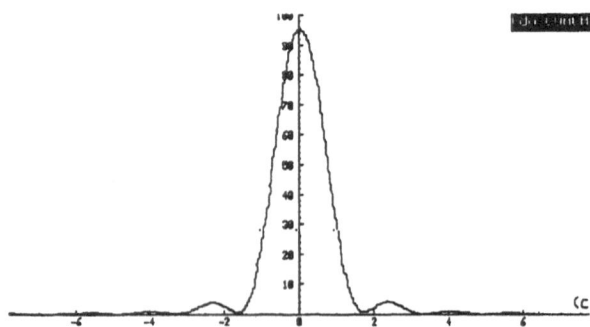

Figure 3   Single slit diffraction pattern

25.3 Conclusions

The main advantages of this technique are that the interference/diffraction patterns can be displayed quickly and the information on intensity and fringe positions can be stored in the computer. Students can change the slit width or slit separation and observe an immediate change in the display. The stored data can be compared with the theoretical predictions, from analytic or dynamic models. To create models, students have to understand physical concepts and develop their own concepts. The models show how the results all come from the same process of adding up waves with phase difference.

The theoretical analysis of light intensity distribution in diffraction patterns extends "observation" of situations which can't be prepared experimentally. For example we can observe change in the intensity distribution when the number of slits changes.

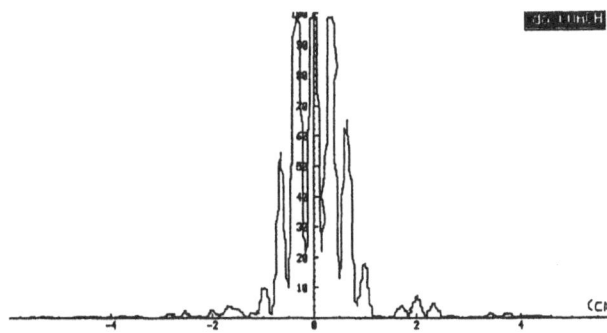

Figure 4 Double slit diffraction pattern

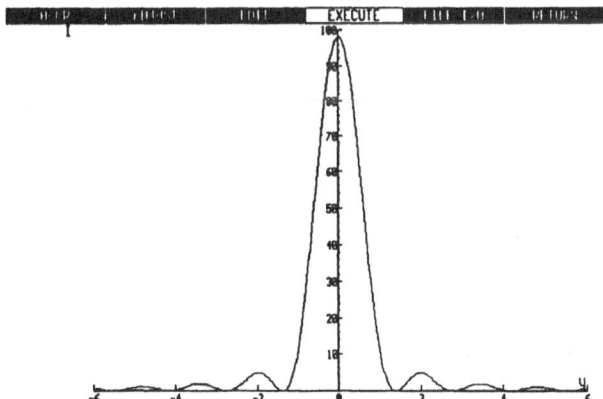

Figure 5 Single slit model

Reducing the drudgery of data collection, time-consuming calculations, and the preparation of graphs and tables allows students to concentrate on the scientific ideas connected to their investigations. Students gain the knowledge from the physical world and construct the theories necessary to understand the physical phenomena for themselves [4][5].

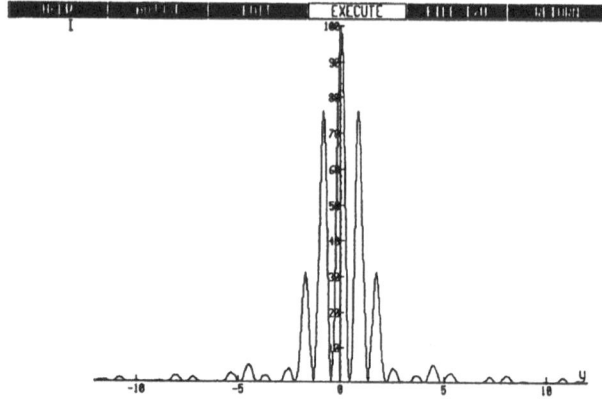

Figure 6 Double slit model

References

1. Manual of IP-Coach.
2. Rogers, L.T.: The computer-assisted laboratory, *Phys. Educ. 22* (1987)
3. Taylor, E.: Comparison of different uses of computers in teaching physics, *Phys. Educ. 22* (1987)
4. Thornton, R.K.: Tools for Scientific Thinking—Microcomputer Based Laboratories for Physics Teaching, *Phys. Educ. 22* (1987)
5. Thornton, R.K.: *Tools for Scientific Thinking: Learning Physical Concepts with Real-time Laboratory Measurement Tools*, Proceedings of the conference Computers in physics instruction, North Carolina 1988

List of Contributors

Emilio Balzano
Dipartimento di Scienze Fisiche
Universita´ di Napoli
Mostra d'Oltremare pad. 20, I 80125
Napoli, Italy

Stephen Bannasch
The Concord Consortium
37 Thoreau Street
Concord, MA 01742 USA
stephen@concord.org

Leslie Beckett
University of London Institute
of Education
20 Bedford Way
London WC 1H 0AL

Boris Berenfeld
TERC, Inc.
2067 Massachusetts Avenue
Cambridge, MA 02140 USA
Boris_Berenfeld@TERC.edu

C. de Beurs
Physics Education Department
University of Amsterdam
Nieuwe Achtergracht 170
1018 WV Amsterdam, The Netherlands
debeurs@phys.uva.nl

Richard Boohan
University of London Institute
of Education
20 Bedford Way
London WC 1H 0AL

Betty Collis
Faculty of Educational Science
and Technology
University of Twente, Postbus 217
7500 AE Enschede, The Netherlands
COLLIS@edte.utwente.nl

V.J. Dorenbos
Physics Education Department
University of Amsterdam
Valckenierstraat 65
NL-1018 XE Amsterdam
The Netherlands
dorenbos@phys.uva.nl

G.H. Dulfer
Basic Science Programme Unit
Centre for Development Cooperation
Services
Vrije Universiteit Amsterdam
The Netherlands
cdcs@dienst.vu.nl

A.L. Ellermeijer
Physics Education Department
University of Amsterdam
Nieuwe Achtergracht 170
1018 WV Amsterdam, The Netherlands
ellermei@phys.uva.nl

Yael Friedler
School of Education
and the Israel Science Teaching Center
Hebrew University of Jerusalem
Givat-Ram, Jerusalem 91904, Israel

Robert G. Fuller
University of Nebraska-Lincoln
Lincoln, NE 68588-0109 USA
rfuller@unlinfo.unl.edu

A.M. Gonçalves
Departamento de Fisica da Faculdade
de Ciências de Lisboa
Ed. C1 4° Piso
Campo Grande
P-1700 Lisboa, Portugal

Frits Gravenberch
National Institute for Curriculum
Development (SLO),
Postbox 2041, 7500 CA Enschede
The Netherlands

Ard Hartsuijker
National Institute for Curriculum
Development (SLO),
Postbox 2041, 7500 CA Enschede
The Netherlands

Dieter Heuer
Lehrstuhl für Didaktik der Physik
Physikalisches Institut der Universität
Würzburg
Am Hubland, D-97074 Würzburg
Germany
heuer@wpax01.Physik.Uni-Wuerzburg.DE

B. Landheer
Physics Education Department
University of Amsterdam,
Nieuwe Achtergracht 170
1018 WV Amsterdam, The Netherlands

Priscilla W. Laws
Department of Physics and Astronomy
Dickinson College
Carlisle, PA 17013 USA
lawsp@dickinson.edu

Marcia C. Linn
Graduate School of Education
University of California
Berkeley, CA 94612 USA
mclinn@violet.berkeley.edu

Angela E. McFarlane
Homerton College
Cambridge, England CB2 2PH
aem14@cus.cam.ac.uk

A.A. Melo
Departamento de Fisica da Faculdade
de Ciências de Lisboa
Ed. C1 4° Piso
Campo Grande
P-1700 Lisboa, Portugal

Ewa Mioduszewska
University of Amsterdam
Physics Education Department
Nieuwe Achtergracht 170
1018 WV Amsterdam
The Netherlands
miodus@phys.uva.nl

P.P.M. Molenaar
Physics Education Department
University of Amsterdam,
Nieuwe Achtergracht 170
1018 WV Amsterdam, The Netherlands

Ricardo Nemirovsky
TERC, Inc.,
2067 Massachusetts Avenue
Cambridge, MA 02140-1363 USA

Örjan Nilsson
Jönköping University
Box 1026
S-551 11 Sweden

Laurence T. Rogers
School of Education
Leicester Universit
21 University Road
Leicester, England LE1 7RF
lto@leicester.ac.uk

Elena Sassi
Dipartimento di Scienze Fisiche
Universita´ di Napoli
Mostra d'Oltremare pad. 20
I 80125 Napoli, Italy
sassi@na.infn.it

Horst P. Schecker
University of Bremen
Institute of Physics Education
Kufsteiner Straße
D-28334 Bremen, Germany
schecker@physik.uni-bremen.de

David R. Sokoloff
Department of Physics
1274 University of Oregon
Eugene, OR 97403-1274 USA
sokoloff@OREGON.UOREGON.EDU

Ivan Stanchev
Faculty of Educational Science
and Technology
University of Twente, Postbus 217
7500 AE Enschede, The Netherlands
stanchev@edte.utwente.nl

Henryk Szydłowski
Institute of Physics
A. Mickiewicz University
ul Umultowska 85
61-614 Poznan, Poland
henryksz@vm.amu.edu.pl

David S.C. Thompson
Educational Consultant, N. Ireland

Ronald K. Thornton
Center for Science
and Mathematics Teaching
Departments of Education and Physics
Tufts University
Medford, MA 02155 USA

Joke M. Voogt
University of Twente
Faculty of Educational Science
and Technology
Enschede, The Netherlands
VOOGT@edte.utwente.nl

Jack M. Wilson
Rensselaer Polytechnic Institute
Troy, NY 12180 USA
wilsoj@rpi.edu

Subject Index

The NATO ASI Series F Special Programme on ADVANCED EDUCATIONAL TECHNOLOGY

NATO ASI Series F

NATO ASI Series F

NATO ASI Series F

Including Special Programmes on Sensory Systems for Robotic Control (ROB) and on Advanced Educational Technology (AET)

NATO ASI Series F

NATO ASI Series F

Including Special Programmes on Sensory Systems for Robotic Control (ROB) and on Advanced Educational Technology (AET)

Vol. 137: Technology-Based Learning Environments. Psychological and Educational Foundations. Edited by S. Vosniadou, E. De Corte and H. Mandl. X, 302 pages. 1994. *(AET)*

Vol. 138: Exploiting Mental Imagery with Computers in Mathematics Education. Edited by R. Sutherland and J. Mason. VIII, 326 pages. 1995. *(AET)*

Vol. 139: Proof and Computation. Edited by H. Schwichtenberg. VII, 470 pages. 1995.

Vol. 140: Automating Instructional Design: Computer-Based Development and Delivery Tools. Edited by R. D. Tennyson and A. E. Barron. IX, 618 pages. 1995. *(AET)*

Vol. 141: Organizational Learning and Technological Change. Edited by C. Zucchermaglio, S. Bagnara and S. U. Stucky. X, 368 pages. 1995. *(AET)*

Vol. 142: Dialogue and Instruction. Modeling Interaction in Intelligent Tutoring Systems. Edited by R.-J. Beun, M. Baker and M. Reiner. IX, 368 pages. 1995. *(AET)*

Vol. 144: The Biology and Technology of Intelligent Autonomous Agents. Edited by Luc Steels. VIII, 517 pages. 1995.

Vol. 145: Advanced Educational Technology: Research Issues and Future Potential. Edited by T. T. Liao. VIII, 219 pages. 1996. *(AET)*

Vol. 146: Computers and Exploratory Learning. Edited by A. A. diSessa, C. Hoyles and R. Noss. VIII, 482 pages. 1995. *(AET)*

Vol. 147: Speech Recognition and Coding. New Advances and Trends. Edited by A. J. Rubio Ayuso and J. M. López Soler. XI, 505 pages. 1995.

Vol. 148: Knowledge Acquisition, Organization, and Use in Biology. Edited by K. M. Fisher and M. R. Kibby. X, 246 pages. 1996. *(AET)*

Vol. 149: Emergent Computing Methods in Engineering Design. Applications of Genetic Algorithms and Neural Networks. Edited by D.E. Grierson and P. Hajela. VIII, 350 pages. 1996.

Vol. 150: Speechreading by Humans and Machines. Edited by D. G. Stork and M. E. Hennecke. XV, 686 pages. 1996.

Vol. 151: Computational and Conversational Discourse. Burning Issues – An Interdisciplinary Account. Edited by E. H. Hovy and D. R. Scott. XII, 202 pages. 1996.

Vol. 152: Deductive Program Design. Edited by M. Broy. IX, 467 pages. 1996.

Vol. 153: Identification, Adaptation, Learning. Edited by S. Bittanti and G. Picci. XIV, 553 pages. 1996.

Vol. 154: Reliability and Maintenance of Complex Systems. Edited by S. Özekici. XI, 589 pages. 1996.

Vol. 156: Microcomputer-Based Labs: Educational Research and Standards. Edited by R.F. Tinker. XIV, 398 pages. 1996.

Springer
and the
environment

At Springer we firmly believe that an international science publisher has a special obligation to the environment, and our corporate policies consistently reflect this conviction.

We also expect our business partners – paper mills, printers, packaging manufacturers, etc. – to commit themselves to using materials and production processes that do not harm the environment. The paper in this book is made from low- or no-chlorine pulp and is acid free, in conformance with international standards for paper permanency.

 Springer